T0140106

Internet of Things

Technology, Communications and Computing

The series Internet of Things - Technologies, Communications and Computing publishes new developments and advances in the various areas of the different facets of the Internet of Things.

The intent is to cover technology (smart devices, wireless sensors, systems), communications (networks and protocols) and computing (theory, middleware and applications) of the Internet of Things, as embedded in the fields of engineering, computer science, life sciences, as well as the methodologies behind them. The series contains monographs, lecture notes and edited volumes in the Internet of Things research and development area, spanning the areas of wireless sensor networks, autonomic networking, network protocol, agent-based computing, artificial intelligence, self organizing systems, multi-sensor data fusion, smart objects, and hybrid intelligent systems.

** Indexing: *Internet of Things* is covered by Scopus and Ei-Compendex **

More information about this series at http://www.springer.com/series/11636

Sanjay Misra • Amit Kumar Tyagi • Vincenzo Piuri
Lalit Garg
Editors

Artificial Intelligence for Cloud and Edge Computing

Editors
Sanjay Misra
Department of Computer Science
and Communication
Ostfold University College
Halden, Norway

Amit Kumar Tyagi
School of Computer Science and
Engineering
Vellore Institute of Technology
Tamil Nandu, India

Vincenzo Piuri
Dipartimento di Informatica
Universita' degli Studi di Milano
Milano, Italy

Lalit Garg
Faculty of Information and Communication
Technology
University of Malta
Msida, Malta

ISSN 2199-1073 ISSN 2199-1081 (electronic)
Internet of Things
ISBN 978-3-030-80823-5 ISBN 978-3-030-80821-1 (eBook)
https://doi.org/10.1007/978-3-030-80821-1

This Springer imprint is published by the registered company Springer Nature Switzerland AG
The registered company address is: Gewerbestrasse 11, 6330 Cham, Switzerland

Preface

In this smart era, artificial intelligence and cloud computing are two significant technologies that are changing the way we live as humans. That is, artificial intelligence (AI) and cloud computing (and edge computing) have emerged together to improve the lives of millions of people. Edge computing is a new advanced version of cloud computing. Today's many digital (smart) devices like Siri, Google Home, and Amazon's Alexa are blending AI and cloud computing in our lives every day. On a larger scale, AI capabilities are employed in the business cloud computing environment to make organizations more efficient, strategic, and insight driven.

Today's cloud and edge computing acts as an engine to increase the scope of an area and impact of AI. AI and cloud (also edge) computing are transforming businesses at every level, especially with a major impact on large-scale businesses. A seamless flow of AI and cloud-based resources make many services a reality. In general, cloud computing offers businesses more flexibility, agility, and cost savings by hosting data and applications in the cloud. A cloud in business (industry) can be public, private, or hybrid, which provides services like SaaS, PaaS, and IaaS to end users. An essential service of cloud computing is infrastructure, which involves the provision of machines for computing and storage. Cloud providers also offer data platform services that span the different available databases.

Some of the benefits of AI and cloud or edge technologies are derivable with access to huge data sets, including refining the data (in an intelligent way), cost-effectiveness, increased productivity, reliability, and availability of advanced infrastructure. For example, an AI-powered pricing module can ensure that a company's pricing will always be optimized. It's not just about making better use of data, it's conducting the analysis and then putting it into action without the need for human intervention. Some other useful benefits of AI integration with cloud and edge computing include powering a self-Managing cloud with AI; improving data management with AI at the cloud level; delivering more value-added service with AI–SaaS integration; and utilizing dynamic cloud-AI integrated services. Thus, many organizations are now improving their commitment to investing in AI-based strategies or cognitive technologies.

Therefore, it is our pleasure to present to you this book: *Artificial Intelligence for Cloud and Edge Computing: Concepts and Paradigms*. The book provides useful in-depth information regarding current research in artificial intelligence on various cybersecurity challenges. The main goal of this book is to conduct analyses, implementation, and discussion of many tools (artificial intelligence, machine learning, and deep learning and cloud computing, fog computing, and edge computing, including concepts of cyber security) towards understanding the integration of these technologies. The book also contains self-assessment problems for increasing the knowledge of readers and case studies about real-world issues, that is, how AI and cloud/edge computing can change business forever. The book comprises 15 chapters from authors around the world. Below is a summary of each chapter.

Queiroz et al. in their chapter titled "An Optimization View to the Design of Edge Computing Infrastructures for IoT Applications" modeled the service placement issue using a multi-objective optimization aiming at minimizing two aspects: the response time for data transmission and processing in the sensors-edge cloud path, and the (energy or monetary) cost related to the number of turned-on edge nodes. They also proposed two heuristics, based on variable neighborhood search and genetic algorithms, and evaluated over a wide range of scenarios, considering a practical smart city application with 100 sensors and up to 10 edge nodes.

Rijal et al., in their chapter titled "AIOps: A Multivocal Literature Review," presented multivocal literature review to define AIOps, the benefits gained from it, the challenges an organization might face, and, finally, what lies in the foreseen future of the AIOps. Their findings revealed that adopting AIOps helps monitor IT work, optimizes time saving, improves human-AI collaboration, makes IT work proactive, and provides faster mean time to recovery (MTTR).

In the chapter titled "Deep Learning-Based Facial Recognition on Hybrid Architecture for Financial Services," Granados and Garcia-Bedoya explored a mechanism for deep learning–based facial recognition to improve customer identification in financial institutions on a hybrid architecture. They proposed a model that integrates traditional classifiers with convolutional neural networks (CNN) to facial recognition involving visual sequences or different image sources as social media as a tool for financial institutions. Thus, a hybrid architecture between edge-cloud computing, which catalyzes these face detection tools in personalized services, was employed.

In the chapter "Classification of Swine Disease Using K-Nearest Neighbor Algorithm on Cloud-Based Framework," Adeniji et al. used supervised learning to conduct real-time classification along with elucidating the device architecture that classifies photographs of diseased animals using k-nearest neighbor (KNN) algorithm. They developed a web-based expert system for the detection of swine diseases in pigs. The developed system is able to predict the likely occurrence of swine disease using the ML algorithm.

Alabi et al., in their chapter titled "Privacy and Trust Models for Cloud-Based EHRs Using Multilevel Cryptography and Artificial Intelligence," proposed a multi-level cryptography approach for achieving security (privacy) at both local and cloud-based electronic medical records (EMR) systems. They also applied subjective logic-based belief or reasoning models for measuring the trustworthiness of the medical personnel.

In the chapter titled "Utilizing an Agent-Based Auction Protocol for Resource Allocation and Load Balancing in Grid Computing," Ali et al. explored resource management challenges for grid systems. They proposed an economic-based approach for effective allocation of resources and scheduling of application. The proposed architecture uses distributed artificial intelligence. The proposed model aims to optimize the usage of resources to satisfy the demands of grid users.

Al-Janabi and Salman, in their chapter titled "Optimization Model of Smartphone and Smart Watch Based on Multi Level of Elitism (OMSPW-MLE)," proposed a model based on two artificial intelligent techniques, namely deterministic selection and ant optimization. The multi-level optimization model was aimed at performing a clustered process of large volume data streams and consisted of five stages which include: data collection, preprocessing of data, finding the most influential nodes in complex networks, finding the best subgraph, and result evaluation using three different measures.

In the chapter titled "K-Nearest Neighbour Algorithm for Classification of IoT-Based Edge Computing Device," Arowolo et al. proposed a method that guarantees the implementation of low-performance ML techniques on hardware in the Internet of Things model, creating means for IoT awareness. The study used the KNN ML algorithm for the implementation, and a confusion matrix in terms of accuracy was used to evaluate the system. The experiment results displayed an accuracy of 85%, outperforming other methods that have been suggested and compared within the literature.

Awotunde et al., in their chapter titled "Big Data Analytics of IoT-Based Cloud System Framework: Smart Healthcare Monitoring Systems," presented a BDA IoT-based cloud system storage for real-time data generated from IoT sensors and analysis of the stored data from the smart devices. The framework is tested using a big data Cloudera platform for database storage and Python for the system design. The framework's applicability is tested in real-time analysis of healthcare monitoring of patients' data for automatic managing of body temperature, blood glucose, and blood pressure. The integration of the system shows improvement in patients' health monitoring situations.

Sudeepa et al., in their chapter titled "Genetic Algorithm-Based Pseudo Random Number Generation for Cloud Security," proposed an innovative, intelligent framework to spawn a pseudo-random number secret key utilizing the combination of the feedback shift register and a genetic algorithm to intensify the key's strength in terms of its length. The proposed system is intended to enhance the sequence's length by the utilization of the genetic algorithm. They carried out various statistical tests to uphold the strength of the generated key sequences.

Khan et al. in "Anomaly Detection in IOT Using Machine Learning" proposed a comparison between different machine learning algorithms that can be used to identify and predict any malicious or anomalous data for a data set of environmental characteristics created by them. The data set is developed from the data exchanged between the sensors in an IoT environment of Bangladesh. The evaluation was carried out to determine which algorithm performs the best in terms of detecting anomalies in the given data set.

In the chapter titled "System Level Knowledge Representation for Edge Intelligence," Maio provided an overview of critical issues from a systems research viewpoint. They also presented novel constructors for AI, the system-level knowledge representation, and its relevance to IoT.

Narasimhan et al., in their paper titled "AI-Based Enhanced Time Cost-Effective Cloud Workflow Scheduling," presented an extension of an earlier work where time-effective scheduling algorithms were discussed. They presented two scheduling algorithms: time-constrained early-deadline cost-effective algorithm (TECA) to schedule these time-critical workflows with minimum cost, and versatile time-cost algorithm (VTCA) which consider both time and cost constraints; these algorithms considerably enhance the earlier algorithms.

In the paper titled "AI-JasCon: An Artificial Intelligent Containerization System for Bayesian Fraud Determination in Complex Networks," Okafor et al. presented AI containerization API system based on JAVA-SQL container (JSR-233) for fraud prediction and prevention in telecommunication networks. Pipeline modeling using Bayesian software implementation paradigm is introduced using AI-JasCon model. A demonstration of how AI engine works with a complex network system for observation of the number of calls, call frequency, and hidden activities for predictive classification (analytics) at the network backend is also discussed.

Finally, in the chapter titled "Performance Improvement of Intrusion Detection System for Detecting Attacks on Internet of Things and Edge of Things," Saheed proposed an improved IDS model for the classification of attacks on IoT and EoT. To protect EoT and IoT appliances and devices and improve IDSs-IoT, implementation of ten different machine learning models is proposed. They carried out normalization using the minimum-maximum (min-max) method. Then they performed dimensionality reduction using principal component analysis (PCA). The light gradient boosting machine, decision tree, gradient boosting machine, k-nearest neighbor, and extreme gradient boosting algorithms were used for classification.

In conclusion, cloud computing utilizing AI is certainly not a radical or progressive change. In numerous regards, it's a transformative one. We require "test and learn" abilities of AI and the cloud. We are sure that the fusion of cloud computing services and AI technology will significantly change the technology industry. In the near future, AI will become a factor of production with large storage capacity. Researchers and scientists will find the topics and examples given in this book very useful and instructive in the domain of AI and cloud and edge computing. Furthermore, investment in AI for cloud and edge computing will improve the profits and productivity of businesses in no small measure. The main goal is to conduct analyses, implementation, and discussion of many tools (artificial intelligence, machine learning, and deep learning and cloud computing, fog computing, and edge computing, including cyber security concepts) to understand the integration of these technologies.

On behalf of all editors,

Ota, Nigeria

Sanjay Misra

Contents

Editors' Biography

Sanjay Misra is a professor at Ostfold University College(HIOF), Halden Norway. Before coming to HIOF, he was a professor in Covenant University (400–500 ranked by THE(2019)) Nigeria for 9 years. He received his PhD. in Inf. & Know. Engg (Software Engg) from the University of Alcala, Spain and M.Tech. (Software Engg) from MLN National Institute of Tech, India. As per SciVal (SCOPUS – Elsevier) analysis (on 01.09.2021), he is the most productive researcher (Number 1) https://t.co/fBYnVxbmiL in Nigeria since 2017 *(in all disciplines)*, in comp science no 1 in the country & no 2 in the whole of Africa. Total more than 500 articles (SCOPUS/WoS) with 500 coauthors worldwide (-110 JCR/SCIE) in the core & appl. area of Soft Engg, Web engineering, Health Informatics, Cybersecurity, Intelligent systems, AI, etc. He got several awards for outstanding publications (2014 IET Software Premium Award(UK)), and from TUBITAK-Turkish Higher Education and Atilim University). He has delivered more than 100 keynote/invited talks/public lectures in reputed conferences and institutes (traveled to more than 60 countries). He is one of the editors of 58 LNCSs, 4 LNEEs, 1 LNNSs, 3 CCISs & 10 IEEE proceedings and 6 books, and editor in chief of *IT Personnel and Project Management* and *Int J of Human Capital & Inf Technology Professionals* – IGI Global and editor in various SCIE journals.

Amit Kumar Tyagi is an assistant professor (senior grade) and senior researcher at Vellore Institute of Technology (VIT), Chennai Campus, India. His current research focuses on machine learning with big data, blockchain technology, data science, cyber physical systems, smart and secure computing, and privacy. He has contributed to several projects such as "AARIN" and "P3-Block" to address some of the open issues related to the privacy breaches in vehicular applications (such as parking) and medical cyber physical systems. He received his Ph.D. degree from Pondicherry Central University, India. He is a member of the IEEE.

Vincenzo Piuri has received his Ph.D. in computer engineering from Politecnico di Milano, Italy (1989). He is full professor of computer engineering at the Università Degli Studi di Milano, Italy (since 2000). He has been associate professor at Politecnico di Milano, Italy, and visiting professor at the University of Texas at

Austin and at George Mason University, USA. His main research interests are: artificial intelligence, computational intelligence, intelligent systems, machine learning, pattern analysis and recognition, signal and image processing, biometrics, intelligent measurement systems, industrial applications, digital processing architectures, fault tolerance, dependability, and cloud computing infrastructures. Original results have been published in more than four hundred papers in international journals, proceedings of international conferences, books, and book chapters.

He is fellow of the IEEE, distinguished scientist of ACM, and senior member of INNS. He is president of the IEEE Systems Council (2020–21) and has been IEEE vice president for technical activities (2015), IEEE director, president of the IEEE Computational Intelligence Society, vice president for education of the IEEE Biometrics Council, vice president for publications of the IEEE Instrumentation and Measurement Society and the IEEE Systems Council, and vice president for membership of the IEEE Computational Intelligence Society.

Prof. Puiri has been editor-in-chief of the *IEEE Systems Journal* (2013–19) and associate editor of the *IEEE Transactions on Cloud Computing*, and has been associate editor of the *IEEE Transactions on Computers*, the *IEEE Transactions on Neural Networks*, the *IEEE Transactions on Instrumentation and Measurement*, and *IEEE Access*. He received the IEEE Instrumentation and Measurement Society Technical Award (2002) and the IEEE TAB Hall of Honor (2019). He is honorary professor at: Obuda University, Hungary; Guangdong University of Petrochemical Technology, China; Northeastern University, China; Muroran Institute of Technology, Japan; and Amity University, India.

Lalit Garg is a senior lecturer in computer information systems at the University of Malta, Malta, and an honorary lecturer at the University of Liverpool, UK. He has also worked as a researcher at the Nanyang Technological University, Singapore, and Ulster University, UK. His supervision experience includes more than hundred ninety masters' dissertations and four PhD thesis. He has edited two books and published over a hundred papers in refereed high-impact journals, conferences, and books, and some of his articles awarded best paper awards. He has delivered more than twenty keynote speeches in different countries, organized/chaired/co-chaired a similar number of international conferences. He was awarded research studentship in healthcare modelling to carry out his Ph.D. research studies in the faculty of computing and engineering at Ulster University, UK. His doctoral research was nominated for the Operational Research Society Doctoral Prize "Most Distinguished Body of Research leading to the Award of a Doctorate in the field of OR." He has also been consulted by numerous public and private organizations for their information system implementations. His research interests are missing data handling, machine learning, data mining, mathematical and stochastic modelling, and operational research, and their applications, especially in the healthcare domain.

An Optimization View to the Design of Edge Computing Infrastructures for IoT Applications

Thiago Alves de Queiroz, Claudia Canali, Manuel Iori, and Riccardo Lancellotti

1 Introduction

A great variety of Internet of Things (IoT) applications have been developed in the last few years and are getting more and more popularity thanks to their capability of providing society and citizens with improved services and lifestyle. Many applications, ranging from autonomous driving, healthcare, smart cities up to industrial environments, increasingly exploit IoT-based services to support decision-making systems and take advantage of data-driven services [2, 45]. IoT applications are based on the presence of geographically distributed sensors to collect heterogeneous kinds of data about the surrounding environment. Such data are typically sent to a cloud computing data center to be processed using machine learning and Artificial Intelligence (AI) algorithms [31]. However, IoT applications are characterized by very different requirements, which will be pointed out in Sect. 2.

For some applications, the traditional cloud-based approach may not represent the most convenient choice. For example, for bandwidth-bound applications, the increasing volume of produced data is likely to make transferring and processing

T. A. de Queiroz
Institute of Mathematics and Technology, Federal University of Catalão, Catalão, Goiás, Brazil
e-mail: taq@ufg.br

C. Canali (✉) · R. Lancellotti
Department of Engineering "Enzo Ferrari", University of Modena and Reggio Emilia, Modena, Italy
e-mail: claudia.canali@unimore.it; riccardo.lancellotti@unimore.it

M. Iori
Department of Science and Methods for Engineering, University of Modena and Reggio Emilia, Reggio Emilia, Italy
e-mail: manuel.iori@unimore.it

S. Misra et al. (eds.), *Artificial Intelligence for Cloud and Edge Computing*,
Internet of Things, https://doi.org/10.1007/978-3-030-80821-1_1

data at a remote cloud data center too expensive or not convenient for network constraints. Indeed, a high network utilization may be unwanted due to costs related to the cloud pricing options, even when the network load does not lead to a performance degradation. Other IoT applications have essential requirements related to latency and response time (e.g., real-time contexts) and cannot cope with a remote cloud data center's high network delays. In these cases, an *Edge Computing* paradigm is likely to represent a preferable solution [25].

The main feature of an edge computing infrastructure is a layer of edge nodes located on the network edge, close to the sensors, to host tasks aimed at pre-processing, filtering, and aggregating the data coming from the distributed sensors. The layer of edge nodes, placed in an intermediate position between sensors and the remote cloud data center, presents a twofold advantage. First, it may reduce the data volume transferred to the cloud through pre-processing and filtering performed on the network edge. Second, the intermediate layer of edge nodes reduces latency and response times for latency-bound applications. However, the increased complexity due to the introduction of the edge nodes layer opens novel issues concerning the infrastructure design.

The allocation of the data flows coming from sensors over the edge nodes and the choice related to the number and location of edge nodes to be activated represent critical aspects for the design and the management process of an edge computing infrastructure. While the problem of an optimized service placement has been largely investigated in terms of resource management within cloud computing data centers [3, 26], it received scarce attention in the edge computing area. This chapter focuses on this critical issue, proposing modeling the service placement based on a multi-objective optimization problem aiming at minimizing two aspects. First, the response time for the data transmission and processing along the path from sensors to edge nodes and then to the cloud data center; second, the (energy or monetary) costs related the number of turned on edge nodes. Due to the complexity related to the optimized design of an edge computing infrastructure that should cope with a non-linear objective function, we propose two approaches based on meta-heuristics, namely Variable Neighborhood Search (VNS) and Generic Algorithms (GA). We evaluate the performance of the two approaches over a range of different scenarios, considering the realistic setting of a smart city application developed in the medium-sized Italian city of Modena. Our experiments show that both heuristics represent promising approaches for designing an edge computing infrastructure, representing effective solutions for supporting IoT applications.

The rest of this chapter is organized as follows. Section 2 describes IoT applications and their classifications based on key requirements. Section 3 focuses on the literature review. Section 4 describes the main performance metrics used to evaluate the behavior of an edge computing infrastructure supporting IoT applications. In contrast, Sect. 5 formalizes the problem of designing and operating an edge computing infrastructure. Section 6 presents the two considered heuristics based on VNS and GA. Section 7 evaluates how the proposed algorithms can cope with the problem of designing and managing an edge computing infrastructure. Finally, Sect. 8 presents some concluding remarks.

2 IoT Applications: Classification and Challenges

IoT applications have recently experienced a remarkable diffusion in very heterogeneous fields, ranging from automotive to industry, health, and smart city applications. The reasons motivating this increasing popularity of applications based on a distributed set of sensors collecting data from the environment are multiple. First of all, thanks to the enhancements of the underlying technologies, the required technical equipment (principally the IoT sensors) is becoming increasingly powerful and efficient in terms of energy consumption, reducing its size at the same time, making possible its use in many contexts before not even conceivable. Simultaneously, the capability of sensors of storing and transmitting data has increased along with the availability of wireless connectivity in many physical environments.

In this section, we initially classify IoT applications based on their field of application. Then, we characterize their essential requirements to understand the challenges from the supporting infrastructure's point of view.

2.1 *A Classification by Field of Application*

2.1.1 Automotive

The global automotive market around artificial intelligence is expected to grow up to $ 8887.6 million by 2025, with a compound annual growth rate of 45% from 2018 to 2025.[1] *Autonomous driving* holds the promise of reducing traffic fatalities, reducing congestion as well as curbing our carbon footprint. According to ABI, a marketing firm, roughly 8 million (10% of global output) vehicles with self-driving capabilities of level 3 or higher, will be shipped in 2025.[2] Connected cars and related city infrastructures also ensure fleet and vehicle health solutions based on collection of data coming from the vehicle, data management at the cloud level, and application of advanced analytics. Furthermore, the concept of *predictive maintenance* is progressively transforming into a practical and straightforward solution integrating vehicles sensors, hardware and software modules, and data transmitters, allowing the tracking of performance and eventual failure factors. Finally, value-added user services and applications are growing in the field of *infotainment*. A wide range of advanced services are offered by companies such as car manufacturers, insurance, service companies, and infotainment providers.

[1] https://www.alliedmarketresearch.com/automotive-artificial-intelligence-market.

[2] https://www.goldmansachs.com/insights/technology-driving-innovation/cars-2025/.

2.1.2 Industry 4.0

The interconnection of machines and IoT sensors able to gather useful data to provide real-time valuable information increases productivity, quality, and worker safety. It is expected that industrial markets will capture more than 70% of IoT value—estimated to exceed $4.6 trillion by 2025.[3] For example, leading institutions and firms in Europe have proposed the Reference Architecture Model Industry 4.0 (RAMI 4.0), describing key concepts and standards for this emerging paradigm [43]. *Industrial IoT* (IIoT) and Artificial Intelligence constitute the fundamental basis for the development and success of the so-called Industry 4.0: IIoT allows to continuously collect data from several and heterogeneous sensors and devices, and to forward the collected data to cloud computing data centers in a secure way. For these reasons, IIoT is able of supporting many Industry 4.0 applications, such as safety and security of industrial processes, quality control, inventory management, optimization of packing, logistics and supply chain, maintenance processes monitoring and effective detection of failures through the use of machine learning predictive techniques [45].

2.1.3 E-Health

The global market size related to IoT-based equipment for healthcare is projected to reach $ 534.3 billion by 2025, with at an annual growth rate of 19.9% over the period 2019–2025.[4] IoT sensors technologies have penetrated various healthcare applications, ranging from monitoring and management of chronic disease, support for home health, education for patients and professionals, up to applications for disaster management. In the field of *medicare interaction*, smart wearable devices are increasingly used to collect data about patients' health status. Significant examples of data that are likely to be collected in these applications are heartbeat, glucose level, and blood pressure. Such data are collected through sensors located on wearable technologies and are then sent to smartphones to be collected and analyzed. In this way, critical information about the medical condition of the patients is collected and afterward transmitted to a remote provider to be analyzed in order to provide the required care support. This approach allows the health care provider to remotely monitor a patient in his/her own home or in a care facility, thus reducing the readmission rates. Furthermore, the interactions between individuals and a provider can be aided with live video streams for consultative, treatment, and diagnostic services. Another sub-field of application is related to *health tracking*, in terms of health practice and education, that has been increasingly supported by mobile devices. E-Health IoT-based applications can range from continuous monitoring of

[3] https://www.forbes.com/sites/louiscolumbus/2017/12/10/2017-roundup-of-internet-of-things-forecasts/.

[4] https://www.reportlinker.com/p05763769/?utm_source=PRN.

health parameters or conditions, to targeted text messages for medicare, up to wide-scale alerts about disease outbreaks.

2.1.4 Smart Cities Applications

IoT applications are enabling Smart City initiatives worldwide. The possibility of collecting data by distributed sensors and analyzing them through artificial intelligence algorithms for predictive purpose or supporting decision-making processes opened a wide range of opportunities for creating innovative services to improve the lives of their residents [2]. Among the many smart cities' applications, we select the following ones as significant examples.

- *Mobility management*. Data collected from distributed sensors reveal patterns of public mobility and transportation. Mobility operators use these data to enhance the traveling experience and eventually improve safety and punctuality. Simultaneously, solutions for personal vehicles can determine the number, location, and speed of moving vehicles. Furthermore, smart traffic lights can automatically manage the lights according to real traffic conditions. Smart services for traffic management may also forecast, based on historical data, traffic evolution over time in order to prevent potential congestion. Finally, smart parking solutions can predict the occupation or the availability of parking spots for creating dynamic parking maps.
- *Energy and waste management*. The sustainable management of energy and waste in our cities is one of the most important field of IoT-based applications. The data collected by smart meters connected to the network can be sent directly to a public utility for further processing and analysis. In this way, utility companies are able to bill for the exact amount of energy, water, and gas consumed by each citizen. Consumption patterns of an entire city can be easily monitored too. Furthermore, waste management applications may facilitate the optimization of waste collection schedules through the accurate tracking of waste levels and analytics allowing route optimization.
- *Public safety*. Applications for public safety for smart cities involve collection, processing, and transmission of a heterogeneous and ever increasing amount of video sources, ranging from home surveillance systems, traffic monitoring, up to individual videos. In general, smart technologies deployment can provide surveillance with the capability to identify different entities, ranging from human beings, objects, events and detect patterns of behavior and anomalies, either real-time or post-events, from large bodies of video and audio streams. Another application is related to Public Warning Systems to alert citizen in a certain area and communicate them to take precautions or actions. In this case, IoT devices are connected to IoT service platforms to convey information to the citizens.

2.1.5 Retail

In the retail industry, IoT and artificial intelligence are applied with the final goal to support, guide, and improve the production of products for stores and businesses. Remote and accurate monitoring, and sales forecast represent the main benefits that IoT-based innovative solutions may provide to the retail business. Tracking and finalizing consumers' purchases and their product history and specific demographics provide retail consumers with heightened convenience. This trend also benefits retailers by achieving an improved understanding of their audience, thanks to the use of artificial intelligence algorithms on the collected data. A typical application in this field is related to *location-aware advertisement and navigation*. Bluetooth beacons provide shoppers with indoor and outdoor localization and information about shops nearby and help vendors focus their marketing campaigns on actual customer behavior. On the same line, IoT-led data tracking allows retail firms to get advanced metrics on the flow of people within the stores and the best selling points for individual customers. An innovative application consists of *smart mirrors*, which can be adopted within fitting rooms for suggesting other items based on what others have bought using an RFID label-scanning system. Smart mirrors are also used in conjunction with augmented reality, allowing customers to dress virtually without physically doing it.

2.1.6 Smart Agriculture

Smart agriculture, also called precision agriculture may significantly help farmers in order to maximize their productivity while minimizing costs and use of resources (e.g., water, seeds, and fertilizer). The deployment of sensors-based fields maps allows the farmers to better understand their farms at a micro-scale level of detail and to reduce the related environmental impacts. The core building block for the development of such smart agriculture are represented by Internet of Things (IoT) and artificial intelligence [8]. Specifically, the role of IoT is fundamental for the automatic collection of data that are subsequently transmitted to cloud data centers for processing. On the other hand, artificial intelligence solutions typically involving artificial neural networks and clustering techniques are then applied to process data to support decision-making. A typical example is related to the identification of the more appropriate amount of water necessary to irrigate fields, that is determined through the analysis of the collected agricultural data. Specifically, an action is taken in order to provide the correct quantity of water as soon as the irrigation need is detected, that is when the field lacks water.

2.2 *Challenges of IoT Applications*

The previously described applications show significantly different characteristics in terms of collected data, type of required processing, and purpose of the data analysis results. The application characteristics imply very different requirements from computational and networking points of view, leading to consequent challenges for the underlying supporting infrastructure. As an example, autonomous driving capabilities require real-time data transmission with strictly controlled latency margins to be safe: a typical requirement of the so-called level 4–5, that is high-level autonomous driving, is that each status message is delivered within 10 ms at maximum to enable appropriate and safe vehicle reactions [38]. On the other hand, medicare applications offering video streaming are more likely to be constrained by bandwidth availability. The near coming transition between closed-world, independent applications, and general-purpose infrastructures capable of supporting diverse applications without intervention requires hardware and software tools capable of finding adaptive trade-offs among different requirements. In the following, we introduce a taxonomy based on four categories, including the majority of IoT applications. Table 1 summarizes the results.

- **Latency-bound**: applications with strong requirements in terms of response times, as in real-time contexts. Examples of applications: mobility management, public surveillance, autonomous driving, and location-aware advertisement and navigation belong to this category.
- **CPU-bound**: applications characterized by computationally heavy tasks with high processing times where the CPU of the processing nodes is likely to become a bottleneck. Examples of applications: public safety, autonomous driving, infotainment systems, smart mirrors, and medicare applications are in this category.

Table 1 IoT applications essential requirements

Application	Latency	CPU	Bandwidth	Reliability	Statefulness
Autonomous driving	✓	✓	✓	✓	–
Predictive maintenance	–	–	–	–	✓
Infotainment	–	✓	✓	–	–
Industrial IoT	–	–	–	–	✓
Medicare interaction	–	✓	✓	✓	–
Health tracking	–	✓	–	✓	–
Mobility management	✓	–	–	–	✓
Utility/Waste	–	–	–	–	–
Public safety	✓	✓	✓	–	–
Location-aware Ads/Nav	✓	–	–	–	–
Smart Mirrors	–	✓	✓	–	✓
Smart agriculture	–	–	–	–	✓

- **Bandwidth-bound**: applications characterized by a significant amount of data to be transferred, where the network bandwidth is an essential requirement for delivering services. Examples of applications: public safety, autonomous driving, infotainment, smart mirrors, and remote medicare interaction.
- **Reliability-bound**: applications where the reliability—intended as integrity, security, privacy, and availability—of data is a crucial requirement. Examples of applications: autonomous driving and e-health applications.
- **Stateful**: applications need to access past information for processing incoming data, hence data flows cannot be distributed over multiple nodes. Typically, applications in which aggregated views of the data are needed. Examples are mobility management, predictive maintenance, and smart mirrors.

3 Literature Review

As discussed in the previous section, an edge computing paradigm may efficiently handle data volume to be processed in IoT applications. In general, these applications require low latency and highly scalable services, motivating edge-based solutions in substituting the standard cloud computing paradigm. There is a vast literature showing the benefits of the edge computing infrastructure, especially when there is a large volume of data coming from geographically distributed devices [33, 35, 42, 44]. One considerable advantage of this architecture is decentralization, which allows pre-process data (for example, filtering and aggregation processing) before they reach the cloud data centers [19, 35, 39].

A survey on edge computing infrastructures was given in [39], who commented on applications and their challenges. Regarding IoT services, Wen et al. [35] discussed concepts and issues that may appear in complex scenarios, as, e.g., smart cities and marine monitoring. These authors handled a generic scenario by using genetic algorithms. Recently, a survey on the integration of edge computing with IoT was given by [28]. The authors discussed many challenges related to the heterogeneity, complexity, and dynamics of applications that need to be taken into account to deliver precise and reliable services. The authors also gave some insights into resiliency and the modeling by game theory. These studies foster the smart city application we are investigating in terms of cost reduction, load balancing, and latency reduction.

In general, the project of edge computing infrastructures is concerned with allocating services over the infrastructure. This was the case of [15], who handled issues related to the power consumption and transmission delay in the edge nodes to cloud data centers communication. On the other hand, the work of [40] modeled the load sharing issues in an edge-to-edge communication. In the context of industrial applications, Foukalas [18] presented a distributed intelligent IoT platform. The author used the cognitive IoT concept, with machine learning classifiers having an overall knowledge of the application, for monitoring and control purposes. Experiments were conducted on a scenario of predictive maintenance in smart factories, demonstrating the effectiveness of the proposed approach. Recently, Caiza et al. [9] discussed the main issues related to architecture, security, latency, and energy consumption in the context of IIoT. The authors commented on the vital role that edge computing plays to deal with these issues, which certainly include the fast processing and real-time storage of large volumes of data. Although our application may carry some similarities with these studies, our initial assumption considers a long-range communication network where multi-hop links exist between sensors. Then, each edge node can serve any sensor.

Our study considers an application to support smart city services and other studies also support this kind of application. For example, Tang et al. [33] proposed a system structured on four layers to handle it. Wang et al. [34] examined the coupling resource management problem, which is related to the multiple requisitions a sensor node is required to attend and then may result in failures of services. With the aim of an edge computing paradigm, the authors could efficiently introduce buffer and controller operations to obtain a sustainable system. In the survey of Zahmatkesh and Al-Turjman [42], the benefits of buffer/caching are discussed when edge computing is used in IoT communications. The authors investigated caching techniques using artificial intelligence and machine learning approaches, showing their potential and future challenges. Concerning the smart traffic monitoring, Dhingra et al. [16] used edge computing in a framework for congestion monitoring and traffic light management. The final results demonstrated a reduction in response time and an increase in the bandwidth. The proposed approach is also flexible, because it does not impose a fixed number of layers and let the number of edge nodes be determined according to the application's requirements.

Determining the number of edge nodes turned on and which sensors each of these nodes will serve is a typical facility location problem. This problem has been extensively explored in the operational research literature, proposing different algorithms to handle it. This problem aims at determining the number and where to open facilities (location decisions), besides which customers each open facility will serve (allocation decisions). The objective is to minimize the location and allocation decisions while satisfying the customers' demand [14].

Facility location problems were surveyed in [1, 13, 17, 22]. Farahani et al. [17] focused their review on multi-criteria problems, including bi-objective, multi-objective, and multi-attribute problems. They also commented on solutions methods. In the review of [22], the problems were classified according to the use of aggregation rules to handle situations with a large number of customers. The authors described aggregation error measurements, besides commenting on conditional location problems. On the other hand, Ahmadi-Javid et al. [1] focused their review on healthcare applications, presenting a framework to classify the problems following the emergency type and location management. The recent survey of Turkoglu and Genevois [13] considers service facility location problems, with a detailed classification of them based on purpose, space, distance, time, parameters, capacity, facilities, objectives, competition, application field, solution method, type of experiment, and other features.

Regarding IoT applications, Klinkowski et al. [24] handled the problem of data center locations with light-path provisioning in elastic optical networks. Silva and Fonseca [32] considered the optimization of human mobility using an edge computing infrastructure. The objective is to reduce the processing in the cloud data centers and the latency of users. Both [24] and [32] started dealing with mathematical models, but due to their inefficiency, they derived heuristic approaches. Recently, Canali, and Lancellotti [10, 11] considered the optimization of a service placement problem with the presence of edge infrastructures. In [10], these authors outlined the problem, while in [11], they presented a GA to examine a smart city application. On the other hand, the GA-based heuristic proposed in this chapter is a clear step ahead concerning these studies, because we introduce the possibility of locating edge nodes to reduce cost and power consumption, besides respecting a service level agreement.

4 Performance Modeling in Edge Computing Infrastructures

In this section, we present the main performance metrics that can be used to evaluate an edge computing infrastructure's behavior supporting IoT applications. We can distinguish two classes of metrics: efficiency metrics, related to the utilization of resources needed to operate, and performance metrics that concern the ability to operate effectively.

4.1 Efficiency Metrics

An edge computing infrastructure needs resources to provide its services. Resources range from hardware components to energy to economic costs related to maintenance and administration. In the following of this analysis, we will refer in general to costs because all the resources needed to support IoT applications can be easily mapped into economic costs. Costs can occur both at deployment time or when running IoT applications on the infrastructure. An example of deployment time is the purchase of the hardware for the infrastructure and its installation. We usually refer to these costs as *Capital Expenditures* (CAPEX). Examples of costs related to the operation of the infrastructure are the energy cost and maintenance. We refer to them as to *Operating Expenditures* (OPEX).

Another way to classify the costs related to the edge computing infrastructure is to consider if the cost is global or related to its infrastructure. For example, costs related to setting up a control center for the edge computing infrastructure are not strongly related to the size of the infrastructure. On the other hand, the cost of the purchase and installation of edge nodes is directly correlated with the number of nodes to install. Similarly, energy costs are directly related to the number of active edge nodes.

The costs occur both when the infrastructure is deployed (usually, in this case, we consider the CAPEX related to the edge nodes that should be installed) and when the infrastructure operates (in this case, the focus is on OPEX such as energy costs for the nodes). It is worth noting that, when the load is subject to variations over time, the infrastructure can be designed to cope with the peak load; however, during off-peak hours, a fraction of the nodes can be turned off to reduce energy costs.

4.2 Performance Metrics

Another class of metrics is related to the performance experienced by the IoT application. These metrics typically can be related to the response time when data flows are processed in the edge infrastructure or can concern the infrastructure's availability.

Among response time related performance metrics, the most common approach is to consider the average response time in a stationary scenario (when there is no transient effect due to sudden variations of the incoming load). An alternative metric is to consider a different estimation of the response time, typically taking into account 90-percentile or 95-percentile, representing a worst-case scenario excluding pathologically long request processing, for example, due to timeouts of transmission errors.

Availability-related metrics can take into account metrics such as service availability over a while (typical values range from 95% availability up to the 5-nines availability for critical services that must be online for 99.999% of the considered period). Another metric to measure a service's ability to process requests is considering the number of requests processed and the number of requests dropped, for example, due to the infrastructure's failures or due to overload in the edge nodes. To this aim, we define the *Throughput* as the number of requests processed in a considered period and the *Goodput* as the number of requests correctly processed in the same time frame. The difference between Throughput and Goodput measures the requests not processed (that is dropped) due to overload or other infrastructure problems. From this metric, we can infer the request drop rate, which is the percentage of requests that cannot be satisfied.

All the considered performance metrics can be used to define some *Service Level Agreement* (SLA) that defines the contract between the provider of the service and its users. An SLA can define a maximum acceptable response time (defined as a limit on the average or on some percentile of the response time), a minimum service availability that must be guaranteed, or a minimum drop rate. Failing to respect an SLA for the service provider usually results in a penalty that reduces the revenues.

5 Problem Formulation

We now formalize the problem of designing and operating an edge computing infrastructure supporting an IoT application. We assume the problem to be both CPU-bound and network-bound, where the network concerns may be either due to network latency or bandwidth issues. Furthermore, we assume our IoT application is a stateful application, where all data from an IoT device (sensor) must be sent to the same edge node for processing. We use as the basis of our model a facility location problem. We aim at (1) identifying a subset of potential edge nodes that should serve requests from IoT devices (referred to as sensors) and (2) mapping data flows from sensors to edge nodes and from edge nodes to cloud data centers to reduce cost and optimize performance.

If the problem concerns the infrastructure (infrastructure deployment) design phase, we consider a set of potential locations where edge nodes can be installed. We know an expected incoming load. Based on a performance model, we determine how many locations should be selected to host an edge node to satisfy SLA requirements while minimizing the infrastructure CAPEX. Similarly, during the edge infrastructure operations (infrastructure management), we can solve the same problem with different input data. In this case, we have a set of edge nodes, and we want to decide the minimum set of nodes to turn on to satisfy the SLA while minimizing the OPEX.

5.1 Model Parameters

To provide a model for the problem, we assume a stationary scenario. Let $\mathcal{S}$ be a set of geographically distributed IoT devices (sensors) producing data at a constant rate. The generic sensor i produces data at a rate λ_i. An intermediate layer of edge nodes will process the data from the sensors. The edge nodes can perform operations such as filtering, aggregation, or anomaly detection with low latency. In our problem, we consider a set $\mathcal{N}$ of potential edge nodes locations. A decision variable E_j is used to decide if edge node j is to be used. If we consider the problem of infrastructure deployment, E_j is used to decide if a potential edge node is to be actually deployed. In the problem of infrastructure management, the variable E_j is used to decide if the edge node is to be turned on or off, based on present load conditions. Each edge node j can process data at a rate μ_j and is associated with a cost c_j that is either a representation of OPEX or CAPEX, depending on the scenario in which the problem is applied. Furthermore, the transmission of data between the sensors and the edge nodes occurs with a delay δ_{ij}, with i being the sensor and j the edge node.

Finally, we consider the third layer of our architecture, that consists in set of cloud data centers $\mathcal{C}$. When data is sent from an edge node j to a cloud data center k, it incurs in a delay δ_{jk}.

Table 2 Summary of notations used in the proposed model

Parameters of the model	
$\mathcal{S}$	Sensors set
$\mathcal{N}$	Edge nodes set
$\mathcal{C}$	Cloud data centers set
λ_i	Outgoing data rate from sensor i
λ_j	Incoming data rate at edge node j; $\lambda_j = \sum_{i \in \mathcal{S}} x_{ij} \lambda_i$
$1/\mu_j$	Processing time at edge node j
c_j	Cost for deploying a potential edge node j (or for keeping an edge node powered on)
δ_{ij}	Communication delay from sensor i to edge node j
δ_{jk}	Communication delay from edge j to cloud k
Indices used in notation	
i	Index of a sensor
j	Index of an edge node
k	Index of a cloud data center
Decision variables	
E_j	Enabling of edge node j
x_{ij}	Communication occurs between sensor i and edge node j
y_{jk}	Communication occurs between edge node j and cloud data center k

To considered problem uses the following binary decision variables.

- E_j, to define if a (potential) edge node located at position j is available to process data from sensors.
- x_{ij}, to define if sensor i is sending data to edge node j.
- y_{jk}, to define if edge node j is sending data to cloud data center k.

The main symbols of the model are summarized in Table 2.

5.2 Objective Functions and SLA

The considered problem is based on the problem in [10] and takes into account two different performance metrics. The first performance metric is related to the infrastructure cost that we aim to minimize. Ideally, the lower is the number of edge nodes that we use, the lower is the global cost. The second performance metric is related to the average response time, and it plays a double role in our problem formulation. On the one hand, we consider an SLA on the average response time to guarantee that the response time remains below a given value. On the other hand, as long as the infrastructure's cost remains the same, we prefer solutions that are characterized by a lower average response time.

Concerning the infrastructure cost, we can define the total cost as

$$C = \sum_{j \in \mathcal{N}} c_j E_j \tag{1}$$

For the second performance metric, we should remember that response time is the sum of three contributions: T_{netSE} that is the network delay related to the sensor-to-edge latency, T_{netEC} that is the delay related to the edge-to-cloud latency, and T_{proc} that is the time due to data processing on the edge nodes.

The network delay components can be modeled as follows:

$$T_{netSE} = \frac{1}{\sum_{i \in \mathcal{S}} \lambda_i} \sum_{i \in \mathcal{S}} \sum_{j \in \mathcal{N}} \lambda_i x_{ij} \delta_{ij} \tag{2}$$

$$T_{netEC} = \frac{1}{\sum_{j \in \mathcal{N}} \lambda_j} \sum_{j \in \mathcal{N}} \sum_{k \in \mathcal{C}} \lambda_j y_{jk} \delta_{jk} \tag{3}$$

For the T_{netSE} definition in Eq. (2), we weight the delays δ_{ij} between each sensor i and each edge node j by the amount of traffic passing through that link, which is $\lambda_{ij} x_{ij}$. The sum is then normalized dividing by the total traffic $\sum_i \lambda_i$. In a similar way, we define T_{netEC} in Eq. (3). The main difference is the use of the term λ_j to refer to the load incoming into each edge node j and forwarded from the edge node to the cloud layer. We can define λ_j as

$$\lambda_j = \sum_{i \in \mathcal{S}} x_{ij} \lambda_i, \quad \forall j \in \mathcal{N} \tag{4}$$

The component concerning the processing time T_{proc} is modeled using the queuing theory considering an M/G/1 system. According to the PASTA theorem [36] and to the Pollaczek-Khintchine formula used to describe the M/G/1 average response time [21], we have

$$T_{proc} = \frac{1}{\sum_{j \in \mathcal{N}} \lambda_j} \sum_{j \in \mathcal{N}} \lambda_j \left(\frac{1}{\mu_j} + \frac{\lambda_j \mu_j V_j^2}{2(\mu_j - \lambda_j)} \right) \tag{5}$$

where V_j^2 is the variance of the service time process. Assuming a variance of the service time close to the one of the inter-arrival time $(2/\mu_j^2)$, as in other example sin literature [4, 11], we can simplify Eq. (5) in

$$T_{proc} = \frac{1}{\sum_{j \in \mathcal{N}} \lambda_j} \sum_{j \in \mathcal{N}} \lambda_j \frac{1}{\mu_j - \lambda_j} \tag{6}$$

It is worth mentioning that we do not consider the cloud layer's details (such as the computation time at the cloud data center level) in the model of our problem. Indeed, this aspect is not meaningful for optimizing the edge infrastructure.

This model for the processing time is used also to describe the SLA for the edge computing infrastructure. Specifically, we expect the average response time to stay below T_{SLA}:

$$T_{SLA} = K\frac{1}{\overline{\mu}} + \overline{\delta}_{SF} + \overline{\delta}_{FC} \tag{7}$$

where K is constant value (in cloud systems it is common to consider $K = 10$ [4]); $1/\overline{\mu}$ is the average service time in an edge node; $\overline{\delta}_{SE}$ and $\overline{\delta}_{EC}$ are the average sensor-to-edge and edge-to-cloud delays, respectively.

5.3 *Optimization Problem*

We can define the model for the edge computing infrastructure problem supporting an IoT application as follows:

Minimize :

$$C = \sum_{j \in \mathcal{N}} c_j E_j \tag{8}$$

$$T_R = T_{netSE} + T_{netEC} + T_{proc} \tag{9}$$

Subject to :

$$T_R \leq T_{SLA} \tag{10}$$

$$\lambda_j < E_j \mu_j, \quad \forall j \in \mathcal{N} \tag{11}$$

$$\sum_{j \in \mathcal{N}} x_{ij} = 1, \quad \forall i \in \mathcal{S} \tag{12}$$

$$\sum_{k \in \mathcal{C}} y_{jk} = E_j, \quad \forall j \in \mathcal{N} \tag{13}$$

$$E_j \in \{0, 1\}, \quad \forall j \in \mathcal{N} \tag{14}$$

$$x_{ij} \in \{0, 1\}, \quad \forall i \in \mathcal{S}, j \in \mathcal{N} \tag{15}$$

$$y_{jk} \in \{0, 1\}, \quad \forall j \in \mathcal{N}, k \in \mathcal{C} \tag{16}$$

The two objective functions, (8) and (9), are related, respectively, to the minimization of: **cost**, which depends on the number of used edge nodes; and **response time**, which is the delay in sensor-edge-cloud data transit expressed through the function introduced in the previous section. The response time objective is subordinated to the cost one, meaning that we aim to minimize (9) as long as the improvement for this objective function does not affect (8).

The model includes the following constraints. Constraint (10) places a limit for the average response time: this value must not violate the *Service Level Agreement* (SLA). Constraints (11) guarantee that no overload occurs on the edge nodes. Hence, the incoming data rate on every node must be lower than the processing rate. For a node that is powered down, no processing occurs. Constraints (12) ensure that exactly one edge node will process the data of each sensor. In a similar way constraints (13) guarantee that exactly one cloud data center receives the data for each edge node. Finally, constraints (14), (15) and (16) describe the binary nature of the decision variables.

6 Heuristics

Several options are available when we aim to solve a problem using a heuristic algorithm. On the one hand, greedy heuristics are usually quite fast. On the other hand, the performance of a greedy algorithms depends on the nature of the problem. Local minima and a non-convex domain, which may hinder their ability to find a solution for the problem. In designing an edge computing platform, the objective function is non-linear. Furthermore, the feasibility domain of the problem is not guaranteed to be convex. For this reason, we explore solutions that differ from the classical greedy approach.

Due to the number of edge nodes and sensors, the problem is characterized by a high dimensionality that may hinder the performance of branch and bound approaches, due to the large solution space to explore.

For these reasons we focus on meta-heuristics that provide a flexible approach to the problem. The ability of these meta-heuristics to solve successfully a broad and heterogeneous set of problems is known in literature [7]. In particular, we consider two approaches, namely Variable Neighborhood Search (VNS) and Generic Algorithms.

6.1 Variable Neighborhood Search

The *Variable Neighborhood Search* (VNS) is a meta-heuristic used to solve hard non-linear and combinatorial optimization problems. Some examples include its application to vehicle routing [37], portfolio selection [5], cutting and packing [30], scheduling [29], among others.

VNS is a single solution-based method, where a solution passes through a shaking and local search phases to become globally optimal concerning all neighborhood structures. These structures represent how to explore the current solution's neighborhood towards new solutions further investigated in the local search phase. When an improved solution is found, VNS comes back to the first neighborhood, or else it continues to the next neighborhood [20, 27].

The VNS iterates through the following general steps, given an initial solution: shaking phase, to obtain a new solution by applying a neighborhood structure; local search phase, to improve the new neighbor solution; and, an acceptance phase, to change the current solution by the one from the previous phase. Once the current solution is changed, the search restarts from the first neighborhood structure; otherwise, it considers the next structure. We present Algorithm 1 with the implemented VNS for the edge computing infrastructure problem. During the search for a solution, we do not accept solutions that violate constraints (10)–(13).

Algorithm 1: VNS for the location-allocation problem

```
1  x ← an initial solution generated by a random constructive heuristic;
2  while the stopping criteria are not reached do
3      k ← 1;
4      while k ≤ K_max do
5          x' ← random solution in the neighborhood structure N_k(x);
6          x'' ← apply the local search on x';
7          if f(x'') < f(x) then
8              x ← x'';
9              k ← 1;
10         end
11         else k ← k + 1
12     end
13 end
14 return x;
```

In Algorithm 1, we code a solution as a matrix of integers. Each matrix's line represents an edge node and contains the sensors it serves. The last cell of each matrix's line keeps the cloud data center that serves the edge node. When there is no sensor in a matrix's line, then that edge node is off. Notice that each sensor will appear in exactly one matrix's line. In the initial solution, we select the closest cloud data center for each edge node to serve it. Similarly, each sensor is served by the closest edge node. In the case an edge node has reached the T_{SLA}, then no other sensor can be served by it. In the latter, the sensor is served by its second-closest edge node, and so on until the edge nodes serve all sensors.

The inner loop of lines 4–12 in Algorithm 1 ends when all neighborhood structures are visited for the current solution. It means this solution is globally optimal with regards to all these structures. In line 5, a new solution is generated randomly on the current solution's neighborhood (i.e., the shaking phase). Next, we apply the local search on this new solution, attempting to improve it. The local search is based on trying all possible allocations of sensors in edge nodes and swaps of sensors in edge nodes. Algorithm 2 describes the local search phase.

A solution has two objectives: (1) the cost associated with the number of edge nodes on; and (2) the delay in sensor-edge-cloud transit of data. We assume the first objective is used to guide the VNS. It means a given solution is better than another if its first objective is smaller, or if their first objectives are equal but its second is smaller. Notice that if the current solution is improved, the search restarts from the first neighborhood structure. Besides that, our VNS considers $K_{max} = 5$ neighborhood structures based on swap and move operations. In particular:

- N_1: select (randomly) an edge node n_1, the farthest sensor s_1 which is served by n_1, the edge node n_2 that is the closest to s_1, and the sensor s_2 which is served by n_2 that is the closest to n_1. Now, let s_1 be served by n_2 and s_2 by n_1.
- N_2: let $\mathcal{N}_{on}$ be the set of edge nodes on. Let $r_j = \lambda_j/\mu_j$ be the load of each edge node j in this set. Calculate the average load of the edge nodes on as $\bar{r} = \left(\sum_{j \in \mathcal{N}_{on}} r_j\right) / |\mathcal{N}|$. Then, select (randomly) $n_1 \in \mathcal{N}_{on}$ whose load $r_1 > \bar{r}$. If one exists, select the farthest sensor s_1 which is served by n_1. Next, select the edge node $n_2 \in \mathcal{N}_{on}$ with the lowest load r_2 and closest to s_1. Now, let s_1 be served by n_2.
- N_3: let $\mathcal{N}_{on}$ be the set of edge nodes on. Select (randomly) an edge node n_1 from this set. Then, compute the average load with all sensors and edge nodes on, except n_1, as $\tilde{r} = \left(\sum_{i \in \mathcal{S}} \lambda_i\right) / \left(\sum_{j \in \mathcal{N}_{on} \setminus \{n_1\}} \mu_j\right)$. If $\tilde{r} < 1$, then for each sensor s_1 which is served by n_1, let s_1 be served by the closest edge node in $\mathcal{N}_{on} \setminus \{n_1\}$.
- N_4: let $\mathcal{N}_{on}$ and $\mathcal{N}_{off}$ be the sets of edge nodes on and off, respectively. If $\mathcal{N}_{off}$ is not empty, select (randomly) an edge node n_1 from this set. On the other hand, select the edge node $n_2 \in \mathcal{N}_{on}$ whose average response time is the highest one. Now, let all sensors of n_2 to be served by n_1.
- N_5: select (randomly) an edge node on and let it be served by the closest cloud data center.

Algorithm 2: Local search phase

1 $x \leftarrow$ an input solution;
2 $k \leftarrow 1$;
3 **while** $k \leq L_{max}$ **do**
4 $\quad x' \leftarrow$ the best solution in the neighborhood structure $M_k(x)$;
5 $\quad$**if** $f(x') < f(x)$ **then**
6 $\quad\quad x \leftarrow x'$;
7 $\quad\quad k \leftarrow 1$;
8 $\quad$**end**
9 $\quad$**else** $k \leftarrow k + 1$
10 **end**
11 **return** x;

The local search phase is presented in Algorithm 2. Given an input solution, it iterates through $L_{max} = 2$ neighborhood structures. At each structure (line 4), the best possible solution in the current solution's neighborhood is chosen. This means that all possible movements defined by the neighborhood structure are tried, so the one whose neighbor solution has the best improvement is considered. The search restarts from the first structure if the current solution is improved. The two neighborhood structures M_k are defined as:

- M_1: for each pair of edges nodes n_1 and n_2, and sensor s_1 which is served by n_1, now let s_1 be served by n_2. Among all these movements/possibilities, choose the one that most reduce the solution cost.
- M_2: for each pair of edges nodes n_1 and n_2, and sensors s_1 which is served by n_1 and s_2 which is served by n_2, now let s_1 be served by n_2 and s_2 by n_1. Among all these movements/possibilities, choose the one that most reduce the solution cost.

6.2 Genetic Algorithm

We now discuss an alternative heuristic for handling the optimization problem based on genetic algorithms (GAs). Evolutionary programming has been used in literature to solve similar problems, like allocating VMs in a cloud data center [41].

In GAs, we model a possible solution as a *population* composed by *individuals*. For each individual, we encode the solution of the problem in the form of a *chromosome*. A generic individual i is represented through its chromosome C^i, which is a sequence of *genes* with a fixed length. We can write that $C^i = \{c^i_j\}$ with each gene representing a parameter characterizing that individual's solution.

The algorithm starts with an initial, randomly generated, population of individuals. The objective function for the optimization problem plays the role of a *fitness function* for the GA. Such function is applied to every individual, to assign a *fitness score* to each chromosome. After this initialization, the population evolves through some *generations*. At each generation, the fitness score of each chromosome is updated. The evolution is carried out applying the following operators to the population:

- **Selection** is an operator that decides if an individual should be preserved in the passage from the Kth generation to the next. The fitness score of each individual is used to discard individuals with undesirable characteristics.
- **Mutation** randomly alters a gene in an individual (more complex mutations may involve more than one gene, for example, swapping their values). In GAs, the role of mutation is to add new genetic material, allowing exploring new areas within the solution space.
- **Crossover** combines two separate individuals creating two new offspring individuals. The offspring is created by exchanging a fraction of the chromosomes of the parents. In GAs, crossover aims to spread positive combinations of genes through the population.

These operations are combined to define an evolutionary strategy. In particular, we adopt the strategy described in [6] as *Simple Strategy*. In every generation, the selection operator is applied to select (and possibly replicate) the fittest individuals. Unfit individuals are likely to be pruned from the genetic pool. After the selection, the new population undergoes the application of the crossover and mutation operators. For each individual, mutation and crossover occurs with a probability defined as P_{mut} and P_{cx}, respectively. For crossover, the offspring (two individuals for each crossover) replaces the parents. In a similar way, mutated individuals replace the originals.

In the following, we discuss the details of the GA applied to our problem. We discuss the chromosome structure and the specific mutation and crossover operators designed for the considered problem.

6.2.1 Problem Definition

If we consider the problem statement, we observe that the decision variables y_{jk} for mapping edge nodes to cloud data centers define a subproblem that can be easily solved by mapping each edge node to the nearest cloud data center. For this reason, we can discard from the genetic algorithm implementation this part of the problem, and we can consider y_{ij} as another problem parameter rather than a decision variable. This reduces the solution space to explore and accelerate the algorithm convergence.

Another design choice aims to simplify the management of the double objective function that characterize our problem. Rather than embedding the ability to change the number of active edge nodes, we rely on constraint (10) to infer this value. We observe that the definition of T_{SLA} in (7) presents three components related to T_{proc}, T_{netSE} and T_{netEC}. We focus in the first part and we assume to have a homogeneous population of nodes where $\mu_j = \mu \ \forall j \in \mathcal{N}$. Furthermore, we assume that the sensors' data rate is the same for every sensor, that is $\lambda_i = \lambda \ \forall j \in \mathcal{S}$. Finally, since we are considering the ideal case that is a lower bound on the processing time, we can impose a perfect load balancing among the edge nodes. We can thus write that:

$$T_{proc} = \frac{1}{NE \cdot \mu - |\mathcal{S}| \cdot \lambda} \leq \frac{K}{\mu} \tag{17}$$

where $NE = \sum_i E_i$ is the number of active edge nodes. From this we can define the minimum number of edge nodes as

$$NE = \left\lceil |\mathcal{S}| \cdot \frac{\lambda}{\mu} \cdot \frac{K+1}{K} \right\rceil \tag{18}$$

To minimize the first objective function, we start the algorithm with some nodes equal to NE and, if no feasible solution is found, we increase NE and re-iterate the genetic algorithm. This approach ensures that the objective function (8) is minimized so that the genetic algorithm can focus on the second objective

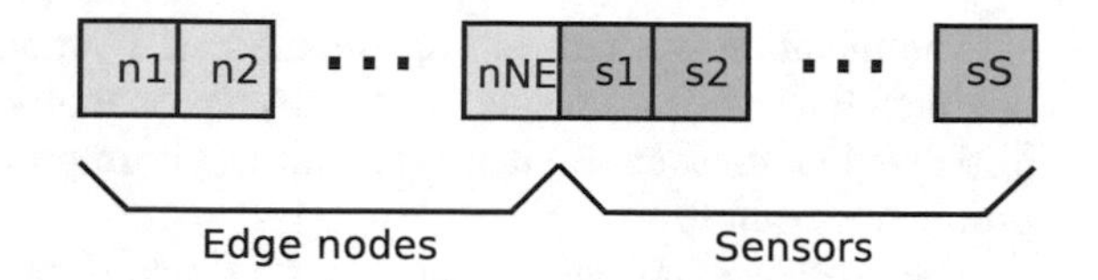

Fig. 1 Solution modeling as a chromosome

function (9). Indeed we can use the definition of T_R in Eq. (9) as the fitness function in our genetic algorithm, with the note that lower response time means better fitness.

6.2.2 Chromosome Encoding

We now consider how to encode a solution in a chromosome. We must embed in the gene sequence two types of information:

- Which edge nodes are active (that is on or selected for deployment) among a set of potential ones. This information represents the E_i decision variables;
- How to map sensors and cloud data centers over the available edge nodes (that would represent the x_{ij} decision variables)

To this aim, we divide the chromosome into two parts (Fig. 1).

The first part contains NE genes $\{n_1, \ldots, n_{NE}\}$ that list the active edge nodes; each generic gene n_i has a value that ranges from 1 to $|\mathcal{N}|$. In this first part of the chromosome, each gene must have a different value from the other, that is $\forall i, j \in [1, NE]$ we have that $n_i = n_j \iff i = j$.

The second part of the chromosome contains $S = |\mathcal{S}|$ genes $\{s_1, \ldots, s_S\}$. Each gene contains a value $\in [1, NE]$ to map the sensors to the list of active edge nodes to data exchange. For the generic sensor $i \in [1, S]$, we have a value e, meaning that sensor i will transmit data to edge node n_e.

The final chromosome will be a concatenation of the two previously described parts: $\{n_1, \ldots, n_{NE}, s_1, \ldots, s_S\}$

6.2.3 Genetic Operators

Given our problem's nature, we have to adapt to the typical genetic operators. First of all, the initialization process must ensure that the basic rules of chromosome encoding are respected. To this aim, the first NE genes must contain no duplicates, and each part of the chromosome must have values in the correct range (that is $[1, |\mathcal{N}|]$ for the first NE genes, $[1, NE]$ for the other S genes).

We can use a standard operator for the selection operator because all the problem-dependent information is embedded in the fitness function. Specifically, we rely on the tournament selection operator. The operator picks K elements randomly within the population and returns the one with the best fitness score.

The mutation operator is implemented as a variation of the uniform mutation operator. Each gene is mutated with a given probability. Again, the mutation, like the initialization, must ensure that the resulting chromosome is feasible for the selected problem encoding.

The crossover operates separately on the two parts of the chromosome. First, we select random parts of the first NE genes, making sure that no duplicates are introduced. Next, we operate on the second part of the chromosome. The particular case to handle is when an edge node n is used in the parent node by a sensor s, but the same node is no longer available in the offspring. In this case, we look among the edge nodes available in the offspring, the node n' that is the closest (from a delay point of view) to n. For each sensor s in the offspring that would refer to n, we use the index of n' to create a feasible chromosome that can inherit some positive feature from the parent.

7 Experimental Evaluation

We now evaluate how the proposed algorithms can cope with designing and managing an edge computing infrastructure. Specifically, we first describe the experimental setup, and then we proceed with a performance comparison between the VNS and GA.

7.1 Experimental Scenario

We refer in our analysis to a realistic IoT application project supported by an edge computing infrastructure. In particular, we consider a smart city project developed into the medium-sized Italian city of Modena (closer to 185.000 inhabitants). The IoT application aims to monitor car, bicycle, and pedestrian traffic using a geographically distributed set of sensors that collect information on the traffic proximity sensors. Depending on the setup, the sensors can also capture low-resolution images. For the considered application, sensors must be placed in the city's main streets. In particular, we identify the sensors' location by mapping the street names using the Open Street Map APIs,[5] using features developed as part of the PAFFI framework [12]. The sensors send the collected data to the edge nodes, which perform pre-processing tasks using AI techniques. In particular, edge nodes can filter and aggregate the proximity sensor readings and, if available, can analyze images from the camera using neural networks to detect cars, bicycles, and pedestrians. For the edge nodes' potential locations, we have a list of buildings belonging to the municipality that could be used to host these nodes. Finally, all the

[5] https://wiki.openstreetmap.org/wiki/API_v0.6.

pre-processed data are then sent to a cloud data center hosted on the municipality data center.

In the sensor prototypes used in this analysis, communication relies on long-range wireless connectivity boards, such as LoRaWAN[6] or IEEE 802.11ah/802.11af [23]. Due to the long-range of these communication technologies, each sensor can potentially communicate with every edge node. On the other hand, the available bandwidth decreases with the sensor-to-edge distance. Consequently, according to other studies in literature [10, 11], the communication delay between sensors and edge nodes is inversely proportional to the physical distance among the communicating entities.

Throughout our experimental evaluation we focus on a scenario where $|\mathcal{S} = 100|$, $|\mathcal{N} = 10|$, and $|C| = 1$. Other experiments with different setup confirm the main findings of our study and are not reported. Instead, we consider a broader analysis of our system's workload scenario to explore the ability of the considered algorithms to adapt to different conditions. In particular, each scenario is defined by three parameters. The first parameter, common to all configurations, is the *sensor data rate* λ, which is the same for all sensors. We consider a preliminary prototype of the smart city applications for traffic monitoring, where each sensor provides a reading every 10 s, meaning that $\lambda_i = \lambda = 0.1, \forall i \in \mathcal{S}$. This setting is suitable when designing an architecture aiming to support a stationary scenario, for example, in the first design phase of the smart city project. An alternative approach (more focused on a dynamic infrastructure management problem and not considered in our experiments) would consider a variable data rate.

The second parameter is the *average utilization* of the system ρ defined as $\frac{\sum_{i \in \mathcal{S}} \lambda_i}{\sum_{j \in \mathcal{N}} \mu_j}$. Since λ is already defined, variations in the system utilization depend mainly on the edge nodes' processing rate μ. Indeed, several computing apparatuses (typically in the form of an embedded board with processing elements, memory, storage, and communications support commonly used in IoT deployments) are available for this task. The CPU frequency, the presence of multiple cores (typically from 1 to 8), and even the support for dynamic voltage and frequency scaling may affect the μ parameter of each node. For the ρ parameter, we consider a wide range of values: $\rho \in \{0.1, 0.2, 0.5, 0.8, 0.9\}$. The last parameter is $\delta\mu$, which is the ratio between average network delay δ, and the processing time $1/\mu$. The parameter can be used to define the *CPU-bound or network-bound nature* of the scenario. For the $\delta\mu$ parameter, we consider values ranging orders of magnitude with $\delta\mu \in \{0.01, 0.1, 1, 10\}$. Our analysis ranges from CPU-bound scenarios (e.g., when $\delta\mu = 0.01$), where computing time is significantly higher than transmission time, to network-bound cases (e.g., when $\delta\mu = 10$) where data transmission dominates the response time. In our setup, this variability of the CPU to network weight refers mainly to the presence and resolution of images transferred over low-bandwidth links.

[6] https://lora-alliance.org/.

The goal of our analysis is to deploy the minimum infrastructure that ensures the respect of SLA as defined in Eq. (7). In our analysis, the constant K is set to 10. To define which nodes are to be deployed, we assume the cost c_j of an edge node at position j is equal to 1, for all $j \in \mathcal{N}$. This assumption is consistent with the observation that all the potential locations for edge nodes already belongs to the municipality. Our problem is thus reduced to minimizing the number of edge nodes used. For the experimental comparison, we evaluate the two considered alternatives described in Sect. 6:

- *Variable Neighborhood Search (VNS)*: described in Sect. 6.1;
- *Genetic Algorithm (GA)*: described in Sect. 6.2

For the VNS, we run it for 300 s or 3000 iterations (the first to reach stops the VNS), while the Genetic algorithm is limited to 300 generations with a population of 200 individuals.

7.2 Experimental Results

We now discuss the most significant results of our performance analysis. In particular, we summarize in Table 3 the scenario parameters and the performance of the considered algorithms. The first two columns contain the scenario-defining parameters ρ and $\delta\mu$. Next, we present the critical value T_{SLA} used to define the maximum acceptable response time. It is worth noting that T_{SLA} changes as a function of both ρ and $\delta\mu$: the dependency from $\delta\mu$ is clearly evident from Eq. (7). The dependency from ρ is because we consider λ as constant, and $1/\mu$ changes as a function of ρ. Finally, we show the values of the two objective functions for the two algorithms: VNS and GA. The first (obj_1) is the number of edge nodes used, while the second (obj_2) is the average response time.

A discussion of the results is provided in the following. First, we observe that for every scenario, each of the proposed algorithms can identify a solution that satisfies the SLA requirement (this is easily verified comparing the columns obj_2 for the two algorithms with the third column T_{SLA}). A second significant observation is that both algorithms find the same minimum number of edge nodes to use to guarantee acceptable performance (to this aim, we can refer to the obj_1 columns in Table 3). Hence, our analysis's first conclusion is that both algorithms are viable to tackle the considered problem.

Concerning the number of nodes, which is the first objective of the optimization problem, we show a plot in Fig. 2 that presents a comparison between the actual number of edge nodes used in the infrastructure and the estimation provided by Eq. (18) (we consider the real value, before applying the $\lceil\cdot\rceil$ operator). We compare the curve of the required number of edge nodes as a function of ρ for two values of $\delta\mu$: a CPU-bound scenario where $\delta\mu = 0.1$ (represented as squares in Fig. 2) and a network-bound scenario $\delta\mu = 10$ (represented as circles). For both considered scenarios, we provide both the estimate from Eq. (18) (in the line with empty circles

Table 3 Experimental results

Parameters			GA		VNS	
ρ	$\delta\mu$	T_{SLA} [s]	obj_1 [#edge]	obj_2 [s]	obj_1 [#edge]	obj_2 [s]
0.1	0.01	1.002	2	0.201	2	0.201
0.1	0.1	1.017	2	0.209	2	0.207
0.1	1	1.172	2	0.293	2	0.268
0.1	10	2.724	2	1.126	2	0.871
0.2	0.01	2.003	3	0.603	3	0.602
0.2	0.1	2.034	3	0.621	3	0.614
0.2	1	2.345	3	0.794	3	0.729
0.2	10	5.449	3	2.525	3	1.867
0.5	0.01	5.009	6	3.072	6	3.070
0.5	0.1	5.086	6	3.121	6	3.096
0.5	1	5.862	6	3.607	6	3.374
0.5	10	13.622	6	8.461	6	6.010
0.8	0.01	8.014	10	4.013	10	4.048
0.8	0.1	8.138	10	4.123	10	4.487
0.8	1	9.380	9	9.343	9	8.724
0.8	10	21.796	9	18.986	9	12.835
0.9	0.01	9.016	10	9.014	10	9.006
0.9	0.1	9.155	10	9.142	10	9.054
0.9	1	10.552	10	10.437	10	9.552
0.9	10	24.520	10	23.175	10	14.496

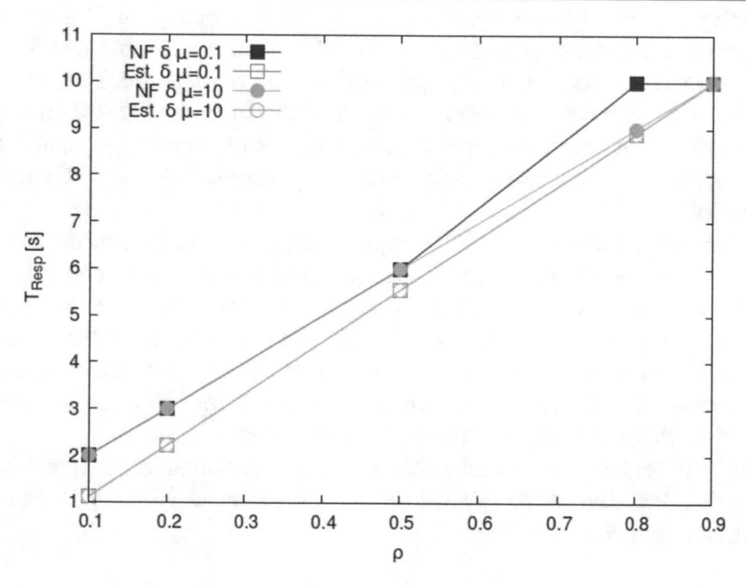

Fig. 2 Number of edge nodes

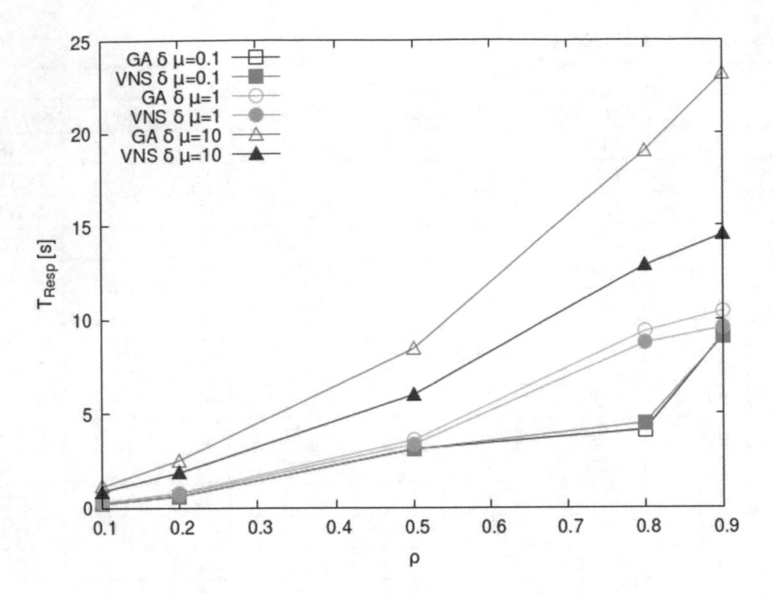

Fig. 3 Response time

and squares) and the objective function (marked with filled circles and squares). We observe that the estimation for the number of edge nodes is a suitable option for a first rough sizing of the infrastructure. The number of edge nodes used is the upper integer compared to the estimation with just one exception. When $\rho = 0.8$, the expected number of nodes is 8.89. We can identify a suitable topology for a network-bound scenario where an optimized linking between the 100 sensors and 9 enabled edge nodes compensates for slightly unbalanced load distribution among the edge nodes (we cannot evenly distribute 100 sensors over 9 edge nodes). On the other hand, in a CPU-bound scenario, finding a feasible solution and 10 edge nodes are required.

Further analysis is the plot of the response time as a function of the load ρ for different $\delta\mu$ scenarios, as in Fig. 3. Specifically, the curves with squares refer to $\delta\mu = 0.1$, the curves with circles refer to $\delta\mu = 1$, and the curves with triangles refer to $\delta\mu = 10$. For every scenario, we compare the performance of the VNS (filled symbols) with the GA (empty symbols). We observe that the two algorithms present similar behavior, with times that (while remaining below the SLA requirement) grow with both the utilization ρ and the $\delta\mu$ parameter.

To fully understand the two algorithms' relative performance, the final comparison shows a heatmap of the relative response time. In particular, we measure the performance gain defined as:

$$R_{\%} = \frac{T_R^{GA} - T_R^{VNS}}{T_R^{VNS}} \cdot 100 \tag{19}$$

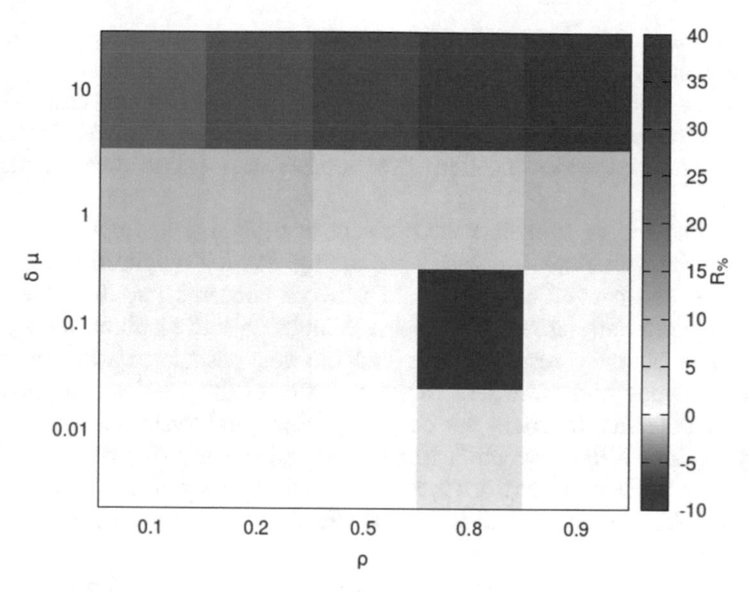

Fig. 4 Comparison of heuristics

When $R_{\%}$ is positive, the VNS is faster than the GA alternative, while when the value is negative, GA provides better performance.

The heatmap in Fig. 4 shows that the VNS outperforms in almost every condition the GA approach. In particular, the performance gain is higher when the network delay plays a significant role and when the load is higher, as testified by the large predominance of red hues in the upper and right part of the graph. Indeed, the VNS iteration is specifically tailored to the considered problem's characteristics, while the GA operators are more generic. Furthermore, the crossover operator is not highly optimized for the considered double meaning of the two parts of the chromosome and may not guarantee the passage of good characteristics from the parents to offspring. This effect is more evident when we need to cope with high network delays.

8 Conclusion and Future Work

Modern IoT applications typically rely on significant amount of data collected by geographically distributed sensors to be processed by artificial intelligence algorithms with different goals. Depending on the field of application and the characteristics of data collected and provided service, the IoT applications may present strong requirements in terms of latency, CPU, bandwidth, reliability, and statefulness. In this scenario, the Edge Computing paradigm may represent a preferable choice with respect to traditional cloud computing systems to reduce network traffic and improve latency by placing computational resources at the edge of the network. However, edge computing opens new issues for the design of the infrastructure and the deployment of IoT applications. Specifically, critical aspects are related to the allocation of data flows coming from sensors over the nodes of the edge layer and the identification of the number and location of edge nodes to be activated.

In this chapter we formalize the problem of designing and operating an edge computing infrastructure supporting, assuming the IoT application to be both CPU-bound and network-bound, where network concerns may be either due to latency or bandwidth issues. We propose a multi-objective optimization problem to minimize both the response time and the cost associated with the number of activated edge nodes. To face the complexity of the problem that involves a non-linear objective function, the chapter presents and evaluates two heuristics, namely Variable Neighborhood Search (VNS) and Generic Algorithms (GA). The performance of the two heuristics is evaluated over a wide range of scenarios based on the realistic setting of a smart city application developed in a medium-sized Italian city.

The experimental evaluation proves that both heuristics can effectively support the design of an edge computing infrastructure. They returned a feasible solution satisfying the service level agreement for all scenario parameters, from CPU-bound to network-bound ones. Each heuristic's solution requires the same number of edge nodes to be active. On the other hand, in terms of relative response time, the VNS approach is overall superior, reaching a performance gain up to almost 40% in scenarios characterized by high network delays and load. This chapter represents an important step along a research line aimed at the identification of promising solutions for the development of future infrastructures supporting IoT applications. In future works, we plan to extend the solutions to handle the possible heterogeneity of the edge nodes and dynamic scenarios where the load can change over time.

References

1. Ahmadi-Javid A, Seyedi P, Syam SS (2017) A survey of healthcare facility location. Comput Operat Res 79:223–263
2. Alavi AH, Jiao P, Buttlar WG, Lajnef N (2018) Internet of things-enabled smart cities: state-of-the-art and future trends. Measurement 129:589 – 606
3. Ardagna D, Ciavotta M, Lancellotti R (2014) A Receding Horizon Approach for the Runtime Management of IaaS Cloud Systems. In: Proceedings of 16th international symposium on symbolic and numeric algorithms for scientific computing (SYNASC), IEEE
4. Ardagna D, Ciavotta M, Lancellotti R, Guerriero M (2018) A hierarchical receding horizon algorithm for QoS-driven control of multi-IaaS applications. IEEE Trans Cloud Comput 9:1–1
5. Bačević A, Vilimonović N, Dabić I, Petrović J, Damnjanović D, Džamić D (2019) Variable neighborhood search heuristic for nonconvex portfolio optimization. Eng Economist 64(3):254–274
6. Back T, Fogel D, Michalewicz Z (2002) Evolutionary computation 1: basic algorithms and operators. CRC Press, Boca Raton

7. Binitha S, Sathya SS, et al (2012) A survey of bio inspired optimization algorithms. Int J Soft Comput Eng 2(2):137–151
8. Bu F, Wang X (2019) A smart agriculture IoT system based on deep reinforcement learning. Future Generation Comput Syst 99:500–507
9. Caiza G, Saeteros M, Oñate W, Garcia MV (2020) Fog computing at industrial level, architecture, latency, energy, and security: a review. Heliyon 6(4):e03706
10. Canali C, Lancellotti R (2019) A fog computing service placement for smart cities based on genetic algorithms. In: Proceedings of international conference on cloud computing and services science (CLOSER 2019), Heraklion
11. Canali C, Lancellotti R (2019) GASP: genetic algorithms for service placement in fog computing systems. Algorithms 12(10):201
12. Canali C, Lancellotti R (2019) Paffi: performance analysis framework for fog infrastructures in realistic scenarios. In: 2019 4th international conference on computing, communications and security (ICCCS), pp 1–8
13. Celik Turkoglu D, Erol Genevois M (2020) A comparative survey of service facility location problems. Annals of Operations Research 292:399–468
14. Cooper L (1963) Location-allocation problems. Oper Res 11(3):331–343
15. Deng R, Lu R, Lai C, Luan TH, Liang H (2016) Optimal workload allocation in fog-cloud computing toward balanced delay and power consumption. IEEE Int Things J 3(6):1171–1181
16. Dhingra S, Madda RB, Patan R, Jiao P, Barri K, Alavi AH (2020) Internet of things-based fog and cloud computing technology for smart traffic monitoring. Internet of Things 14:100175
17. Farahani RZ, SteadieSeifi M, Asgari N (2010) Multiple criteria facility location problems: A survey. Appl Math Model 34(7):1689–1709
18. Foukalas F (2020) Cognitive IoT platform for fog computing industrial applications. Comput Electr Eng 87:106770
19. Gill SS, Tuli S, Xu M, Singh I, Singh KV, Lindsay D, Tuli S, Smirnova D, Singh M, Jain U, Pervaiz H, Sehgal B, Kaila SS, Misra S, Aslanpour MS, Mehta H, Stankovski V, Garraghan P (2019) Transformative effects of IoT, Blockchain and Artificial Intelligence on cloud computing: evolution, vision, trends and open challenges. Int Things 8:100118
20. Hansen P, Mladenović N, Moreno Pérez JA (2010) Variable neighbourhood search: methods and applications. Ann Oper Res 175(1):367–407
21. Harrison PG, Patel NM (1993) Performance modeling of communication networks and computer. Addison-Wesley, Boston
22. Irawan C, Salhi S (2015) Aggregation and non aggregation techniques for large facility location problems - a survey. Yugoslav J Oper Res 25:313–341
23. Khorov E, Lyakhov A, Krotov A, Guschin A (2015) A survey on IEEE 802.11 ah: an enabling networking technology for smart cities. Comput Commun 58:53–69
24. Klinkowski M, Walkowiak K, Goścień R (2013) Optimization algorithms for data center location problem in elastic optical networks. In: 2013 15th international conference on transparent optical networks (ICTON), pp 1–5
25. Liu F, Tang G, Li Y, Cai Z, Zhang X, Zhou T (2019) A survey on edge computing systems and tools. Proc IEEE 107(8):1537–1562
26. Marotta A, Avallone S (2015) A Simulated Annealing Based Approach for Power Efficient Virtual Machines Consolidation. In: Proceedings of 8th international conference on cloud computing (CLOUD), IEEE
27. Mladenović N, Hansen P (1997) Variable neighborhood search. Comput Oper Res 24(11):1097–1100
28. Moura J, Hutchison D (2020) Fog computing systems: state of the art, research issues and future trends, with a focus on resilience. J Netw Comput Appl 169:102784
29. Queiroz TAd, Mundim LR (2020) Multiobjective pseudo-variable neighborhood descent for a bicriteria parallel machine scheduling problem with setup time. Int Trans Oper Res 27(3):1478–1500
30. Santos LFM, Iwayama RS, Cavalcanti LB, Turi LM, de Souza Morais FE, Mormilho G, Cunha CB (2019) A variable neighborhood search algorithm for the bin packing problem with compatible categories. Expert Syst Appl 124:209–225

31. Shanthamallu US, Spanias A, Tepedelenlioglu C, Stanley M (2017) A brief survey of machine learning methods and their sensor and IoT applications. In: 2017 8th international conference on information, intelligence, systems applications (IISA)
32. Silva RAC, Fonseca NLS (2019) On the location of fog nodes in fog-cloud infrastructures. Sensors 19(11):2445
33. Tang B, Chen Z, Hefferman G, Wei T, He H, Yang Q (2015) A hierarchical distributed fog computing architecture for big data analysis in smart cities. In: Proceedings of the ASE BigData & socialInformatics 2015, ACM, New York, ASE BD&SI '15, pp 28:1–28:6
34. Wang T, Liang Y, Jia W, Arif M, Liu A, Xie M (2019) Coupling resource management based on fog computing in smart city systems. J Netw Comput Appl 135:11–19
35. Wen Z, Yang R, Garraghan P, Lin T, Xu J, Rovatsos M (2017) Fog orchestration for internet of things services. IEEE Int Comput 21(2):16–24
36. Wolff RW (1982) Poisson arrivals see time averages. Oper Res 30(2):223–231
37. Xu Z, Cai Y (2018) Variable neighborhood search for consistent vehicle routing problem. Expert Syst Appl 113:66–76
38. Yamanaka N, Yamamoto G, Okamoto S, Muranaka T, Fumagalli A (2019) Autonomous driving vehicle controlling network using dynamic migrated edge computer function. In: 2019 21st international conference on transparent optical networks (ICTON), Angers
39. Yi S, Li C, Li Q (2015) A survey of fog computing: Concepts, applications and issues. In: Proceedings of the 2015 workshop on mobile big data, ACM, New York, Mobidata '15, pp 37–42
40. Yousefpour A, Ishigaki G, Jue JP (2017) Fog computing: Towards minimizing delay in the internet of things. In: 2017 IEEE international conference on edge computing (EDGE), pp 17–24
41. Yusoh ZIM, Tang M (2010) A penalty-based genetic algorithm for the composite SaaS placement problem in the cloud. In: IEEE congress on evolutionary computation, pp 1–8
42. Zahmatkesh H, Al-Turjman F (2020) Fog computing for sustainable smart cities in the IoT era: Caching techniques and enabling technologies - an overview. Sustainable Cities Soc 59:102139
43. Zezulka F, Marcon P, Vesely I, Sajdl O (2016) Industry 4.0—an introduction in the phenomenon. IFAC-PapersOnLine 49(25):8–12; 14th IFAC conference on programmable devices and embedded systems PDES 2016
44. Zhang C (2020) Design and application of fog computing and internet of things service platform for smart city. Future Gener Comput Syst 112:630–640
45. Zheng P, Wang H, Sang Z (2018) Smart manufacturing systems for industry 4.0: Conceptual framework, scenarios, and future perspectives. Front Mech Eng 13:137–150

AIOps: A Multivocal Literature Review

Laxmi Rijal, Ricardo Colomo-Palacios, and Mary Sánchez-Gordón

1 Introduction

Information technology (IT) has transformed almost every industry and created an impact from business to everyday life. As technology is becoming a crucial part of society, IT installations are becoming larger and more complex, especially for large-scale data centers [1]. The increasing number of systems and applications is creating new challenges, and the communication among these applications makes the interconnectivity so narrow that applications become an inseparable and complicated living ecosystem [2]. At the same time, IT operators are challenged to complete the complex task manually without additional automation and assistance [3]. Moreover, support services are challenged due to improper incident management, problem management, and service level management [4]. In order to minimize these challenges, IT service management (ITSM) plays an important role as it delivers the quality of the IT services in the best possible way by implementing, managing, and delivering the quality product to meet the business needs [5]. ITSM ensures the proper mix of people, processes, and technology which helps to reduce the risk of losing business opportunities and client trust [5, 6].

In the last years, there is a move toward a continuous improvement of ITSM [7] in which, following standards like ITIL or CMMI Service, there is a continuous approach in getting maturity, including personnel aspects as one of the cornerstones of the approach [8]. Other approaches include the implementation of service management offices [9], co-creation approaches [10], or gamification [11], citing some of the current developments. A broad review of the new approaches could be found in the work of Marrone and Hammerle [12]. Traditionally, ITSM facilitates cost savings, reduced occurrences of incidents, and increased customer satisfaction

L. Rijal · R. Colomo-Palacios (✉) · M. Sánchez-Gordón
Østfold University College, Halden, Norway
e-mail: laxmi.rijal@hiof.no; ricardo.colomo-palacios@hiof.no; mary.sanchez-gordon@hiof.no

S. Misra et al. (eds.), *Artificial Intelligence for Cloud and Edge Computing*,
Internet of Things, https://doi.org/10.1007/978-3-030-80821-1_2

[6]. However, the shift to multi-cloud environments, DevOps, microservices architectures, and rapid data growth is increasing IT complexity [13] which IT operators are increasingly unable to deal with [14]. Thus, it has become clear that the IT business itself needs a digital transformation to cope with the increasing operational uncertainty and its costs [3, 14]. This digital transformation must be constructed upon, among other factors, artificial intelligence.

Given the growing interest and investment in this process, Gartner introduced the concept of Algorithmic IT Operations back in 2016. Later on, it was changed to Artificial intelligence for IT operations (AIOps) based on public opinion [15]. AIOps explores the use of artificial intelligence (AI) to control and optimize IT services [3]. It uses big data, machine learning, and other advanced computational tools to develop IT operations directly and indirectly [16]. Moreover, it provides strategic insights and suggestions to minimize errors, boost mean time to recovery (MTTR), and effectively distribute computing resources [14, 17–19] in the look for lowering also personnel costs [20], an aspect crucial for modern IT [21].

Besides, AIOps has started getting more industry attention in literature as well as in research [14, 22]. So far, this specific branch of the IT industry is seeing massive growth, as new products, open-source projects, and service provider companies are emerging. In fact, Gartner predicts that 40% of the large business will combine machine learning and big data to replace legacy services by 2022; however, only 5% was using it by 2018 [15]. Tools in the AIOps panorama include AppDynamics, BigPanda, SL1, Instana, Dynatrace, Moogsoft, PagerDuty, SysTrack, Optanix, and DataDog or Splunk, naming just some of the most important tools available in the market early 2021 [23].

Despite of the importance of AIOps, to the best of the authors' knowledge, there is a lack of work devoted to review and provide insights on the use of AI for IT operations. Therefore, this study aims to identify the definition of AIOps, its benefits and opportunities, as well as its challenges by conducting a Multivocal Literature Review (MLR). The search string is applied to two databases (Google Scholar and Google Search). In fact, no prior literature review on AIOps, with the same objective as the one proposed in this study, is available. However, there is a systematic mapping study conducted by Notaro et al. [3] to identify the past research in AIOps. These authors considered data-driven approaches based on ML and data mining for searching for and identifying relevant studies. They performed the searches in three database libraries (IEEE Xplore, ACM Digital Library, and arXiv). The result of the study is a taxonomy in which the majority of papers are associated with failure-related tasks (62%), i.e., anomaly detection and root cause analysis. This work will complement the work performed by these authors adding also insights from gray literature as well as more recent works on the topic.

The paper is organized as follows. The next section presents the background about AI and AIOps. Section 3 presents the research methodology including research questions and the data collection procedure. Section 4 presents results of the research questions and the limitation of the study. Finally, the conclusions and future work are presented in Sect. 5.

2 Background

2.1 Artificial Intelligence

Nilsson et al. [24] define artificial intelligence (AI) as "that activity devoted to making machines intelligent... [where] intelligence is that quality that enables an entity to function correctly and with foresight in its environment."

AI is not relatively a new term [25]. The concept of modern-day AI was created in 1955, by Mr. John McCarthy along with Marvin Minsky, Nathan Rochester, and Claude Shannon in a conference at Dartmouth by submitting a proposal named "A Proposal for the Artificial Intelligence Summer Research Project in Dartmouth" [26]. Although AI was introduced back then, currently the discipline has gained momentum both in popularity and real repercussion. Over time, AI is making some impact in society, and it is often associated with the term "machine learning" (ML) or "deep learning" [25, 27].

AI is the blend of various advanced technologies having the capabilities of replicating and/or improving different human tasks and cognitive capabilities such as image and speech recognition, planning, and learning [28]. More precisely, AI is a technological domain with core components such as machine learning (ML), deep learning, natural language processing (NLP) platforms, predictive application programming interfaces (APIs), and image and speech recognition tools [29]. More importantly, the reason electronic devices and machines are assumed to be crossing the boundaries is due to the blend of technologies, knowledge, and materials. In fact, most groundbreaking elements lie in the AI-equipped machines to change their actions and alter their objective based on the previous experience as well as in response to changing environment [30, 31].

While AI has become an integral part of common applications, a study by McKinsey Global Institute predicts that the use of AI in industries will result in a USD 13 trillion global value-added contribution by 2030 [32]. In addition, the International Data Corporation (IDC) predicts that global spending on AI and ML will be double, rising from $50.1 billion in 2020 to over $110 billion by 2024 [33]. Thus, this trend also should be addressed seriously which might also have a direct or indirect impact on AIOps.

2.2 Artificial Intelligence for IT Operations (AIOps)

As mentioned earlier, AIOps is the combination of artificial intelligence for IT operations. In particular, Gartner states that "AIOps platforms utilize big data, modern machine learning and other advanced analytics technologies to directly and indirectly enhance IT operations (monitoring, automation and service desk) functions with proactive, personal and dynamic insight. AIOps platforms enable the concurrent use of multiple data sources, data collection methods, analytical (real-time and deep) technologies, and presentation technologies" [34].

AIOps evolved from the need to monitor and analyze the activities performed in an IT environment (both hardware and software), such as processor use, application response times, API usage statistics, and memory loads [18]. AIOps offers the information needed to filter out the info required for faster and safer decisions with intelligent data correlation and dynamic pattern analysis, techniques that are not possible by means of classic methods [22].

In recent years, AIOps has evolved, and now it offers a wide variety of tools for different applications from resource and complexity management to task scheduling, anomaly detection, and recovery [22, 35, 36]. Besides, OpsRamp [37] conducted a survey named "The OpsRamp State of AIOps Report" of 200 IT managers throughout the United States to understand their experience with AIOps. The result of the study showed that 85% of responses were for automating the tedious tasks, followed by 80% suppression/de-duplication/correlation of alerts and 77% reduction in open incident tickets. In fact, a study by Digital Enterprise Journal shows that, since 2018, there has been an 83% increase in the number of companies implementing or looking to deploy AIOps capabilities in their IT operations [38].

3 Research Methodology

To address the goal of this study, a Multivocal Literature Review (MLR) was conducted based on the guideline provided by [39]. A MLR is a form of systematic literature review that allows to include primary, secondary, as well as gray literature (e.g., blog posts, videos, and white papers) [39]. Such an approach is gaining interest in academic literature [39], and it diminishes the gap by combining the knowledge of the state of the art and practice. As a result of its usefulness, there are many recent studies in computing at large including aspects on microservices [40], function as a service [41], or software as a service [42], mentioning just some of the most recent publications.

In what follows, a brief description of the research procedure is presented including a description of the three research questions for this study and an explanation of the search strategy adopting, mentioning, and discussing aspects on data sources, search string used, search process, and search execution.

3.1 Research Questions

The goal of this MLR is to get a state of the art and practice related to AIOps by defining AIOps, the benefits gained from it, and the challenges organizations adopting AIOps might face. Based on the above goal, three research questions (RQs) are formulated:

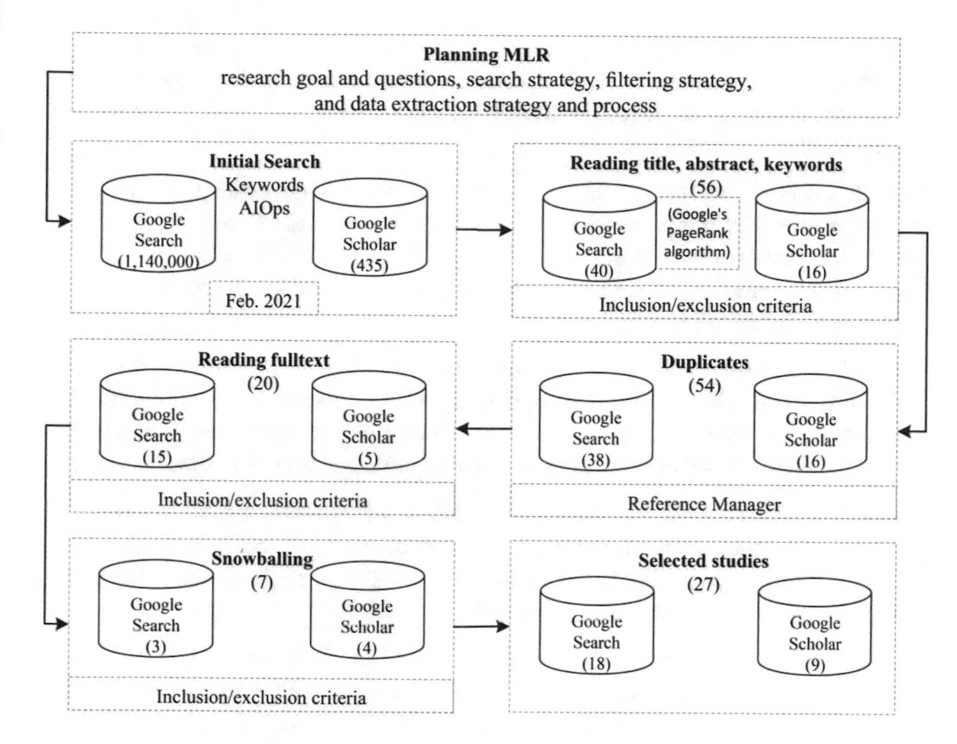

Fig. 1 Overview of the search process

RQ_1: How does the literature define AIOps?
RQ_2: What are the reported benefits of AIOps?
RQ_3: What are the reported challenges of AIOps?

3.2 *Search Strategy*

The scope of this step is to characterize the search and evaluation strategy for identifying the primary studies. This allows a thorough search of the available literature needed to answer the proposed research questions. Figure 1 depicts an overview of the process.

Data Source To conduct a literature review, data on AIOps were collected by using two search engines: Google Scholar (http://scholar.google.com/) and Google Search (http://www.google.com/). For this study, both engines were estimated to be adequate since they cover all major publishing venues (i.e., IEEE, ScienceDirect, Springer, ACM, or Wiley). Google Scholar was used in collecting the scientific literature, while gray literature was collected using Google Search.

Search Term The search string is constructed in order to retrieve the most relevant literature on AIOps. As AIOps is relatively a new term "AIOps" was identified as the search string to have the broader scope of the results and to answer the aforementioned research questions

Search Process The search process allows the authors to select primary studies from the scientific literature (Google Scholar). The process is comprised of four phases that follow a test-retest approach to reduce bias in the selection process. The same process was also conducted to identify gray literature on Google Search. The four phases are as follows:

Phase 1. Initial Search The search string was applied to the search engines in order to identify the literature related to topic under review. Searches are limited to title, abstract, and keywords. In terms of timeline, this study was conducted in February 2021, and thus the authors included the papers published until that time.

Phase 2. Remove Duplicates Studies identified during phase one of the selection process will be checked in order to remove the duplicates. If duplication is identified, papers providing detailed information such as an abstract or the full text of the paper, complete references of the publication will be selected.

Phase 3. First Selection Process Studies selected in phase two will be evaluated with inclusion and exclusion criteria. In this phase, the title and abstract of each paper will be reviewed. If the papers are out of the inclusion criteria, the papers will be completely discarded; however, if the papers fall under the inclusion criteria, the papers will be selected for the next phase.

Phase 4. Second Selection Process Studies selected during phase three will be reviewed thoroughly. This stage will be done to ensure that publication contains the relevant information for the study under review. This approach helps in omitting irrelevant literature.

Search Criteria The search criteria aim at identifying those studies that provided direct empirical evidence about the research questions. In order to narrow down the initial search results, a general set of inclusion and exclusion criteria were established (see Table 1).

Search Execution The search procedure was carried out using the method described above. However, it is worth mentioning that search space was restricted using the relevance ranking approach (e.g., Google's PageRank algorithm) while

Table 1 Summary of inclusion/exclusion criteria

Inclusion criteria	Exclusion criteria
Studies discuss the concept of AIOps	Studies are not relevant to AIOps
Studies that highlighted benefits of AIOps	Studies are inaccessible
Studies that highlighted challenges of AIOps	Studies contained a summary only
Studies written in English language	Studies that are duplicated/repeated

Table 2 Paper selected from full reading and snowballing

	Google Search		Google Scholar		Total
Full reading	15	[18, 34, 44–56]	5	[1, 3, 14, 22, 57]	20
Snowballing	3	[37, 58, 59]	4	[60–63]	7

Table 3 Sources and their relevance to research questions

RQs	Google Scholar		Google Search	
RQ_1 (definition)	2	[22, 57]	4	[18, 44, 45, 58]
RQ_2 (benefits)	6	[1, 3, 14, 60, 61, 63]	10	[18, 34, 46–50, 56, 58, 59]
RQ_3 (challenges)	2	[22, 62]	6	[37, 51–55]

using Google Search. In this case, the above search string was applied to Google Search and returned 1,140,000 results. However, after observation, it is found that only the first few pages were relevant for the study. Therefore, in this work, authors adopted an approach to proceed further only if needed as proposed in [43]. In other words, $(n + 1)$th page was checked only if the result on n^{th} page was found relevant. In Google Scholar, the same search string was applied and returned 435 results. In this case, all papers retrieved from Google Scholar were reviewed by researchers.

Figure 1 shows a summary of the search results. First, the search strings were applied in Google Scholar and Google Search. The initial results include 1,140,435 results (Google Scholar returned 435 results and Google Search returned 1,140,000). In the first phase, 1,140,387 results were excluded after reviewing the title, keywords, and abstract, and this resulted in 56 articles, in Phase 2. All the duplicate papers were removed, which left 54 papers. Then, those papers were checked with inclusion/exclusion criteria again, and the total papers were 31. Afterward, full texts were analyzed, and 20 studies were selected. Finally, a snowballing approach was carried out, and seven papers were selected. Table 2 shows the papers selected from full reading (20) and snowballing (7) approaches.

All 27 papers were sorted in a reference manager tool, namely, Zotero. To ensure the inclusion of all relevant papers, forward and backward snowballing approaches were used as recommended by MLR guidelines, on the set of sources already in the pool. Forward snowballing is identifying articles that have cited the articles found in the search, and backward snowballing is identifying articles from the reference lists.

All the selected sources were used to answer the three research questions listed in Sect. 3.1. Table 3 presents the search results according to the research questions and the search engines. The first column presents the RQs, while the second and third columns present the number of papers collected from Google Scholar and Google Search, respectively, to answer the specific research question. Based on the select studies, it is observed that 8 out of 27 (30%) sources are related to challenges, while 16 (60%) are related to benefits. Therefore, the selected sources reported more challenges than benefits in the AIOps scope.

4 Results

In the following sections, the results of the MLR with regard to the RQs, limitation, and trends of the study are presented. The results of the search process are analyzed taking into account the selected sources retrieved from both the gray and formally published literature and the RQs formulated.

4.1 Trends

As mentioned above, a total of 27 sources are included in this MLR (see the bibliographic details in Appendix A). To visually see the growth of the field (AIOPs), the authors report next a summary on the trends in the final pool of sources, based on two aspects: number of sources by source type and number of sources per year (growth of attention in this area) by literature type (formally published versus gray literature).

Figure 2 shows the number of sources by source type in each of the two categories: formally published versus gray literature. In the formally published literature category, there were six conference papers followed by two journal papers and one book chapter. In the gray literature category, there were ten blog posts and eight web pages. It is not surprising that the gray literature in this area has surpassed the formally published literature due to the contextual fact that the AIOps term and discussions have their origins in industry.

Figure 3 shows the cumulative number of sources per year. As one can see, sources in both literature categories have been on a steady increasing trend from

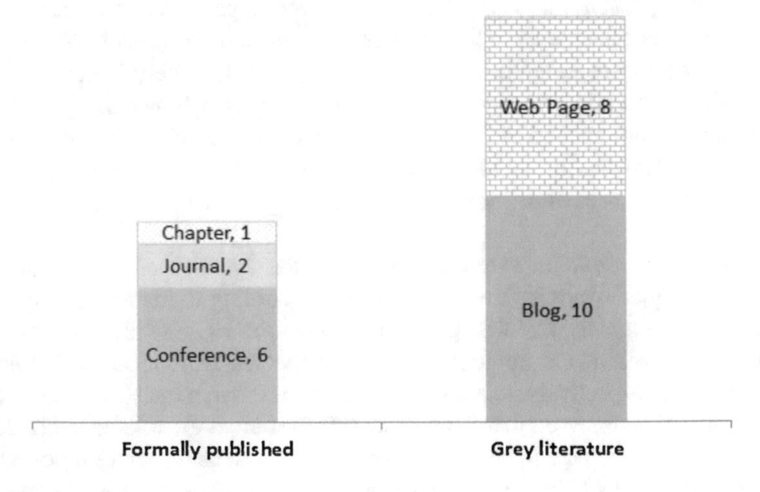

Fig. 2 Number of sources by year

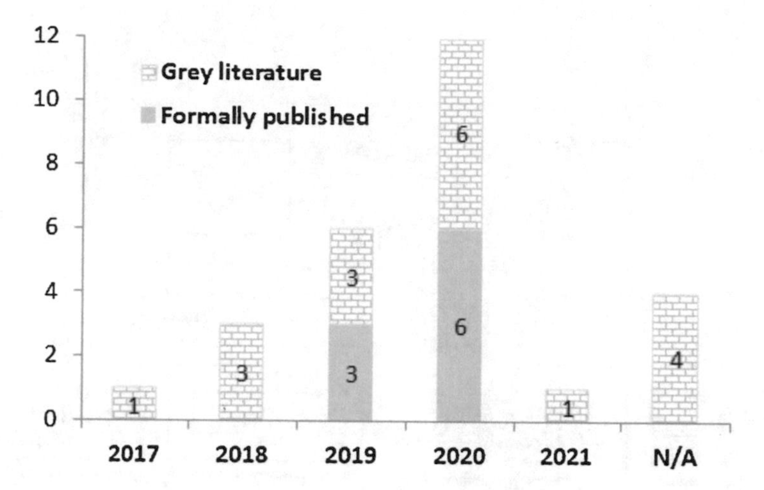

Fig. 3 Cumulative number of sources by year. N/A no available

Fig. 4 Overview of topics shown by the word cloud of all source titles

2017 to 2020. However, there is a group of web pages (4 out of 27 sources) in which their year of publication is not available (N/A), while there is one web page in 2021. The formally published literature in this area seems to start around year 2019, denoting the increasing attention of scholars on this important topic.

Before analyzing the paper, the authors got an overview of the topics covered in the sources according to their titles. Word clouds are a suitable tool for this purpose. Figure 4 shows a word cloud of all sources (the authors used the online tool https://www.wordclouds.com/).

Finally, Fig. 5 shows an overview of the results as observed in the selected studies. Businesses are turning to big data, ML, or artificial intelligence for IT operations (AIOps). As one can see, there are a set of benefits (see Sect. 4.3) and challenges (see Sect. 4.4) related to AIOps. In fact, previous research identified

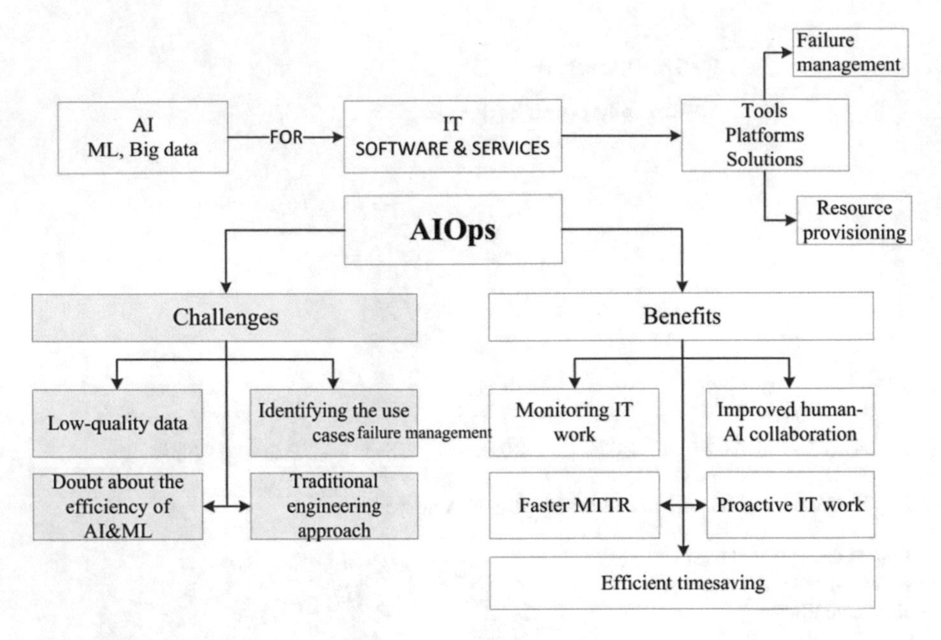

Fig. 5 Overview of the results as observed in the selected studies

AIOps contributions in failure management and resource provisioning [3]. The first one is related to how to deal with undesired behavior in the delivery of IT services, while the second one is about allocation of energetic, computational, storage, and time resources for the optimal delivery of IT services.

4.2 *Definition of AIOps (RQ₁)*

Although there is still not a generally accepted definition of AIOps, in order to answer this question, six papers were selected of which two are scientific literature and the remaining four are gray literature. The scientific literature [22] highlights that AIOps seeks to empower software and service engineers to use AI and ML techniques to develop software and services that are easy to handle in an accessible and productive way. On the other hand, the gray literature defined AIOps as a new approach for automating and enhancing IT operations through ML and analytics in order to identify and respond to IT operational issues in real-time [18, 45, 58]. According to [57], a leading company in Cambridge Massachusetts named Forrester has defined AIOps as follows:

> AIOps primarily focuses on applying machine learning algorithms to create self-learning—and potentially self-healing—applications and infrastructure. A key to analytics, especially predictive analytics, is knowing what insights you're after.

Similarly, Techopedia dictionary defines [44]:

AIOps is a methodology that is on the frontier of enterprise IT operations. AIOps automates various aspects of IT and utilizes the power of artificial intelligence to create self-learning programs that help revolutionize IT services.

4.3 *Benefits of AIOps (RQ$_2$)*

AIOps is relatively young and far from a mature technology, even so, it already has reported some potential benefits. To answer this question, 16 papers were selected: 6 are scientific literature, while the remaining 10 are gray literature. The right side of Fig. 5 shows the identified benefits grouped in five categories: (i) monitoring IT work, (ii) efficient time saving, (iii) improved human-AI collaboration, (iv) proactive IT work, and (v) faster MTTR.

Monitoring IT Work AIOps solutions monitor and analyze the activities performed in an IT environment (both hardware and software), e.g., processor use, application response times, API usage statistics, and memory loads [18, 61]. These analytics and ML capabilities allow AIOps to perform powerful root cause analysis that speeds up troubleshooting and solution to difficult and unusual problems [47, 59], e.g., if the workload traffic exceeded a normal threshold by a certain percentage, the AIOps platform could add resources to the workload or migrate it to another system or environment much like a human admin does [56].

Efficient Time Saving When AIOps platforms are set up properly, the time and effort of IT professionals can be spent on more productive tasks in their jobs [1, 34, 47], meaning less time to spend with routine request and system monitoring every day [14, 46]. Therefore, IT teams can work on innovative tasks to add value to the business [50, 61].

Improved Human-AI Collaboration Collaboration and workflow activities between IT teams and other business units can be enhanced by AIOps [14, 49]. Given that AIOps learns from the input data, they can automate the process requiring less human effort. In this sense, customized reports and dashboards can help teams to understand their tasks and requirements faster and better and to communicate with others without having to understand the complexity of the various areas in detail [18, 60]. Moreover, the IT operations team can focus on tasks with greater strategic value to the business [60]. As an example, if AI teams get specific alerts to meet the service level threshold instead of being loaded with an alert from the various environment, IT teams can respond to them more quickly and can possibly stop the slowdown and outages of the services with less effort [18].

Proactive IT Work AIOps reduces the operational burden of IT systems and facilities with constructive actionable dynamic insight by utilizing big data, machine learning, and other advanced analytics technologies to boost IT operations [47, 58]. This means that AIOps platforms can provide predictive warnings that allow potential issues to be solved by IT teams before they lead to slowdowns or outages. In fact, a survey from 6000 global IT leaders about AIOps revealed that 74% of the IT professionals want to use proactive monitoring and analytics tools [48]. However, 42% of them are still using monitoring and analytic tools reactively to detect and fix technological challenges and issues.

Faster MTTR MTTR is the average time taken to resolve an outage and restore service to end-users. AIOps assists the IT operators in finding the root causes and assists in finding the solutions quicker and more effectively than humanly possible [3]. Infrastructure failure must be addressed at ever-increasing speeds. According to [18], it saves millions of dollars by avoiding direct (fines, opportunity costs) and indirect (customer dissatisfaction and lost reference) costs in IT operation. In fact, one study highlights that MTTR can be reduced from 60 minutes to 30 seconds with the help of AIOps [63].

4.4 Challenges of AIOps (RQ$_3$)

Despite that AIOps shows real promise as a path to success, it upholds some challenges from both technical and nontechnical perspectives [22]. To understand the challenges, a total of eight papers were identified: two papers are scientific literature, while the remaining six papers are gray literature. The left side of Fig. 5 shows the identified challenges grouped into four categories: (i) doubt about the efficiency of AI and ML, (ii) low-quality data, (iii) identifying the use cases, and (iv) traditional engineering approaches.

Doubt About the Efficiency of AI and ML AIOps solutions' basic approach is to learn from experience to predict the future and to recognize trends from huge volumes of data [22]. However, IT professionals who are already working in the field for a while are questioning the efficiency of analytics and ML, even after realizing the need for digital transformation [53, 62]. One possible explanation is their previous experience on piloted projects or attempted analytic projects in-house or with other suppliers which resulted in failed or mixed responses [22]. Therefore, it seems that businesses require more time to develop trust in the validity and reliability of recommendations from AIOps [37].

Low-Quality Data The performance of the AIOps highly depends on the quality of the data [53]. While major cloud providers capture terabytes and even petabytes of telemetry data every day/month today, there is still a shortage of representative and high-quality data for developing AIOps solutions [22]. It is simply becoming too complex for manual reporting and analysis. In this scenario, current issues are noisy data, irregular or inadequate reporting frequencies, and even inconsistent naming convention [51, 53]. Besides, essential pieces of information are "unstructured" types of data presenting poor data quality [53]. Therefore, a constant improvement of data quality and quantity is essential, taking into account that AIOps solutions are based on data [22].

Identifying the Use Cases Use cases in the AIOps is the process of analyzing and identifying the challenges and opportunities across the IT operation environment [51], in addition to building the models to solve these problems and monitoring the performance of the developed model [55]. Companies believe using AI- and ML-related features will increase the efficiency of the current development within the organization [52]. However, without identifying the underlying issue, AIOps implementation might not be effective [51], as AIOps solutions require analytical thought and adequate comprehension of the whole problem space such as market benefit and constraints, development models, and considerations of system and process integration [22]. Therefore, the organization should start examining underlying systems, applications, and processes from the top level and decide the integration of AIOps to have the greatest leverage [51].

Traditional Engineering Approach Successful AIOps implementation requires significant engineering efforts [22]. As it is relatively young and far from mature technology, only limited AIOps engineers are available [22]. Therefore, instead of focusing on building new AIOps initiative, reshaping the existing approach and processes in the organizations is important for the new realities of digital business [54, 55]. These works indicate that traditional approaches do not work in dynamic, elastic environments. However, ideal practice/principles/design patterns are yet to be established in the industry [53].

4.5 Limitations and Potential Threat to Validity

In order to make sure that this review is repeatable, the systematic approach is described in Sect. 3.2. However, despite that search engines, search terms, and inclusion/exclusion criteria are carefully defined and reported, there are some limitations in the selection process that can lead to incomplete set of primary studies.

As a single term has been used in the search string (AIOps), the main threat to the validity of this study is that the literature regarding AIOps is still scarce. A significant part of the available information about AIOps comes from informal publication channels as blogs. In order to mitigate risk of finding all relevant studies, the authors included publications that apparently are not peer reviewed but that they consider being of high enough quality or that have already been cited by peer-reviewed publications.

For controlling threats due to search engines, the authors have included an academic database "Google Scholar" and a general web search database "Google Search." Moreover, a snowballing process of the selected studies was done to ensure that other relevant studies had been included. While this introduces a subjective quality assessment step that has the risk of being biased, it gives the opportunity to provide a definition of AIOps according to how the term is currently being used by scholars and practitioners.

In addition, applying inclusion/exclusion criteria can suffer from researchers' bias. To reduce researcher biases, two authors were actively involved in the search process, while the remaining author supervised all the process. The authors also limited themselves to publications written in English so that relevant studies in other languages are missed out. Therefore, although the authors recognized that additional relevant published studies may have been overlooked, they believe that their MLR provides an adequate overview of the field. Finally, it is worth to note that their findings are within the IT field, particularly AIOps. Beyond this field, they had no intention to generalize their results, but they believe that the value of their MLR should not be undermined.

5 Conclusions and Future Work

This study aims to identify the definition of AIOps, the benefits gained from it, and the challenges an organization might face. It is expected that this study will help to achieve a deeper and wider understanding of this field from a practitioner's point of view based on literature.

In this study, a MLR was conducted based on Google Scholar and Google Search. As a result, 27 sources were identified after applying the inclusion and exclusion criteria (see Table 2). Among them, only 9 were academic literature, and the remaining 18 consist of gray literature. The findings reveal that AIOps is defined as an approach for automating and enhancing IT operations through ML and analytics to identify and respond to IT operational issues in real-time. AIOps evolves from the need to track and manage the highly demanding big data and advanced analytics strategies with adequate means. In other words, AIOps can be thought of as continuous integration and deployment (CI/CD) for core IT functions.

Moreover, a set of benefits and challenges related to AIOps were identified. As AIOps is not yet fully developed, it upholds challenges such as IT organizations questioning the efficiency of AI and ML, poor quality of data affecting the results, and lack of engineering effort to think strategically about reshaping the approach, processes, and organizations to account for the new realities of digital business. On the other side, adopting an AIOps solution will not only improve human-AI collaboration, but AIOps will also monitor the IT work, analyze the root cause, and speed up troubleshooting by saving the time of IT teams. In addition, AIOps could foster proactive IT work, improve the MTTR, and provide a solution to difficult and unusual problems in IT operation.

Given the increasing attention that AIOps has gained from practitioners, further research is needed to better understand how AIOps provides human augmentation to enhance human productivity in terms of senses, cognition, and human action. Taking this into account, authors aim at continuing this work by investigating specific aspects on the connection of AIOps with DevOps and DevSecOps environments, including needed competences and pipelines toward the integration of security aspects in the automation picture. A second proposed line of research is the one devoted to investigating aspects on IT governance and IT compliance and their cascading over AIOps tools and processes. Finally, the authors would like to work on the integration of ITSM methods and frameworks (e.g., ITIL) in AIOps-enabled IT service operations.

A.1 Appendix A: List of Sources Included in the MLR

	Reference	Year	Authors	Title	Type
1	[1]	2020	Gulenko, A., Acker, A., Kao, O., Liu, F.	AI-Governance and Levels of Automation for AIOps-Supported System Administration	Conference
2	[3]	2020	Notaro, P., Cardoso, J., Gerndt, M.	A Systematic Mapping Study in AIOps	Conference
3	[14]	2019	Levin, A., Garion, S., Kolodner, E.K., Lorenz, D.H., Barabash, K., Kugler, M., McShane, N.	AIOps for a Cloud Object Storage Service	Conference
4	[18]	2020	IBM Cloud Education	AIOps	Web Page
5	[22]	2019	Dang, Y., Lin, Q., Huang, P.	AIOps: Real-World Challenges and Research Innovations	Conference
6	[34]	2017	Lerner, A	AIOps Platforms	Blog
7	[37]	2019	OpsRamp	The OpsRamp State of AIOps Report	Web Page
8	[44]	Not available	Techopedia	What Is AIOps? – Definition from Techopedia	Web Page
9	[45]	2020	Sagemo, I.	What Is AIOps?	Web Page
10	[46]	2018	Oats, M.	What Is AIOps? The Benefits Explained	Web Page
11	[47]	2020	Oehrlich, E.	What Is AIOps? Benefits and Adoption Considerations	Web Page

(continued)

	Reference	Year	Authors	Title	Type
12	[48]	2018	Jacob, S.	The Rise of AIOps: How Data, Machine Learning, and AI Will Transform Performance Monitoring | AppDynamics	Blog
13	[49]	2019	Mercina, P.	The Benefits of AIOps	Blog
14	[50]	Not available	Moogsoft	What Is AIOps | A Guide to Everything You Need to Know About AIOps	Web Page
15	[51]	2019	Rogers, P.	Four Problems to Avoid in Order to Have a Successful AIOps Integration	Blog
16	[52]	2019	OPTANIX	AIOps Solutions Concerns Considered by IT Leaders	Blog
17	[53]	2018	Paskin, S.	Concerns and Challenges of IT Leaders Considering AIOps Platforms – BMC Blogs	Blog
18	[54]	2020	CloudFabrix	Top 5 Practical Challenges and Considerations with AIOps | Our Latest Blog Posts | CloudFabrix Buzz	Blog
19	[55]	2020	Analytics Insight	AIOps: Understanding the Benefits and Challenges in IT Landscape	Blog
20	[56]	Not available	Bigelow, S. J.	What Is AIOps (Artificial Intelligence for IT Operations)? – Definition from WhatIs.com	Web Page
21	[57]	2019	Masood, A., Hashmi, A.	AIOps: Predictive Analytics and Machine Learning in Operations	Chapter
22	[58]	Not available	AISERA	AIOps Platforms: A Guide to What You Should Know About AIOps	Blog
23	[59]	2021	Sacolick, I.	What Is the AI in AIOps?	Blog

(continued)

	Reference	Year	Authors	Title	Type
24	[60]	2020	Banica, L., Polychronidou, P., Stefan, C., Hagiu, A.	Empowering IT Operations Through Artificial Intelligence – A New Business Perspective	Journal
25	[61]	2020	Gheorghiță, A.C., Petre, I.	Securely Driving IoT by Integrating AIOps and Blockchain	Journal
26	[62]	2020	Kostadinov, G., Atanasova, T., Petrov, P.	Reducing the Number of Incidents in Converged IT Infrastructure Using Correlation Approach	Conference
27	[63]	2020	Shen, S., Zhang, J., Huang, D., Xiao, J.	Evolving from Traditional Systems to AIOps: Design, Implementation, and Measurements	Conference

References

1. Gulenko A, Acker A, Kao O, Liu F (2020) AI-governance and levels of automation for AIOps-supported system Administration. In: 2020 29th International conference on computer communications and networks (ICCCN). IEEE, pp 1–6
2. Astakhova L, Medvedev I (2020) The software application for increasing the awareness of industrial enterprise workers on information security of significant objects of critical information infrastructure. In: 2020 Global smart industry conference (GloSIC). IEEE, pp 121–126
3. Notaro P, Cardoso J, Gerndt M (2020) A systematic mapping study in AIOps. ArXiv Prepr. ArXiv201209108
4. Jäntti M, Cater-Steel A (2017) Proactive management of IT operations to improve IT services. JISTEM-J Inf Syst Technol Manag 14:191–218
5. Galup SD, Dattero R, Quan JJ, Conger S (2009) An overview of IT service management. Commun ACM 52:124–127
6. Iden J, Eikebrokk TR (2013) Implementing IT service management: a systematic literature review. Int J Inf Manag 33:512–523
7. Heikkinen S, Jäntti M (2019) Studying continual service improvement and monitoring the quality of ITSM. In: Piattini M, Rupino da Cunha P, García Rodríguez de Guzmán I, Pérez-Castillo R (eds) Quality of information and communications technology. Springer International Publishing, Cham, pp 193–206. https://doi.org/10.1007/978-3-030-29238-6_14
8. Dávila A, Janampa R, Angeleri P, Melendez K (2019) ITSM model for very small organisation: an empirical validation. IET Softw 14:138–144. https://doi.org/10.1049/iet-sen.2019.0034
9. Lucio-Nieto T, Gonzalez-Bañales DL (2021) Implementation of a service management office into a world food company in Latin America. Int J Inf Technol Syst Approach IJITSA 14:116–135

10. Winkler TJ, Wulf J (2019) Effectiveness of IT service management capability: value co-creation and value facilitation mechanisms. J Manag Inf Syst 36:639–675. https://doi.org/10.1080/07421222.2019.1599513
11. Orta E, Ruiz M, Calderón A, Hurtado N (2017) Gamification for improving IT service incident management. In: Mas A, Mesquida A, O'Connor RV, Rout T, Dorling A (eds) Software process improvement and capability determination. Springer International Publishing, Cham, pp 371–383. https://doi.org/10.1007/978-3-319-67383-7_27
12. Marrone M, Hammerle M (2017) Relevant research areas in IT service management: an examination of academic and practitioner literatures. Commun Assoc Inf Syst 41. https://doi.org/10.17705/1CAIS.04123
13. Larrucea X, Santamaria I, Colomo-Palacios R, Ebert C (2018) Microservices. IEEE Softw 35:96–100. https://doi.org/10.1109/MS.2018.2141030
14. Levin A, Garion S, Kolodner EK, Lorenz DH, Barabash K, Kugler M, McShane N 2019 AIOps for a cloud object storage Service. In: 2019 IEEE international congress on big data (BigDataCongress). IEEE, pp 165–169
15. Prasad P, Rich C (2018) Market guide for AIOps platforms. Retrieved March 12, 2020
16. Wang H., Zhang H (2020) AIOPS prediction for hard drive failures based on stacking ensemble model. In: 2020 10th Annual computing and communication workshop and conference (CCWC). IEEE, pp 0417–0423
17. Goldberg D, Shan Y (2015) The importance of features for statistical anomaly detection. In: 7th USENIX workshop on hot topics in cloud computing (HotCloud 15)
18. IBM Cloud Education. AIOps, https://www.ibm.com/cloud/learn/aiops. Last accessed 05 Feb 2021
19. Liu X, Tong Y, Xu A, Akkiraju R (2020) Using language models to pre-train features for optimizing information technology operations management tasks
20. Dařena F, Gotter F (2021) Technological development and its effect on IT operations cost and environmental impact. Int J Sustain Eng 0:1–12. https://doi.org/10.1080/19397038.2020.1862342
21. Casado-Lumbreras C, Colomo-Palacios R, Gomez-Berbis JM, Garcia-Crespo A (2009) Mentoring programmes: a study of the Spanish software industry. Int J Learn Intellect Cap 6:293–302
22. Dang Y, Lin Q, Huang P (2019) AIOps: real-world challenges and research innovations. In: 2019 IEEE/ACM 41st international conference on software engineering: companion proceedings (ICSE-Companion). IEEE, pp 4–5
23. Sen A (2021) DevOps, DevSecOps, AIOPS-paradigms to IT operations. In: Singh PK, Noor A, Kolekar MH, Tanwar S, Bhatnagar RK, Khanna S (eds) Evolving technologies for computing, communication and smart world. Springer, Singapore, pp 211–221. https://doi.org/10.1007/978-981-15-7804-5_16
24. Nilsson NJ (2009) The quest for artificial intelligence. Cambridge University Press, Cambridge
25. Long D, Magerko B (2020) What is AI literacy? Competencies and design considerations. In: Proceedings of the 2020 CHI conference on human factors in computing systems, pp 1–16
26. Rajaraman V (2014) John McCarthy—father of artificial intelligence. Resonance 19:198–207
27. Touretzky D, Gardner-McCune C., Martin F, Seehorn D (2019) Envisioning AI for k-12: what should every child know about AI? In: Proceedings of the AAAI conference on artificial intelligence, pp 9795–9799
28. Martínez-Plumed F, Tolan S, Pesole A, Hernández-Orallo J, Fernández-Macías E, Gómez E (2020) Does AI qualify for the job? A bidirectional model mapping labour and AI intensities. In: Proceedings of the AAAI/ACM conference on AI, ethics, and society, pp 94–100
29. Martinelli A, Mina A, Moggi M (2021) The enabling technologies of industry 4.0: examining the seeds of the fourth industrial revolution. Ind Corp Change 30:161–188
30. Fanti L, Guarascio D, Moggi M (2020) The development of AI and its impact on business models, organization and work. Laboratory of Economics and Management (LEM), Sant'Anna School of Advanced ...

31. Brynjolfsson E, Mitchell T, Rock D (2018) What can machines learn, and what does it mean for occupations and the economy? In: AEA papers and proceedings, pp 43–47
32. Bughin J, Seong J, Manyika J, Chui M, Joshi R (2018) Notes from the AI frontier: modeling the impact of AI on the world economy. McKinsey Global Institute
33. IDC Corporate USA (2020) International data corporation spending guide. https://www.idc.com/getdoc.jsp?containerId=prUS46794720. Last accessed 23 Jan 2021
34. Lerner A (2017) AIOps platforms. Gart. Blog August
35. Bogatinovski J, Nedelkoski S, Acker A, Schmidt F, Wittkopp T, Becker S, Cardoso J, Kao O (2021) Artificial intelligence for IT operations (AIOPS) workshop white paper. ArXiv Prepr. ArXiv210106054
36. Li Y, Jiang ZM, Li H, Hassan AE, He C, Huang R, Zeng Z, Wang M, Chen P (2020) Predicting node failures in an ultra-large-scale cloud computing platform: an AIOps solution. ACM Trans Softw Eng Methodol TOSEM 29:1–24
37. OpsRamp (2019) The-OpsRamp-State-of-AIOps-Report.pdf. https://www.opsramp.com/wp-content/uploads/2019/04/The-OpsRamp-State-of-AIOps-Report.pdf. Last Accessed 6 Feb 2021
38. Simic B (2020) Strategies of top performing organizations in deploying AIOps. https://www.dej.cognanta.com/2020/05/04/the-aiops-maturity-research-study-key-findings/. Last accessed 09 Feb 2021
39. Garousi V, Felderer M, Mäntylä MV (2019) Guidelines for including grey literature and conducting multivocal literature reviews in software engineering. Inf Softw Technol 106:101–121
40. Pereira-Vale A, Fernandez EB, Monge R, Astudillo H, Márquez G (2021) Security in microservice-based systems: a multivocal literature review. Comput Secur 103:102200. https://doi.org/10.1016/j.cose.2021.102200
41. Scheuner J, Leitner P (2020) Function-as-a-service performance evaluation: A multivocal literature review. J Syst Softw 170:110708. https://doi.org/10.1016/j.jss.2020.110708
42. Saltan A, Smolander K (2021) Bridging the state-of-the-art and the state-of-the-practice of SaaS pricing: a multivocal literature review. Inf Softw Technol 133:106510. https://doi.org/10.1016/j.infsof.2021.106510
43. Garousi V, Mäntylä MV (2016) A systematic literature review of literature reviews in software testing. Inf Softw Technol 80:195–216
44. Techopedia (2019) What is AIOps? – Definition from Techopedia. http://www.techopedia.com/definition/33321/aiops. Last accessed 06 Feb 2021
45. Sagemo I (2020) What is AIOps? https://www.aims.ai/resources/what-is-aiops. Last accessed 07 Feb 2021
46. Oats M (2018) What is AIOps? The benefits explained. https://www.intellimagic.com/resources/aiops-benefits-explained/. Last accessed 05 Feb 2021
47. Oehrlich E (2020) What is AIOps? Benefits and adoption considerations. https://enterprisersproject.com/article/2020/3/what-is-aiops. Last accessed 05 Feb 2021
48. Jacob S (2018) The rise of AIOps: how data, machine learning, and AI will transform performance monitoring | AppDynamics. https://www.appdynamics.com/blog/news/aiops-platforms-transform-performance-monitoring/. Last accessed 06 Feb 2021
49. Mercina P (2019) The benefits of AIOps. https://www.parkplacetechnologies.com/blog/the-benefits-of-aiops/. Last accessed 05 Feb 2021
50. Moogsoft M (2021) What is AIOps | A guide to everything you need to know about AIOps. https://www.moogsoft.com/resources/aiops/guide/everything-aiops/. Last accessed 09 Feb 2021
51. Intelligent CA (2019) Four problems to avoid in order to have a successful AIOps integration. https://www.intelligentcio.com/africa/2019/03/12/four-problems-to-avoid-in-order-to-have-a-successful-aiops-integration/. Last accessed 10 Feb 2021
52. OPTANIX (2019) AIOps solutions concerns considered by IT leaders. https://www.optanix.com/aiops-solutions-challenges-and-concerns/. Last accessed 10 Feb 2021

53. Paskin S (2018) Concerns and challenges of IT leaders considering AIOps platforms – BMC blogs. https://www.bmc.com/blogs/concerns-and-challenges-of-it-leaders-considering-aiops-platforms/. Last accessed 10 Feb 2021
54. CloudFabrix (2020) Top 5 practical challenges & considerations with AIOps | Our latest blog posts | CloudFabrix Buzz. https://cloudfabrix.com/blog/aiops/top-5-practical-challenges-considerations-with-aiops/. last Accessed 10 Feb 2021
55. Analytics Insight (2020) AIOps: understanding the benefits and challenges in IT landscape. https://www.analyticsinsight.net/aiops-understanding-benefits-challenges-landscape/. Last accessed 10 Feb 2021
56. Bigelow SJ (2019) What is AIOps (artificial intelligence for IT operations)? Definition from WhatIs.com, https://searchitoperations.techtarget.com/definition/AIOps. Last accessed 21 Feb 2021
57. Masood A, Hashmi A (2019) AIOps: predictive analytics & machine learning in operations. In: Cognitive computing recipes. Springer, pp 359–382
58. AISERA (2020) AIOps Platforms: A guide to what you should know about AIOps, https://aisera.com/blog/aiops-platforms-a-guide-to-aiops/. Last accessed 06 Feb 2021
59. Sacolick I (2021) What is the AI in AIops? https://www.infoworld.com/article/3603953/what-is-the-ai-in-aiops.html. Last accessed 13 Feb 2021
60. Banica L, Polychronidou P, Stefan C, Hagiu A (2020) Empowering IT operations through artificial intelligence–a new business perspective. KnE Soc Sci 412–425
61. Gheorghiță AC, Petre I (2020) Securely driving IoT by integrating AIOps and blockchain. Romanian Cyber Secur J 2
62. Kostadinov G, Atanasova T, Petrov P (2020) Reducing the number of incidents in converged IT infrastructure using correlation approach. In: 2020 International conference automatics and informatics (ICAI). IEEE, pp 1–4
63. Shen S, Zhang J, Huang D, Xiao J (2020) Evolving from traditional systems to AIOps: design, implementation and measurements. In: 2020 IEEE international conference on advances in electrical engineering and computer applications (AEECA). IEEE, pp 276–280

Deep Learning-Based Facial Recognition on Hybrid Architecture for Financial Services

Oscar Granados and Olmer Garcia-Bedoya

1 Introduction

The interaction of big data, blockchain, edge–cloud computing, AI, and other emerging technologies has emerged rapid changes to the financial industry [1–3]. The use of old and new AI tools has grown in different activities such as loan applications, sales and trading, credit reports, compliance processes, identity confirmation, fraud prevention, anti-money laundering compliance, and marketing, to list a few. Several technologies as face recognition have a transformable role in the future of the financial industry because it has been the relevant biometric technique for identity authentication [4]. The biometric data and face recognition algorithms could be linked to client financial data for a new approach to the banking records, credit scores, and payment services [5]. Some studies proposed facial recognition for preventing identity fraud [6], implementing payment transactions [7], providing services [8], and authenticating system access [9] on physical platforms or mobile devices [10, 11]. While some studies propose to include the history of employment or social media to achieve better accuracy and digital identity clients, others argue that these tools affect privacy and could be generated demographic differentials or errors by gender, age, or social and ethnic group [12, 13].

Different scholars have implemented deep learning methods for facial recognition in different ways, especially convolutional neural networks (CNNs), but the high computational costs affect these kinds of proposals and their perfor-

O. Granados (✉)
Department of Economics and International Trade, Universidad Jorge Tadeo Lozano, Bogotá, Colombia
e-mail: oscarm.granadose@utadeo.edu.co

O. Garcia-Bedoya
Department of Industries and Digital Technologies, Bogotá, Colombia
e-mail: olmer.garciab@utadeo.edu.co

S. Misra et al. (eds.), *Artificial Intelligence for Cloud and Edge Computing*, Internet of Things, https://doi.org/10.1007/978-3-030-80821-1_3

mance. In this case, several scholars have proposed architectures to consolidate the performance of CNN frameworks. Some focused on energy efficiency [14], infrastructure efficiency [15], performance optimization and accelerators [16, 17], accelerators with parallel computing based on streaming architecture [18] or parallel computing to optimize the CNN process [19], data processing efficiency with edge–cloud computing [20, 21], distributed computing for data security and privacy [22], wireless networks and communication systems efficiency [22, 23], or custom hardware [24], to list a few.

Our proposal tries to provide some facial anthropometric and aesthetic analysis tools to financial institutions that could help them to consolidate the client information. The result of this tool is to obtain specific facial measures to fight against growing crimes like financial fraud, tax evasion, money laundering, identity fraud, and cybercrime that help authenticating users and recognize the online and offline identity of current and future financial institutions' customers. This framework based on several image sets analyzed in real-time needs an architecture that extracted maximum performance and reliability with computational infrastructure optimization and energy consumption reduction. Because in the input process of the facial recognition framework exists several required steps before processing the image sets that demand resources beyond of traditional information systems of financial institutions. To achieve better performance and efficiency, we integrate to the CNN-based facial recognition framework an edge–cloud computing architecture. Hence, this chapter aims to explore a mechanism for deep learning-based facial recognition for customer identification in financial institutions with scalable deep learning hybrid architecture that facilitates the framework operability and optimizes the infrastructure resources.

We organized the chapter as follows: Sect. 2 evaluated several methods to create a framework to use facial measures as a tool for facial recognition from social media and other image sets in a first subsection and related work about edge–cloud computing. In Sect. 3, we proposed a facial recognition model combining three methods and hybrid architecture model. In Sect. 4, we presented the experimental results and finally Sect. 5 that concluded and identified the future work.

2 Related Work

2.1 Facial Recognition

Different scholars have studied the anthropometry of the head and face with different computational, mathematical, and visualization tools in two or three dimensions as a part of medical metrics [25], clinical diagnosis [26], facial attractiveness [27], geometry structure [28], and cephalometry [29]. Other studies define those measures to apply at different issues like population and ethnicity structure [30–34], face perception [35], and sports [36, 37] among others.

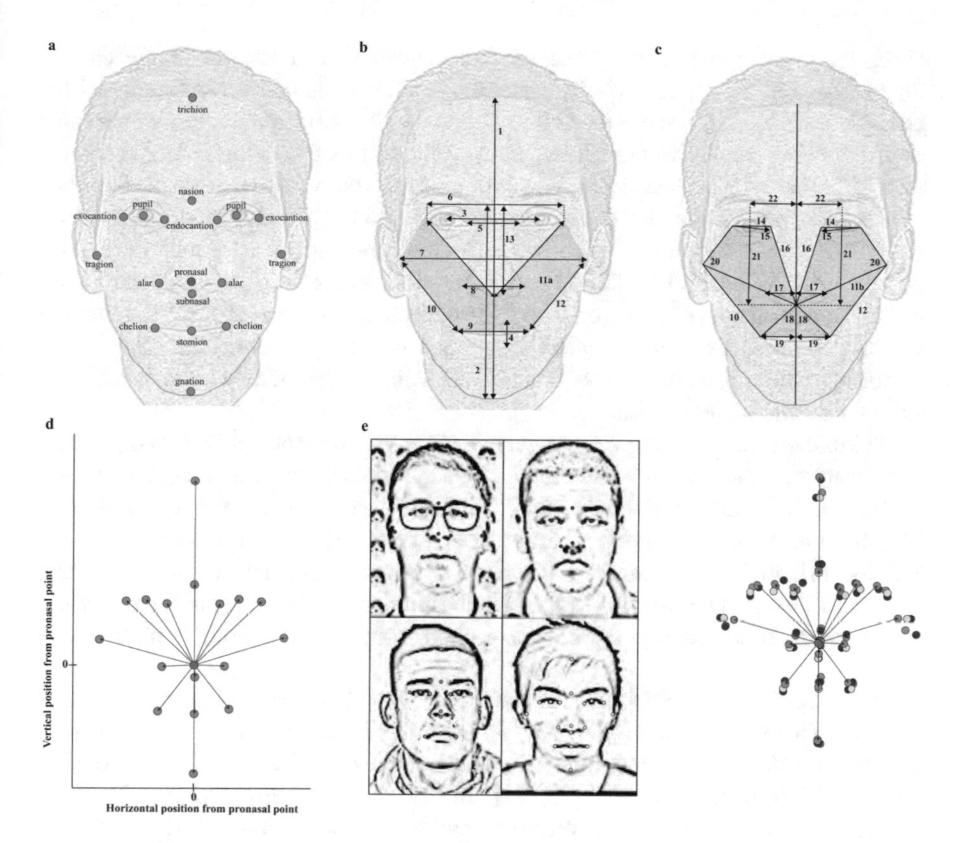

Fig. 1 Facial measures. (**a**) References points. (**b**) Anthropometric parameters of frontal face. 1. Physiognomic face height, 2. average nose height, 3. nose height, 4. buccal mucosa height, 5. internal interocular width, 6. external interocular width, 7. biauricular width, 8. nose width, 9. mouth width, 10. external naso-ocular angle, 11a. senses area (face), 12. senses perimeter, and 13. biocular width. (**c**) Face symmetry, 11b. senses area (face), 14. palpebral width, 15. eye inclination, 16. eye separation angle, 17. pronasal–alagenium distance, 18. naso-buccal angle, 19. stomion–chelion distance, 20. naso-ocular and oto naso-ocular angle, 21. pupil–subnasal height, and 22. pupil–face axis distance. (**d**) Vector representation in a bidimensional space. (**e**) Vector representation in a group of persons with different facial measures and features

The face structure has the reference points (Fig. 1a) to identify the anthropometric parameters (Fig. 1b) and face symmetry (Fig. 1c). Those points are essential to define those measures and differences between people because they create a new set of measures with a vector representation in a bidimensional space (Fig. 1d). However, measures bring about highly micro-variation among population groups, including the similar features groups (Fig. 1e). Additionally, several facial poses or positions affect the landmark detection. To solve this problem, we used facial features around facial key points to complement deep learning methodologies.

It exists a large group of classifiers to use for facial recognition [38–40] that could help us to develop a current or potential client recognition for financial institutions.

First, the local binary pattern (LBP) is a concept that a texture image can be decomposed into a set of essential small units called texture units and modeled the background that could be clutter or low illuminate [41–43]. Second, the Weber local descriptor (inspired by Weber's Law from psychology - LDs) defines that the change of a stimulus will be just noticeable is a constant ratio of the original stimulus. When the change is smaller than this, constant ratio of the original stimulus could recognize it as background noise rather than a valid noise, i.e., consists of two components: differential excitation and orientation, where the differential excitation component is computed based on the ratio between the two terms: one is the relative intensity differences of a current pixel against its neighbors, and the other is the intensity of the current pixel. The differential excitation component extracts the local salient patterns in the input image [44].

Third, the principal component analysis (PCA) is an extracted feature algorithm that examines the dataset collectively, which implies that the image does not evaluate individually, but which extracts a feature based on evaluating the dataset [45, 46]. Fourth, the histogram of oriented gradients (HOGs) counts occurrences of gradient orientation in localized portions of an image with a uniform grid. Namely, this classifier computes image gradients that capture a small number of features from a very large set of potential features, especially contour and silhouette information [47–49].

Other options are Support Vector Machine (SVM) and Relevance Vector Machine (RVM) classifiers [50, 51] or their combination with abovementioned classifiers [46]; also, Convolutional Neural Network (CNN) [52], Generative Adversarial Networks (GANs) [53, 54], and Long Short Term Memory Network (LSTM), this latter method evolved to a combination of CNN and LSTM [55–57]. The supervised convolutional neural network (CNN) models and their variations have been used on image recognition and, more recently, on processing video [52, 58, 59] and real-time decision-making for sustainable development plans [60]. Those models allow processing of variable-length input sequences as well as outputs, which could help institutions to include different sources of images for their facial recognition systems [61]. Several of those works show a standard data process to facial recognition (Fig. 2a) that it complemented with different processes that use a machine learning or other methodologies to recognize facial features (Fig. 2b). Those processes identify the sequence of our proposed model in the Methods section.

2.2 Edge–Cloud Computing

Cloud computing has emerged as a framework with a wide range of virtualized services where data or applications are accessed from any location over the internet [62–65]. Although those services are consolidated as scalable and uninterrupted architecture with a simplified and power structure, they do not guarantee the quality of experience (QoE) and that the orchestration always works. Thus, versatility and

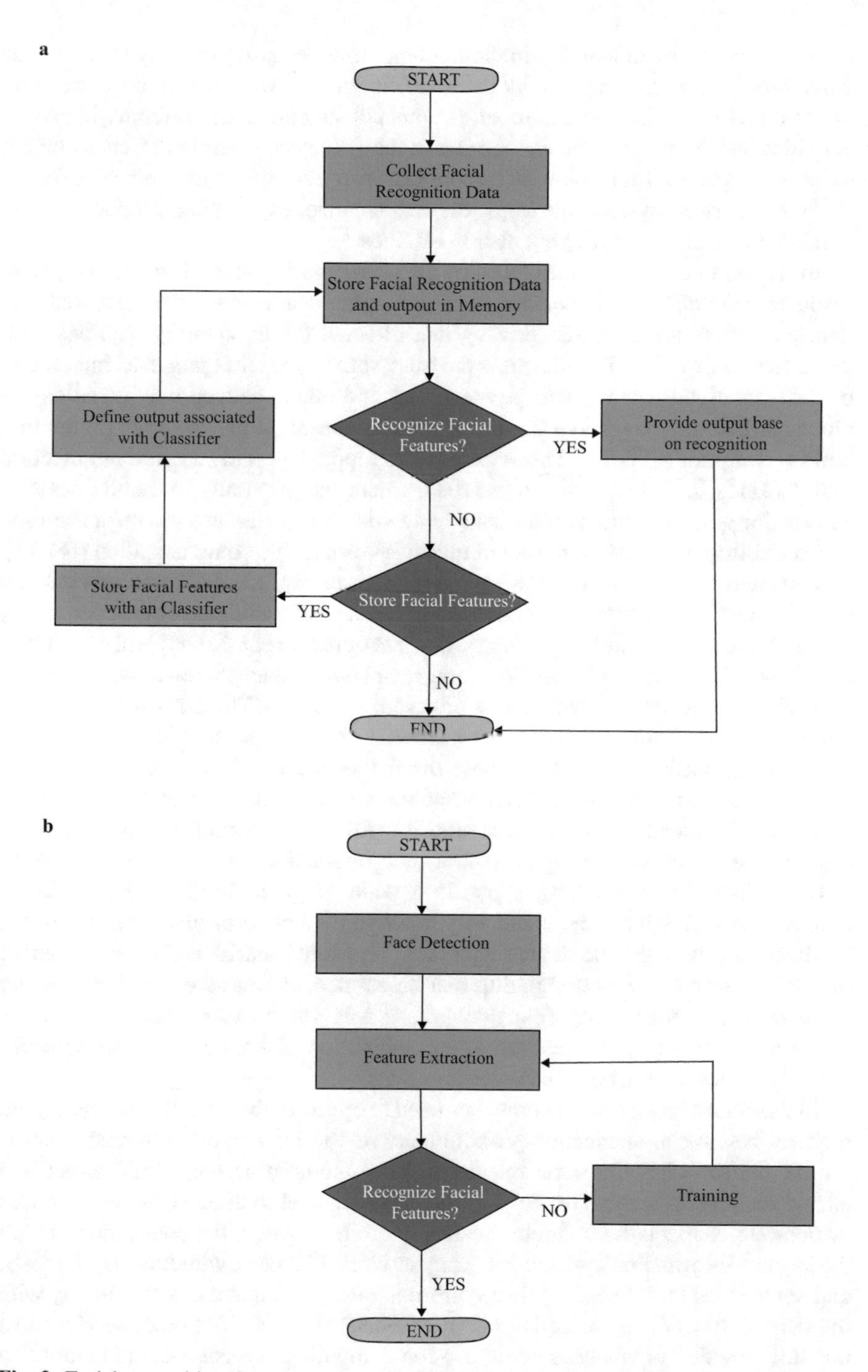

Fig. 2 Facial recognition data process. (**a**) Collect facial recognition data. (**b**) Face recognition system

elasticity are the main features in cloud computing because uses may have diverse variations in a wide range of industries [66]. In this way, the cloud computing architecture is an instrument to allow the consolidation of different industrial activities and business services, because it may connect several artifacts to obtain or provide data and information not only in everyday life ecosystem [67, 68] or in the business ecosystem like logistics, maintenance, exploration, production, i.e., industrial infrastructure [69] but also in services.

In the middle 2010s, some tasks in cloud computing started to demand more resources that affected its performance, and some challenges were presented like data protection, security, and privacy that affected the trustworthy qualities with cloud technology [70]. To robustness the trust, some researches proposed interaction of different clouds (public and private) [71], and others defined new variations of cloud architecture to resolve the infrastructure consumption like edge computing and fog computing [72, 73]. Those variations support especially the new architecture that demands the internet of things (IoT), where the proximity to mobile devices, sensors, or general artifacts with data centers is crucial because the response time and scalability play a relevant role in machine-to-machine communication [68, 74].

Although the two frameworks have a relevant utility, some differences could be better for one or other business activities. In the case of traditional cloud computing, some operations as capture and storage data are optimal (Fig. 3a). Hybrid computing could optimize capture, processing, storage, and output data processes when a group of artifacts is involved and interacted repeatedly (Fig. 3b). Those architectures have simple variations, but the results are different depending on activities.

In several business activities where the IoT or mobile devices are the artifacts to gather data or information, fog computing could accelerate and optimize those processes. Principally, because it creates a distributed environment that results in devices near to users, and supports real-time interaction, i.e., reduces the latency [74]. Besides, fog computing supports a wide range of artifacts that facilitate connections and reduces the mandatory homogeneity of some platforms where the hardware admits only one device type (e.g., registered machines in a supermarket or banking branch). Fog computing can be connected from the cloud computing network to the edge computing network, and in some cases where the process needs high computational resources, a combination of both is a key to optimize the information interactions in business activity.

In financial services, where branches need to optimize the time, the answer period requires better communication. Additionally, if the interaction with customers is online, the time has the same relevance. Thus, some processes could affect that period for several reasons especially, the same channel to attend different services or dedicate channels for a single service, but in both cases, the connection is with the central information system. Banking services like cash withdrawals, deposits, and some investments need a private information system that connects directly with the central network (monolithic system or cloud) (Fig. 3a). But other services that are still provided in branches could use fog computing and edge–cloud computing framework (Fig. 3b), and they could consolidate a similar or equal process in online and branch platforms.

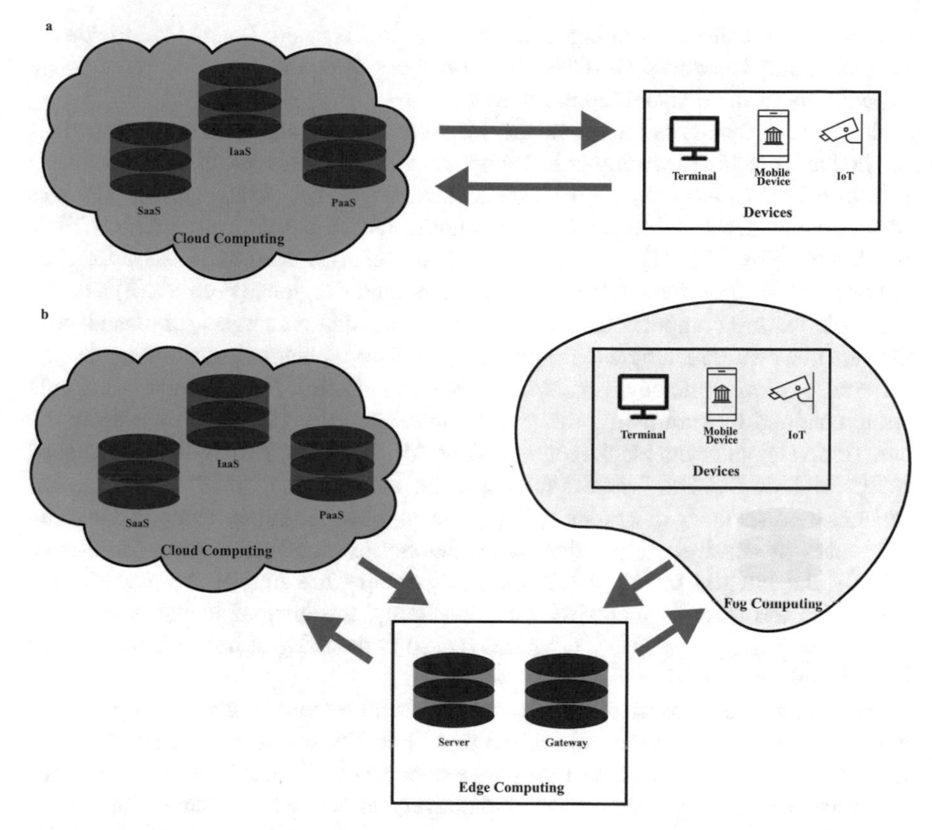

Fig. 3 Edge–cloud computing scenarios. (**a**) Cloud computing. (**b**) Hybrid computing

3 Methods

3.1 Facial Recognition Methods

Facial recognition has to analyze as a dynamic process because people have facial changes during their life and some suspicious agents use those changes and their magnification for fraud and financial crime; namely, the face recognition is a pipeline that consists of four common stages: detect, align, represent, and verify.

The convolutional neural networks have feature learning capabilities, and the combination with LSTM helps to obtain information from image sequences or image in different contexts with several facial expressions, improving the accuracy of recognition. However, in some cases, the quality of pictures or the face position affect the results because faces are not frontal, neutral expression, or standard illumination [75]. We created a framework that tried to reduce those anomalies, especially with a combination of a traditional classifier and a deep learning method. In this case, we did not use the activity recognition as a goal, but as a tool to identify the facial structure combining convolutional layers, i.e., Long-term Recurrent Convolutional Networks (LRCNs) because the sequential dynamics reduces the dependency of fixed visual representation.

In consequence, first, we use the Haar-cascade frontal face detection. This method is a machine learning-based approach where a cascade function (classifier) is trained from a group of positive and negative images, i.e., positive images (images of faces) and negative images (images without faces), which used different kinds of features [76, 77]: (1) the value of a two-rectangle feature is the difference between the sum of the pixels within two rectangular regions (Fig. 4a), (2) a three-rectangle feature computes the sum within two outside rectangles subtracted from the sum in a center rectangle (Fig. 4b), and (3) a four-rectangle feature computes the difference between diagonal pairs of rectangles (Fig. 4c). Each feature is a single value obtained by subtracting the sum of pixels under a white rectangle from the sum of pixels under the black rectangle (Fig. 4d). However, rectangle features need a representation of the image knowing as an integral image to identify relevant features because using this procedure, any rectangular sum can be computed in array references successively depending on the number of rectangle features (Fig. 4e–f); namely, the integral image at location x, y contains the sum of the pixels above and to the left of x, y, inclusive. Mathematically, the integral image is given by $ii(x, y) = \sum_{x' \leq x, y' \leq y} i(x', y')$, where $ii(x, y)$ is the integral image and $i(x, y)$ is the original image [77].

Second, we implemented CNN to capture facial features where the input is the detected face in gray scale and resized to 128×128 pixels or its equivalence in frames or points that result from the above step. The CNN architecture comprises a sequence of layers that start with the input layer, following by a convolutional layer that consists of a kernel of a fixed size to extract features. The next layer rectifies linear units, and the pooling layer (max pooling) is responsible for down-sampling and dimensionality reduction also has a kernel to extract dominant features that are rotational and positional invariant. The flatten layer convert the two-dimensional features into one dimension that we used in the third step. Finally, the FC layer connects each neuron in the input to each neuron in the output. This layer is responsible for computing the score of a particular class, where N denoted the categories to classify [56, 58]. Figure 5 shows the network structure. Other models that we use, including the state-of-the-art (SOTA) deep neural network models as VGG-Face [78], FaceNet [61], OpenFace [79], and DeepFace [80], have different input and output shapes.

Third, we used the flatten layer to combine CNN with LSTM. When we have an image in a vector form using any of the CNN methods explained above, we can compare the similarities to identify images from previous observations.

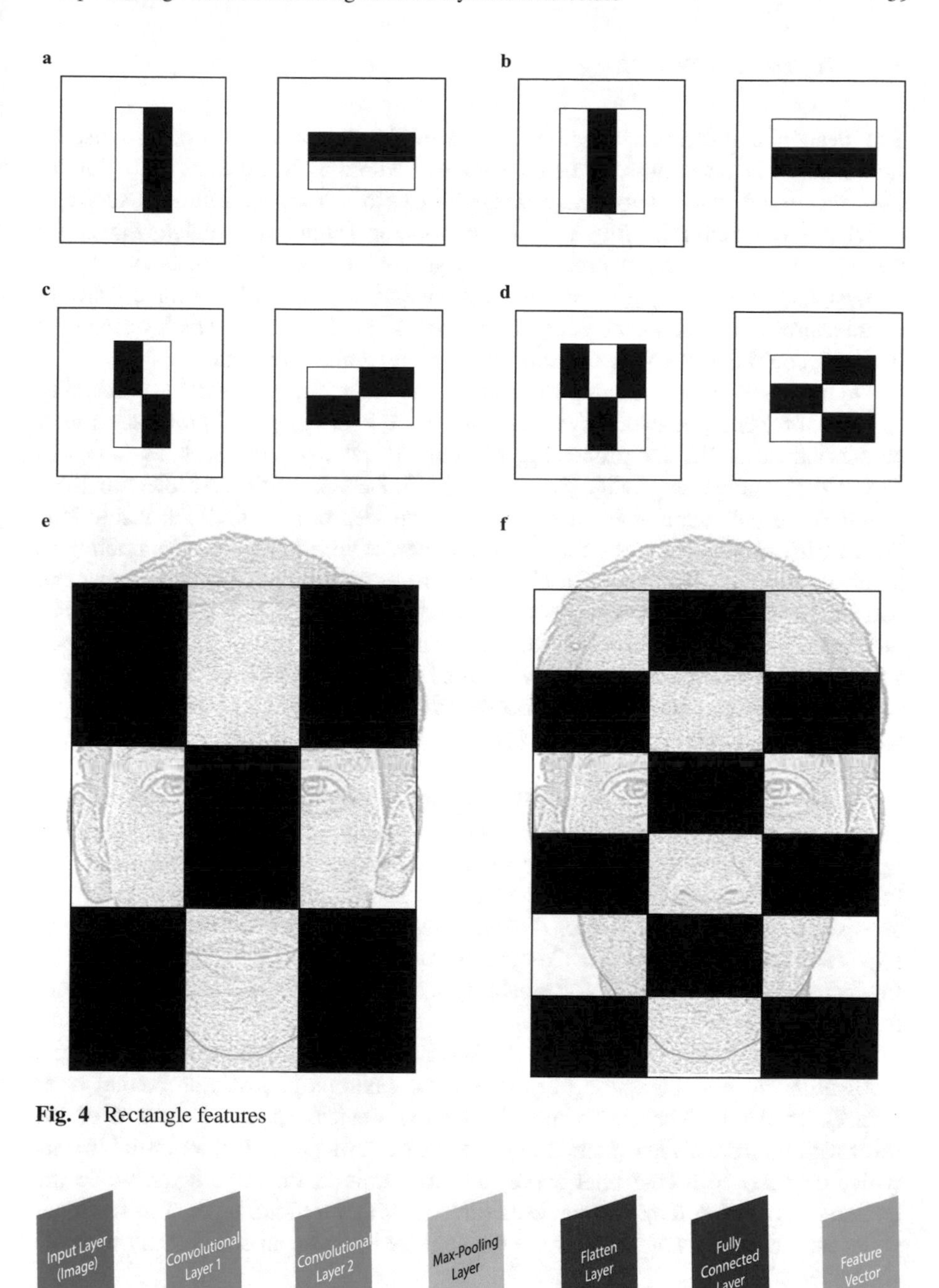

Fig. 4 Rectangle features

Fig. 5 CNN structure

3.2 Hybrid Architecture

The deep neural networks method described in the previous section must be supported by infrastructure on the edge–cloud computing to create an IoT solution. This separation would allow easy integration with financial institutions services, which requires high reliability in data networks for banking transactions that could be compromised by cloud processing images in real time. Those devices must be designed with highly efficient, reliable, secure, and scalable machine learning architectures [81]. The facial recognition process based on CNN, which we showed in Fig. 5, could have two ways in our hybrid architecture proposal.

On the one hand, we propose that in the edge, there must be processing capacity of neural network systems such as GPUs (graphics processor units) or specialized ASIC for processing of deep neural networks such as Google's TPU (tensor processing unit) [82] or NVIDIA Deep Learning Accelerator [83]. Figure 6 presents the architecture of the general idea of how financial institutions through IoT devices or mobile devices can interact with edge–cloud computing to identify customers. This hybrid architecture has the advantage of energy saving and bandwidth saving because we could send an encoded image over the network after the target identification on the image. For example, a high definition image of 720p compressed will require around 200 kb to be sent by the network and many times require several seconds, but if we processed the image in the edge, the model output will be required to send a sparse vector of fewer than 10,000 registers, only when a new face is detected.

Having images' identification at the edge allows working with files and not with databases of models and required encodings. This scheme creates challenges on how the update processes of device firmware perform to ensure that the models' version is compatible in different devices for the face classification process to be consistent. The authors [84] discuss advantages and trials in orchestrating cloud and edge resources in an "Osmotic Computing" paradigm. This paradigm is driven by the increase in resource capacity/capability at the network edge, along with support for data transfer protocols.

On the other hand, we identify the distribution of deep learning-based facial recognition in three computing layers: the fog layer, edge layer, and cloud layer (Fig. 7). In the first layer, we include the IoT devices (e.g., the camera of the information terminal) and mobile devices. These artifacts collect an initial image, which captures in a controller server and transmits to the edge layer, while the banking transaction terminal connects directly with the cloud layer. The fog layer waits for instructions or results from the edge layer that connects directly with the controller server.

The second layer receives the image, and it processes the three first parts of the CNN framework in a first sublayer. This sublayer has a storage capacity and computing resources to execute those parts, i.e., they cached in this sublayer. After that, the second sublayer implements CNN's max pooling that demands a greater computational capacity because it is responsible for down-sampling and

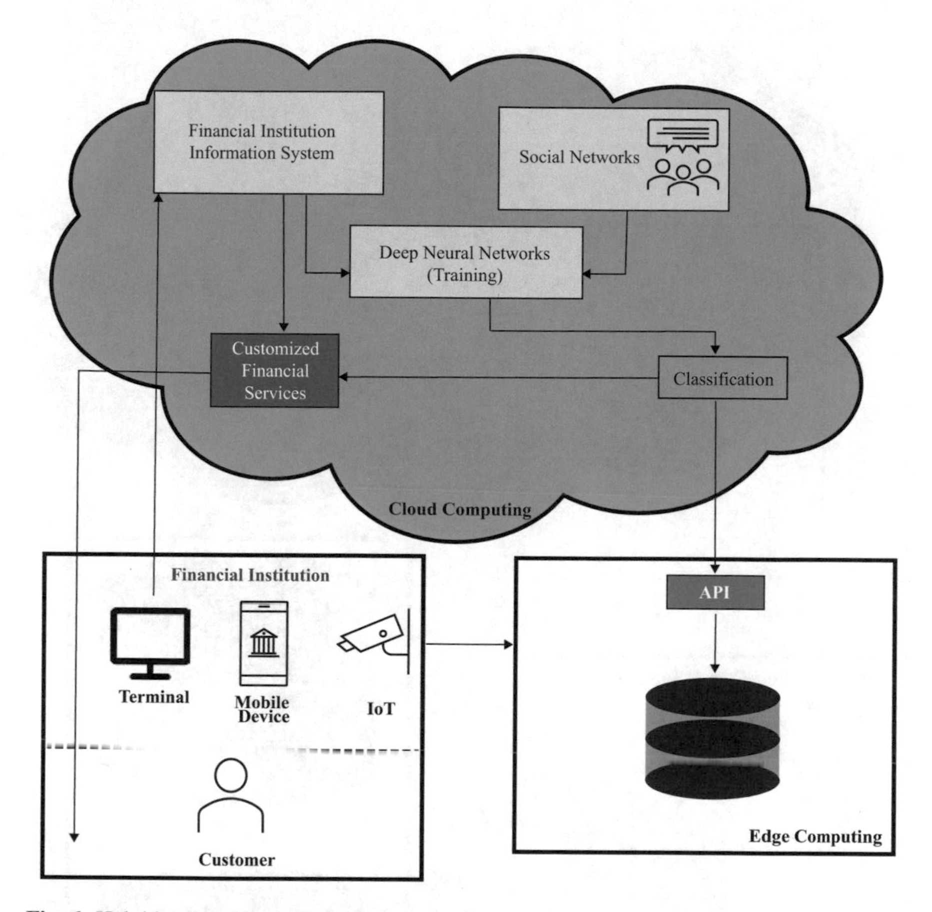

Fig. 6 Hybrid architecture including edge–cloud computing

dimensionality reduction of image. Meanwhile, the image pooling logs are saved for future processes, and after that, this sublayer implements the flatten layer to create a vector that sends it to process in the cloud layer. This sublayer connects with the cloud layer. Additionally, the edge layer can manage other functions that facilitate the communication process, security and privacy protection, or the customized services (second option). These services result during the facial recognition process.

The third layer is responsible for the CNN algorithms training and manages the data from social media. Additionally, this layer has other responsibilities: manages the information system of the financial institution, supports sales and CRM services, and creates advanced analytical tools for optimizing the facial recognition process. Finally, we connected the financial transactions system directly with the cloud layer because it consolidates the security of financial transaction information.

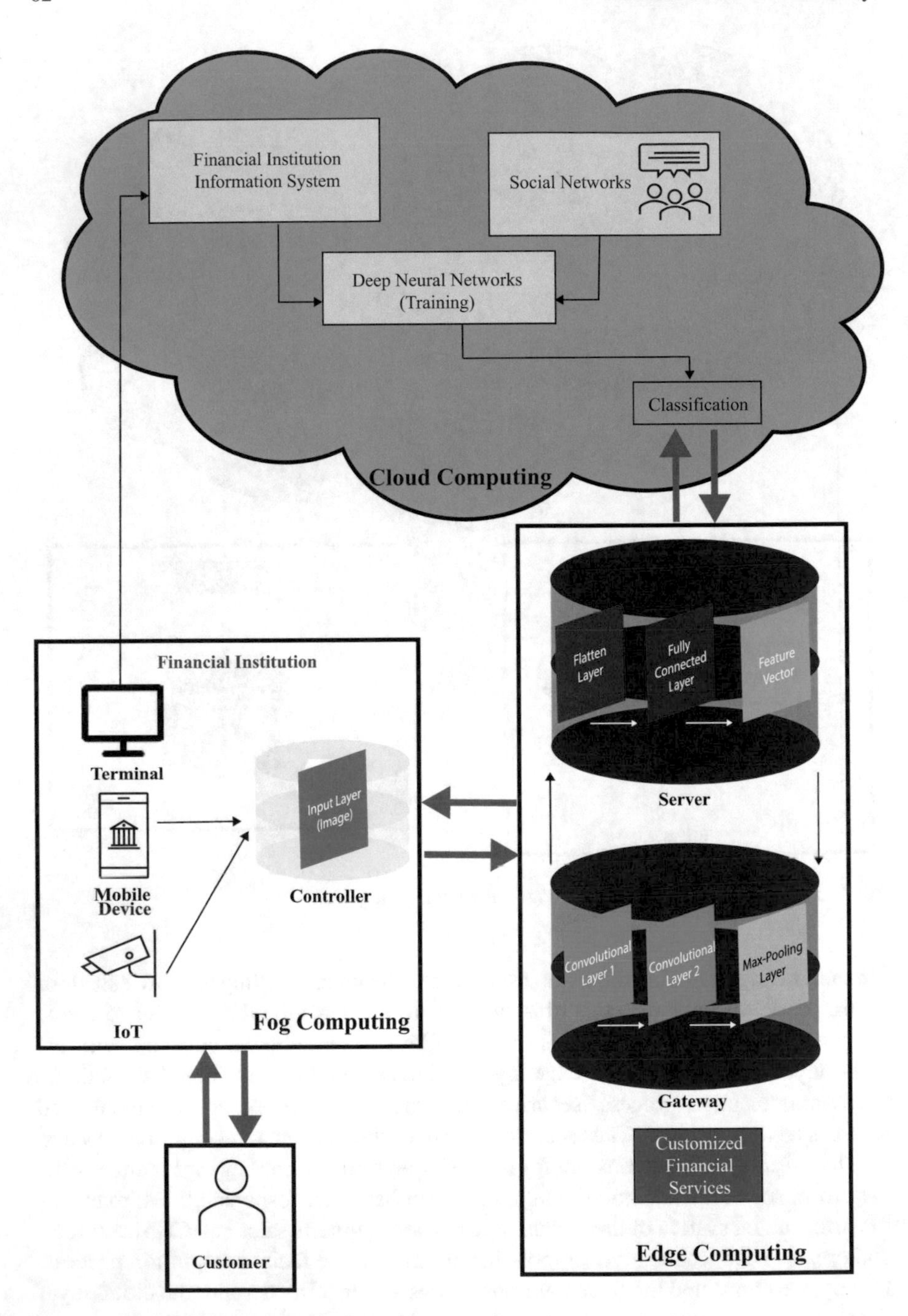

Fig. 7 Hybrid architecture including fog computing and edge–cloud computing

4 Experimental Results

We present an approach using image processing techniques to improve face recognition accuracy based on a combination of different models. In this section, we give the experiment results from the model performance through that combination. The performance of the proposed face recognition method was carried out using a public dataset of persons in different activities, which has almost 2500 pictures recollected from several internet sources. Other open datasets as AffectNet, Cohn-Kanade, and Affectiva-MIT, to list a few, could help us to consolidate the results. However, the data processing with our dataset was enough to detect the face in all images using the Haar-cascade frontal face detection. We used a variation of the Haar-cascade and single shot multibox detector (SSD), knowing as OpenCV and Dlib, that use a histogram of oriented gradients (HOGs) to extract features. In both cases, the methods are fast, but they need robust CNN tools to improve the accuracy (we recommend using GPU-CUDA). However, DeepFace implements this part in a function, which detects the faces. This method can be implemented in face recognition using *face_recognition.api.face_landmarks*, which generate *left_eye* and *right_eye* features.

The results of these experiments are summarized with an example in Fig. 8. As it can be inferred from the figure, the highest rate of recognition is achieved with the full face with the recognition rate of 100% using each classifier. However, the recognition rate starts to drop down with some methodologies and metrics.

To verify the feasibility of our CNN proposal, we evaluate four deep learning models with different metrics and compare their results. However, the most common approach is the Euclidean or cosine distance that helps us to tune the decision of which image of the dataset is more similar to the new image. The diagonal line of the confusion matrix is the ratio of the predicted result to the actual result that reveals that the darker color has to concentrate along this diagonal because this feature confirms the accuracy of the model (Fig. 9).

The VGG-Face that learns on a large-scale database and extracts the activation vector of the fully connected layer in the CNN architecture shows recognition rates with different levels especially, a 100% accuracy rate for Euclidean L2 metric. Other metrics had some data that were misjudged in seven and twelve situations for the cosine and the Euclidean, respectively (Fig. 9a). The FaceNet that creates embeddings directly rather than extracting them as other models showed a better performance of the four models. The score was 100% in the three metrics (Fig. 9b). While the OpenFace and DeepFace presented some misjudged in different metrics. OpenFace misjudged four times for the cosine metrics, but in the case of Euclidean and Euclidean L2, the results are not significant because the misjudges have a considerable distance from the diagonal line to the matrix (Fig. 9c). The DeepFace presents similar not significant results (Fig. 9d), but the reason could be the training process.

Finally, hyper-parameter identification is crucial to the face recognition process and architecture efficiency. This identification could change the hardware required

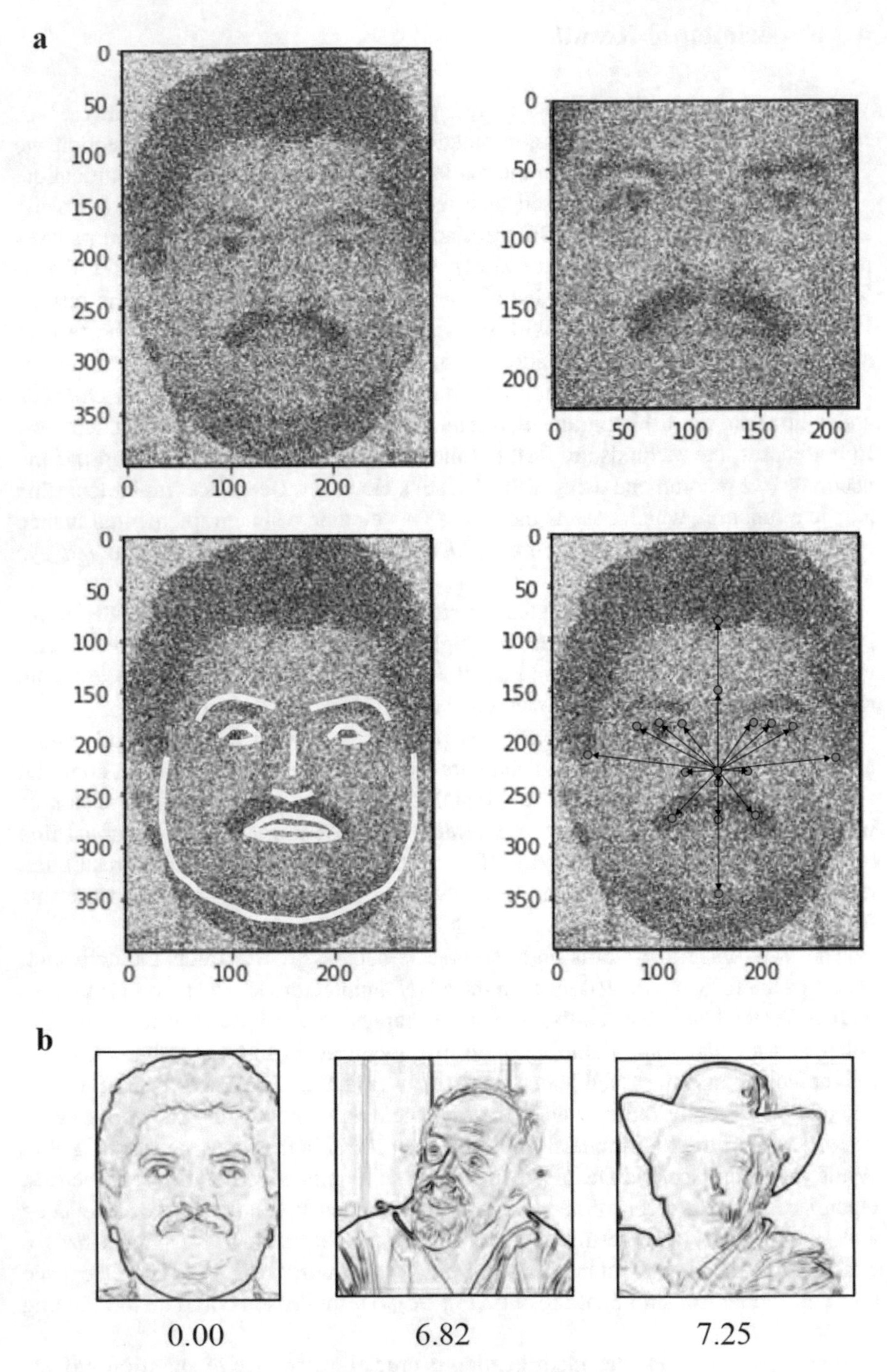

Fig. 8 An example of face recognition results. (**a**) Using OpenFace model back-end cosine distance. (**b**) Using FaceNet model back-end Euclidean L2 distance

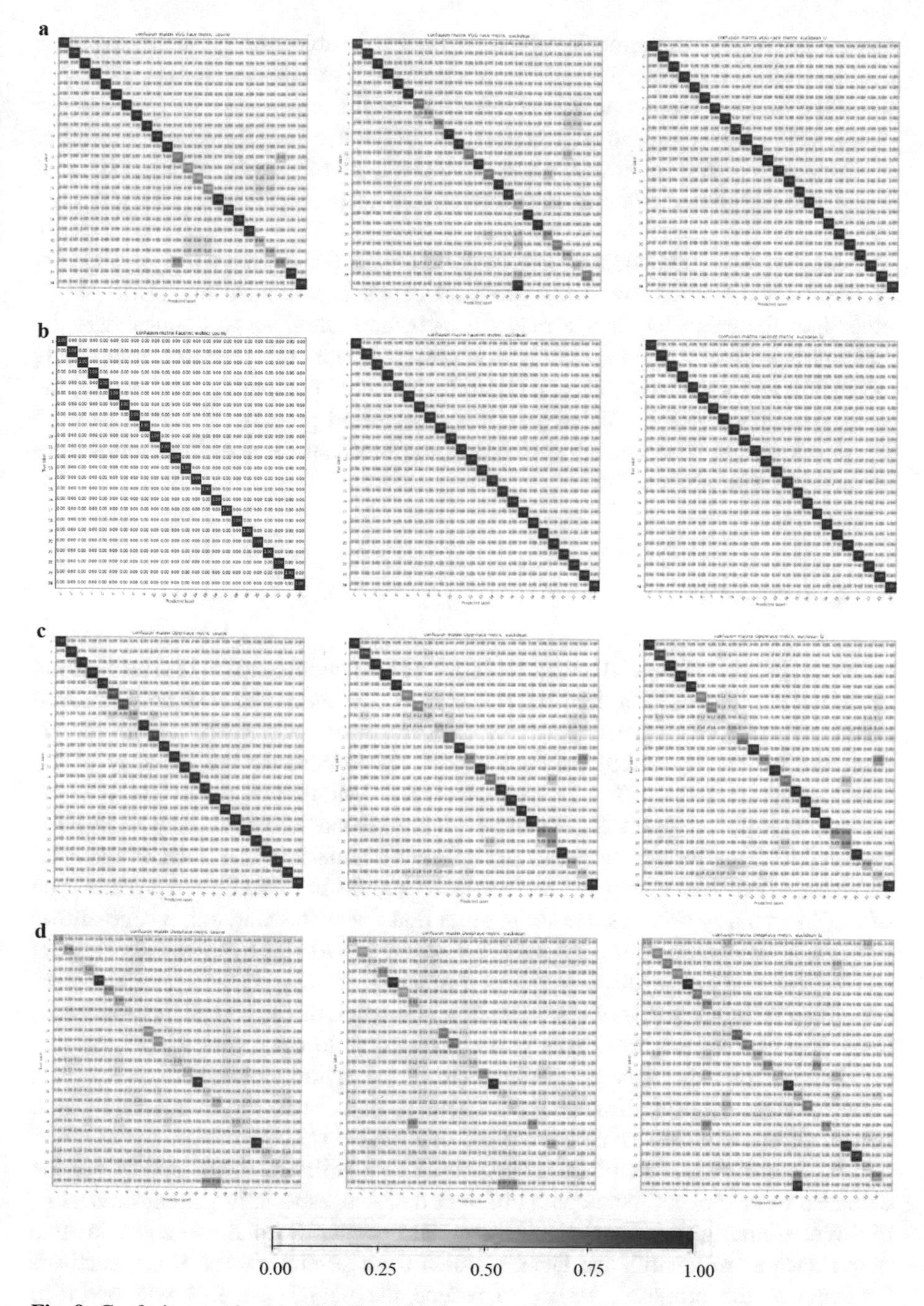

Fig. 9 Confusion matrix. (**a**) VGG-Face. (**b**) FaceNet. (**c**) OpenFace. (**d**) DeepFace

and the network's requirements, which are the two enablers of this technology that could be efficiently used in real-world scenarios. For edge computing example, the algorithm's selection to detect the face in the image has to consider the parallelism opportunities in edge computing. In the hybrid architecture, the choice of computation of similarity (cosine, Euclidean, or L2) could be an enabler of distributed computing, which lets to decentralize face recognition that we proposed above.

Although this methodology has discussed face detection to use in financial services' identification, we could use the DeepFace pretrained models to obtain other face features such as age range, gender, and emotions that consolidate the customize process of financial services. However, to work this process is necessarily the hybrid architecture because, in an exclusive cloud computing architecture, the latency could affect the efficiency of this customized process especially, because customers do not have enough time. Also, the financial services institutions have the moment of truth in sales.

5 Conclusion

In this chapter, we evaluated several tools of artificial intelligence to face recognition systems. We proposed a framework using facial measures and convolutional neural networks. The performance of the framework and CNN architecture is evaluated by tuning various parameters of CNN to enhance the accuracy of the facial recognition tool. This type of intelligent systems that try to use state-of-the-art algorithm requires large computational resources, which may need large bandwidths if performed in the cloud. In transactional services such as financial services, where the quality of service (QoS) should not be affected by this type of complementary services, the proposal is that the processing at the edge allows minimizing integration problems. Hence, tests made over this hybrid architecture let us conclude that bandwidth reduction and energy efficiency enable first, an increase of customized financial services, second, a mitigation of the compliance risk, and, third, the possibilities to distribute the processing to open and to increase the efficiency of financial services. To summarize, the hybrid architecture on financial services opens opportunities to generate benefits in a distributed architecture where banks' different processes integrate in a decentralized way.

The future directions of this work can be as follows. First, models can be extended to new experiments with different datasets, especially with customers of different ethnicity, emotional signals, age, and gender. Second, we could include new measures to identify the facial features from several angles or perspectives. To improve the proposal, we try to reduce the model size that will optimize the technological infrastructure. We will also look into ways of improving the processing time because the financial services need a fast decision to implement solutions or strategies for financial users, especially, in the case of imperfect facial data.

Acknowledgments We thank the Colombian Ministry of Science and technology for the resources to implement this methodology in a real case and Todosistemas STI engineers for their insights on face recognition and architecture to integrate on its risk software.

References

1. Zheng Xl, Zhu My, Li Qb, Chen Cc, Tan Yc (2019) Front Inf Technol Electr Eng 20(7):914. https://doi.org/10.1631/FITEE.1700822
2. Boobier T (2020) AI and the future of banking. Wiley, Hoboken. https://doi.org/10.1002/9781119596165
3. Sinha S, Al Huraimel K (2020) Reimagining businesses with AI. Wiley, Hoboken. https://doi.org/10.1002/9781119709183
4. Wang M, Deng W (2021) Neurocomputing 429:215. https://doi.org/10.1016/j.neucom.2020.10.081
5. Zhong Y, Oh S, Moon HC (2021) Technol Soc 64:101515. https://doi.org/10.1016/j.techsoc.2020.101515
6. Delgrosso D, Orr F (2004) System and method for preventing identity fraud. United States Patent number 20040258281A1, Assignee US Biometrics Corp, 05012004, Springfield, IL
7. Agarwal P, Annamalai P (2014) Systems and methods for implementing transactions based on facial recognition. United States Patent number 10043184B2, Assignee PayPal Inc, 05302014, San Jose, CA
8. Chandrasekaran S, Ho D, Kalenichenko D, Chitilian V, Zwiebel TR, Hashal JA (2016) Identifying consumers via facial recognition to provide services. United States Patent number 20160321633A1, Assignee Google, 04292016, Mountain View
9. Tussy KA (2015) Facial recognition authentication system including path parameters. United States Patent number 9953149B2, Assignee Facetec Inc, 08282015, San Diego
10. Hendrikse R (2019) Biometric Technol Today 2019(4):5. https://doi.org/10.1016/S0969-4765(19)30053-0
11. Ríos-Sínchez B, Silva DCd, Martín-Yuste N, Sánchez-Ávila C (2020) Biometrics 9(3):109. https://doi.org/10.1049/iet-bmt.2019.0093
12. Andrade NNGd, Martin A, Monteleone S (2013) IEEE Security Privacy 11(3):21–28. https://doi.org/10.1109/MSP.2013.22
13. Wright E (2019) Fordham Intell Prop Media Ent LJ 29(2):611
14. Chen Y, Emer J, Sze V (2016) 2016 ACM/IEEE 43rd annual international symposium on computer architecture (ISCA), pp 367–379. https://doi.org/10.1109/ISCA.2016.40
15. Caulfield AM, Chung ES, Putnam A, Angepat H, Fowers J, Haselman M, Heil S, Humphrey M, Kaur P, Kim J, Lo D, Massengill T, Ovtcharov K, Papamichael M, Woods L, Lanka S, Chiou D, Burger D (2016) 2016 49th annual IEEE/ACM international symposium on microarchitecture (MICRO), pp 1–13. https://doi.org/10.1109/MICRO.2016.7783710
16. Park J, Sharma H, Mahajan D, Kim JK, Olds P, Esmaeilzadeh H (2017) Proceedings of the 50th annual IEEE/ACM international symposium on microarchitecture. Association for Computing Machinery, New York, MICRO-50 '17, pp 367–381. https://doi.org/10.1145/3123939.3123979.
17. Gui CY, Zheng L, He B, Liu C, Chen XY, Liao XF, Jin H (2019) J Comput Sci Technol 34(2):339. https://doi.org/10.1007/s11390-019-1914-z
18. Zhu Y, Qian L, Wang C, Ding L, Yang F, Wang H (2019) In: Fahrnberger G, Gopinathan S, Parida L (eds) Distributed computing and internet technology. Springer International Publishing, Cham, pp 150–158. https://doi.org/10.1007/978-3-030-05366-6_12
19. Ting Z (2020) Microproces Microsyst 103456. https://doi.org/10.1016/j.micpro.2020.103456
20. Hossain MS, Muhammad G (2019) Inf Sci 504:589. https://doi.org/10.1016/j.ins.2019.07.040

21. Wang W, Lin H, Wang J (2020) J Cloud Comput 9(1):27. https://doi.org/10.1186/s13677-020-00172-z
22. Song C, Li T, Huang X, Wang Z, Zeng P (2019) In: Sun X, Pan Z, Bertino E (eds) Artificial intelligence and security. Springer International Publishing, Cham, pp 283–292. https://doi.org/10.1007/978-3-030-24274-9_25
23. Gupta BB, Agrawal DP, Yamaguchi S (2019) J Amb Intell Human Comput 10(8):2907. https://doi.org/10.1007/s12652-018-0919-8
24. Moolchandani D, Kumar A, Sarangi SR (2020) J Syst Architect 101887. https://doi.org/10.1016/j.sysarc.2020.101887
25. Farkas L (1994) Anthropometry of the head and face. Raven Press, New York
26. Naini FB (2011) Facial aesthetics: concepts and clinical diagnosis. Wiley, Hoboken
27. Bashour M (2007) Is an objective measuring system for facial attractiveness possible? Universal-Publishers, Boca Raton
28. George RM (2007) Facial geometry: graphic facial analysis for forensic artists. Charles C Thomas Publisher, Springfield
29. Swennen GR, Schutyser F, Hausamen JE (2005) Three-dimensional cephalometry: a color atlas and manual. Springer, Berlin
30. Hosoi S, Takikawa E, Kawade M (2004) Sixth IEEE international conference on automatic face and gesture recognition, 2004. Proceedings, pp 195–200. URL 10.1109/AFGR.2004.1301530
31. Lin H, Lu H, Zhang L (2006) 2006 6th world congress on intelligent control and automation, vol 2, pp 9988–9991. URL 10.1109/WCICA.2006.1713951
32. Rehnman J, Herlitz A (2006) Memory 14(3):289. https://doi.org/10.1080/09658210500233581.
33. Berretti S, Del Bimbo A, Pala P (2012) ACM Trans Intell Syst Technol 3(3):1. https://doi.org/10.1145/2168752.2168759.
34. Camargo Díaz JL, PÃ E, SotaquirÃ¡ GutiÃ M, GutiÃ De Aguas R (2011) Ingeniería y Desarrollo 29:127
35. Fuentes CT, Runa C, Blanco XA, Orvalho V, Haggard P (2013) PLOS ONE 8(10). https://doi.org/10.1371/journal.pone.0076805
36. Asli KH, Yavarmasroor S, Zidashti ZH (2014) Computational modeling for anthropometry. CRC Press, Boca Raton
37. Perret-Ellena T, Skals SL, Subic A, Mustafa H, Pang TY (2015) Procedia Eng 112:98. 'The Impact of Technology on Sport VI' 7th Asia-Pacific Congress on Sports Technology, APCST2015. https://doi.org/10.1016/j.proeng.2015.07.182
38. Zhao W, Chellappa R, Phillips PJ, Rosenfeld A (2003) ACM Comput Surv 35(4):399. https://doi.org/10.1145/954339.954342
39. Tolba AS, El-Baz A, El-Harby A (2008) Int J Comput Inf Eng 2(7):2556. https://publications.waset.org/vol/19
40. Kaur P, Krishan K, Sharma SK, Kanchan T (2020) Med Sci Law 60(2):131.PMID: 31964224. https://doi.org/10.1177/0025802419893168.
41. Wang L, He DC (1990) Pattern Recog 23(8):905. https://doi.org/10.1016/0031-3203(90)90135-8
42. Ojala T, Pietikäinen M, Harwood D (1996) Pattern Recog 29(1):51. https://doi.org/10.1016/0031-3203(95)00067-4
43. Bouwmans T, Silva C, Marghes C, Zitouni MS, Bhaskar H, Frelicot C (2018) Comput Sci Rev 28:26. https://doi.org/10.1016/j.cosrev.2018.01.004
44. Chen J, Shan S, He C, Zhao G, Pietikäinen M, Chen X, Gao W (2010) IEEE Trans Pattern Analy Mach Intell 32(9):1705. https://doi.org/10.1109/TPAMI.2009.155
45. Min WY, Romanova E, Lisovec Y, San AM (2019) 2019 IEEE conference of Russian young researchers in electrical and electronic engineering (EIConRus), pp 2208–2212. https://doi.org/10.1109/EIConRus.2019.8657240
46. Sharma S, Bhatt M, Sharma P (2020) 2020 5th international conference on communication and electronics systems (ICCES), pp 1162–1168. https://doi.org/10.1109/ICCES48766.2020.9137850

47. Freund Y, Schapire RE (1995) In: Vitányi P (ed) Computational learning theory. Springer, Berlin, pp 23–37
48. Viola P, Jones MJ (2004) Int J Comput Vision 57(2):137. https://doi.org/10.1023/B:VISI.0000013087.49260.fb
49. Dalal N, Triggs B (2005) 2005 IEEE computer society conference on computer vision and pattern recognition (CVPR'05), vol 1, pp 886–893 vol 1. https://doi.org/10.1109/CVPR.2005.177
50. Zhang S, Qiao H (2003) IEEE international conference on robotics, intelligent systems and signal processing, 2003. Proceedings. 2003, vol 2, pp 726–730. https://doi.org/10.1109/RISSP.2003.1285674
51. Karthik HS, Manikandan J (2017) 2017 IEEE international conference on consumer electronics-Asia (ICCE-Asia), pp 26–30. https://doi.org/10.1109/ICCE-ASIA.2017.8307832
52. Wu X, He R, Sun Z, Tan T (2018) IEEE Trans Inf Foren Security 13(11):2884. https://doi.org/10.1109/TIFS.2018.2833032
53. Bayramli B, Ali U, Qi T, Lu H (2019) In: Gedeon T, Wong KW, Lee M (eds) Neural information processing. Springer International Publishing, Cham, pp 3–15. https://doi.org/10.1007/9783030367084_12
54. Rong C, Zhang X, Lin Y (2020) IEEE Access 8:68842. https://doi.org/10.1109/ACCESS.2020.2986079
55. Donahue J, Hendricks LA, Rohrbach M, Venugopalan S, Guadarrama S, Saenko K, Darrell T (2017) IEEE Trans Pattern Analy Mach Intell 39(4):677. https://doi.org/10.1109/TPAMI.2016.2599174
56. Li TS, Kuo P, Tsai T, Luan P (2019) IEEE Access 7:93998. https://doi.org/10.1109/ACCESS.2019.2928364
57. An F, Liu Z (2020) Visual Comput 36(3):483. https://doi.org/10.1007/s00371-019-01635-4
58. K B P, J M (2020) Procedia Comput Sci 171:1651. Third International Conference on Computing and Network Communications (CoCoNet'19). https://doi.org/10.1016/j.procs.2020.04.177.
59. Zhang M, Khan S, Yan H (2020) Pattern Recog 100:107176. https://doi.org/10.1016/j.patcog.2019.107176
60. Emmanuel O, M A, Misra S, Koyuncu M (2020) Sustainability 12(24):10524. https://doi.org/10.3390/su122410524
61. Schroff F, Kalenichenko D, Philbin J (2015) 2015 IEEE conference on computer vision and pattern recognition (CVPR), pp 815–823. https://doi.org/10.1109/CVPR.2015.7298682
62. Jadeja Y, Modi K (2012) 2012 international conference on computing, electronics and electrical technologies (ICCEET), pp 877–880. https://doi.org/10.1109/ICCEET.2012.6203873
63. Buyya R, Yeo CS, Venugopal S, Broberg J, Brandic I (2009) Future Gener Comput Syst 25(6):599. https://doi.org/10.1016/j.future.2008.12.001
64. Gill SS, Tuli S, Xu M, Singh I, Singh KV, Lindsay D, Tuli S, Smirnova D, Singh M, Jain U, Pervaiz H, Sehgal B, Kaila SS, Misra S, Aslanpour MS, Mehta H, Stankovski V, Garraghan P (2019) Int Things 8:100118. https://doi.org/10.1016/j.iot.2019.100118
65. Mansouri Y, Babar MA (2021) J Parallel Distributed Comput 150:155. https://doi.org/10.1016/j.jpdc.2020.12.015
66. Al-Dhuraibi Y, Paraiso F, Djarallah N, Merle P (2018) IEEE Trans Serv Comput 11(2):430. https://doi.org/10.1109/TSC.2017.2711009
67. Soliman M, Abiodun T, Hamouda T, Zhou J, Lung C (2013) 2013 IEEE 5th international conference on cloud computing technology and science, vol 2, pp 317–320. https://doi.org/10.1109/CloudCom.2013.155
68. Al-Fuqaha A, Guizani M, Mohammadi M, Aledhari M, Ayyash M (2015) IEEE Commun Surveys Tutor 17(4):2347. https://doi.org/10.1109/COMST.2015.2444095
69. Qiu T, Chi J, Zhou X, Ning Z, Atiquzzaman M, Wu DO (2020) IEEE Commun Surv Tutorials 22(4):2462. https://doi.org/10.1109/COMST.2020.3009103

70. Saxena VK, Pushkar S (2016) 2016 international conference on electrical, electronics, and optimization techniques (ICEEOT), pp 2583–2588. https://doi.org/10.1109/ICEEOT.2016.7755159
71. Mishra SD, Dewangan BK, Choudhury T (2020) Cyber security in cloud platforms. CRC Press, Boca Raton, pp 159–180
72. Gubbi J, Buyya R, Marusic S, Palaniswami M (2013) Future Gener Comput Syst 29(7):1645. https://doi.org/10.1016/j.future.2013.01.010
73. Satyanarayanan M (2017) Computer 50(1):30. https://doi.org/10.1109/MC.2017.9
74. Jain A, Singhal P (2016) 2016 international conference system modeling advancement in research trends (SMART), pp 294–297. https://doi.org/10.1109/SYSMART.2016.7894538
75. Masud M, Muhammad G, Alhumyani H, Alshamrani SS, Cheikhrouhou O, Ibrahim S, Hossain MS (2020) Comput Commun 152:215. https://doi.org/10.1016/j.comcom.2020.01.050
76. Papageorgiou CP, Oren M, Poggio T (1998) Sixth international conference on computer vision (IEEE Cat. No.98CH36271), pp 555–562. https://doi.org/10.1109/ICCV.1998.710772
77. Viola P, Jones M (2001) Proceedings of the 2001 IEEE computer society conference on computer vision and pattern recognition. CVPR 2001, vol 1, pp I–I. https://doi.org/10.1109/CVPR.2001.990517
78. Parkhi OM, Vedaldi A, Zisserman A (2015) British machine vision conference
79. B. Amos, B. Ludwiczuk, M. Satyanarayanan, Openface: A general-purpose face recognition library with mobile applications. Technical report, CMU-CS-16-118, CMU School of Computer Science (2016)
80. Taigman Y, Yang M, Ranzato M, Wolf L (2014) 2014 IEEE conference on computer vision and pattern recognition, pp 1701–1708. https://doi.org/10.1109/CVPR.2014.220
81. Shafique M, Theocharides T, Bouganis C, Hanif MA, Khalid F, Hafız R, Rehman S (2018) 2018 design, automation test in Europe conference exhibition (DATE), pp 827–832. https://doi.org/10.23919/DATE.2018.8342120
82. ass S (2019) IEEE Spectrum 56(5):16
83. Zhou G, Zhou J, Lin H (2018) 2018 12th IEEE international conference on anti-counterfeiting, security, and identification (ASID), IEEE, pp 192–195
84. Villari M, Fazio M, Dustdar S, Rana O, Ranjan R (2016) IEEE Cloud Comput 3(6):76. https://doi.org/10.1109/MCC.2016.124

Classification of Swine Disease Using K-Nearest Neighbor Algorithm on Cloud-Based Framework

Emmanuel Abidemi Adeniyi, Roseline Oluwaseun Ogundokun, Babatunde Gbadamosi, Sanjay Misra, and Olabisi Kalejaiye

1 Introduction

In many countries around the world, agriculture plays a key role in the economy, thus being the backbone of developed countries like Nigeria [1]. In the lifetime of rearing animals, diseases are inevitable. In the commercial rearing of animals, for instance, in pigs, it is found out that almost 60% of the animals were infected with diseases which is the cause of wastage of time and money to the farmers rearing them [2, 3]. A lot of animals have suffered and even died from diseases that could have been easily cured if they were discovered earlier [4]. Swine diseases are illnesses that affect pigs, a very common example is the African swine fever popularly known as ASF [5]. It was first discovered in Africa in 1921, in a small town called Montgomery in Kenya. The first case ever of ASF was in Portugal, near Lisbon, in 1957, where death was caused by African swine fever with over 100% of the pigs killed. Three years later in 1960, ASF reappeared in Portugal after an epidemiological breakout spreading like wildfire throughout the entire Iberian Peninsula. Since then, ASF has remained present in Spain and Portugal for more than 20 years until its elimination has been accomplished [6]. Another popular disease common among pigs is porcine parvovirus which causes reproductive failure in pigs. It is said that porcine parvovirus has two genetic lineages from which it originated from: one from Germany and the other from Asia. It was first discovered and isolated in Germany in 1965 when a herd of cattle was found dead. Traces of

E. A. Adeniyi · R. O. Ogundokun (✉) · B. Gbadamosi · O. Kalejaiye
Department of Computer Science, Landmark University, Omu Aran, Nigeria
e-mail: ogundokun.roseline@lmu.edu.ng

S. Misra
Department of Computer Science and Communication, Ostfold University College, Halden, Norway
e-mail: sanjay.misra@hiof.no

S. Misra et al. (eds.), *Artificial Intelligence for Cloud and Edge Computing*,
Internet of Things, https://doi.org/10.1007/978-3-030-80821-1_4

the disease were also found in South Korea where it brought a huge reduction to the swine industry [7]. Other very popular swine diseases include mastitis which is a swine disease in which the mammary glands of the pigs are infected, coccidiosis which is a bacterial disease that affects the intestines of piglets and could cause death between 7 and 21 days if not noticed, etc. Predicting these diseases would help reduce the rate at which pigs die which would be a very big advantage to farmers.

Predictive analytics was first discovered as far back as the nineteenth century when the government began using early computers, although it had existed for decades. Predictive analytics is used in making forecasts about behaviors and patterns. It uses several techniques ranging from statistics, data mining, machine learning, artificial intelligence, and so on to analyze data and make predictions. It can be used in both the business world and the medical world to predict potential dangers and opportunities for people all over the world [8]. The growing advancement of global businesses has encouraged the dissemination of ailments. Most of the time, these diseases have the capability to transverse nations' boundaries and transmit speedily. Averting and detecting these ailments have aroused to be a community medical difficulty of great importance. Swine diseases have hunted and tormented farmers for many years, and a solution should be provided to stop these losses that farmers experience. Therefore, a system should be created to predict these swine diseases to prevent pigs from dying. It should allow the farmers to input the symptoms that they notice once a pig starts showing signs of sickness, and then the system would predict the disease that the pig is likely to have so that preventive measures can be carried out and also to prevent the spread of such disease on the farm.

Therefore, this study examines the way predictive algorithms can be used in the cloud-based database to prevent swine disease and help farmers take quick preventive measures once the diseases have been predicted to reduce the death rates of the pig. The study aims to design a disease prediction application that detects the likely occurrence of swine disease using predictive algorithms, allow farmers to input the symptoms noticed in the sick pigs, derive correct symptoms from a pig suffering a particular illness, and lastly to prevent swine diseases from affecting other healthy pigs on the farm. This study also intends to do an actual classification using supervised learning, together with the explanation of the machine architecture that classifies objects of the diseased animals, using the k-nearest neighbor (KNN) algorithm. A web cloud-based expert system was developed for the detection of swine diseases in pigs. MySQL was used for the cloud-based database which is at the backend of the website developed. PHP was used for the coding aspect of the system; the KNN algorithm was embedded into the PHP version 7.1.30.

The article's remaining part is structured as thus: Section 2 discussed the literature review on cloud computing and related works in the area of ML and AI. Section 3 discussed the materials and methods. The ML procedure used for the execution of the system in the study was conferred in the unit as well. Section 4 discussed the execution and testing of the system. Here, the findings discovered are also conferred in this section. The study was concluded in Sect. 5.

2 Background and Literature Review

Information and communication processes are invisibly rooted in the world surrounding us in the IoT model. This would result in the development of an immense volume of data that must be created, interpreted, and delivered in a smooth, effective, and easily interpretable manner. According to [9], cloud computing is the newest concept to evolve, offering high reliability, scalability, and independence. In reality, it provides pervasive entry, resource provisioning exploration, and composability needed for future IoT applications. This portal serves as a data transmitter for ubiquitous sensors, a device for the processing and visualization of data, and a provider for the comprehension of Internet visualizations. Not only does the cloud minimize the expense of installing ubiquitous software, but it is also extremely elastic. Sensing content providers can use a database cloud to access the network and deliver their results, computational technology engineers can provide their development software, machine learning professionals can provide their data processing and machine learning resources, and computer animation designers can eventually provide a variety of visualization techniques. These resources may be offered to the IoT by cloud computing as infrastructures, networks, or applications where the full capacity of human imagination can be utilized. In this context, the data generated, the equipment used, and the method of creating complicated visualizations are concealed [10]. Cloud computing connects a large number of computing, storage, and information tools to form a large public global resource pool from which people can buy utilities like hydropower. Due to the rapid popularization of cloud computing applications, cloud computing has entered diverse areas, such as scientific research, manufacturing, education, consumption, and entertainment [11]. Cloud computing mainly provides three service delivery models which are infrastructure as a service (IaaS), platform as a service (PaaS), and software as a service (SaaS) [11]. Many of these providers offer computer services on request, such as storage and data processing. Cloud computing also focuses on the complex optimization of pooled resources by multiple users, in addition to providing the services listed above. For instance, Western users are assigned a cloud storage utility (such as email) depending on their time zone. The same resource is allocated to Asian users by cloud storage, often dependent on their time zone.

Cloud computing has recently arisen and evolved rapidly as a versatile and preordained innovation. Cloud computing infrastructure provides extremely flexible, accessible, and programmable virtual servers, space, computational resources, and digital communication, according to customer requirements. It will also offer a solution kit for the electronic information transition if it is optimized for IoTs and combined with modern technologies for data collection, delivery, and preservation. In comparison, the user creates information very easily as long as the storage capacity allows [12]. Most of the information can also be used by the client, who wants to be quickly accessible. Media storage is one of the main facets of cloud computing, as the cloud allows vast volumes of electronic content to be stored, handled, and exchanged. This functionality will play a crucial role for digital media linked to IoTs. Many other multimedia services for people on the move, such as smartphones, tablets, and desktop users, ad hoc vehicle networks, and various ambulance and emergency operations, will be possible in the future. Cloud computing is planning to run a very strong function for those services with important position in the administration of programs and infrastructure. The server will be used more commonly, particularly with the expanded network, fog computing [13], also identified as edge computing, or micro data center (MDC). Cloud computing is a handy solution for managing content in distributed systems. It offers seamless access to information without the burden of managing massive storage and processing tools. Exchanging a vast volume of online media is another function cloud service offers. Apart from social networks, conventional cloud computing offers extra functionality for sharing and content management. Similarly, if the information is to be exchanged, it is not possible to import individual files one after the other. Cloud storage solves this problem since all information can be viewed at once by all people with which the information is distributed. Also, more context-aware services can be offered by cloud storage, as IoT and sensor networks are not dense enough in assets to execute those activities. Data stored in the cloud could also be fully processed to develop more personalized and usable applications [14].

One of the most exciting recent artificial intelligence technologies is machine learning. Intelligent agent is the name of an artificial intelligence (AI) program. AI has been used for the prediction and detection of diseases [15–17]. For various subjects, for instance, smart cities, computer vision, medical care, self-propelled, and machine learning algorithms are used [1, 18, 19]. There are numerous instances of what way machine learning may be applied on edge gadgets in these areas [20]. The smart agent can coordinate with the world. The agent can identify the state of an environment through its sensors, and then, through its actuators, it can influence the state. The key AI factor is the control strategy of the system, which implies how the signals received from the sensors are transmitted to the actuators, i.e., how the sensors are connected to the actuators, which is achieved by the feature within the operator. AI's ultimate aim is to achieve human-like computer intelligence. However, through learning algorithms that attempt to mimic how the human brain learns, such a dream can be achieved. AI systems do the most intriguing things, though, such as web search or photo tagging or anti-spam email. So, as a modern capability for computers, machine learning was created, and today it reaches several sectors of business and basic science. There is autonomous robotics, and there is machine biology. In the last 2 years, about 90% of the world's data has been created on its own, and the inclusion of the machine learning library known as Mahout into the Hadoop ecosystem has made it possible to face the challenges of big data, especially unstructured data. The focus is mostly on selecting or designing an algorithm and doing tests based on the algorithm in the field of machine learning science. This inherently distorted perspective limits the effect of applications in the real world.

In this article, the machine learning procedures that can be used in resource-constrained edge computing situations are discussed. The machine learning techniques presented in the following subsections are the utmost widely used and present the challenge of bringing artificial intelligence to hardware-resource-constrained applications. To store training data and processing resources to train massive models, machine learning models currently need adequate memory. New device models have been developed to operate on edge equipment effectively by using shallow models to be used on IoT devices that require sufficiently low processing power. Alternatively, reducing the size of model inputs for applications for classification would increase the speed of learning [21]. The k-nearest neighbors (KNN) algorithm is a technique used to classify patterns based on identical artifact characteristics to the one measured. This approach is being used for both grouping and regression concerns. Many adapted KNN variants help to operate the algorithm on hardware-restricted machines, the most revolutionary being the ProtoNN. It is a KNN-based algorithm. The main cloud computation KNN problems are the size of the training data (the algorithm uses the whole dataset to make predictions), the time of prediction, and the choice of distance metrics [22].

The k-nearest neighbors (KNN) algorithm is an algorithm used to classify patterns based on identical artifact characteristics to the one considered. This technique is used for both classification and regression problems. Various modified KNN variants help to execute the algorithm on hardware-constrained computers, with the most innovative being ProtoNN. It is a KNN-dependent algorithm. The major issues with KNN for edge computing are the scale of the training data (the algorithm uses the whole dataset to create predictions), the assumption of speed, and location metrics [23]. Tree-based ML algorithms are used for the learning algorithm, a very standard occurrence in the IoT field. Nevertheless, due to the finite assets of the machines, the regular tree algorithms could not be extended to them. A changing algorithm is Bonsai. The tree algorithm is specifically built for heavily resource-restricted IoT devices and retains predictive accuracy while reducing model complexity and predictive pace. It first found that a single compact tree decreases the scale model and then allows nonlinear representations of the inner nodes and the leaf ones. The sparse matrix is increasingly being learned by Bonsai, transforming all knowledge into a low-dimensional space where the tree is taught. This helps the algorithm to be used on handheld computers such as those of the IoT [24].

Among the most commonly used ML techniques at the integrated stage is the SVM. SVM is a supervised learning technique that can be used for both classification and regression concerns. The methodology segregates among two or more categories of data by defining an optimal hyperplane that separates all classes. Support vectors are the data nearest to the hyperplane that redefines the hyperplane itself if it is deleted. For these purposes, the essential elements of the data collection shall be considered. The loss function used by the method is usually the loss of the hinge, and the descent of the gradient form is the optimization function [25].

2.1 Related Works

A significant amount of research has been carried out in this field, which has been unique to particular animal disease and has also been generalized for different single animal diseases as well as diseases affecting multiple animals.

Some of the related works on disease predictions are discussed in this section.

Adebayo Peter Idowu et al. [26] projected a scientific method for the prediction of immunizable diseases that happen to children between the ages of 0 and 5 years. This model was deployed and tested for use within six local areas in Osun state. They made use of MATLAB's ANN toolbox, a statistical toolbox for regression and classification, and a classifier called Naive Bayes for developing their model. The data mining techniques used for the implementation of this model brought about the discovery of various trends and patterns which helped a lot in the prediction of these diseases in each location in Osun State.

Sohail et al. [27] talked about a clinical assessment for predicting diabetes mellitus in Nigeria. They made use of machine learning techniques for implementing over a data mining platform by using some rule classifications such as a decision table for measuring accurateness and regression on patients that have suffered from diabetes mellitus and other deadly diseases. They collected data in Nigeria from December 2017 to February 2019, and the rule classifiers used realized a mean accuracy of 98.75%.

Sellappan and Rafiah [28] developed a porotype for intelligence heart disease prediction and made use of a couple of data mining techniques including decision tree, Naive Bayes, and neural networks. The system could answer questions like "what if" which the traditional decision support system could not answer. It made use of some parameters such as age, sex, blood pressure, and the amount of sugar level in the blood to predict the likelihood of someone getting a heart disease. It was a web-based application and was implemented on a network platform. Chronic diseases were also predicted in hospitals using electronic health records.

Brisimi et al. [29] predicted a chronic disease using electronic health records. They focused on heart disease and diabetes which were two chronic diseases affecting old people. Also, they developed data-driven methods to predict a patient's hospitalization due to heart diseases and diabetes. Patient's medical history from electronic health records was used, and they formulated a prediction problem as a binary classification problem and made use of various machine learning techniques. Some of the techniques are kernelized and sparse vector machines, logistic regression, and random forests. They made use of two methods to achieve accuracy and interpretability which are K-LRT and joint clustering and classification.

Idowu et al. [30] predicted a deadly disease called HIV/AIDS. HIV/AIDS is commonly known as human immunodeficiency virus. Acquired immunodeficiency disease syndrome is also a deadly epidemic that is common in Nigeria. The naïve Bayes approach was used in Nigeria in the prediction of HIV/AIDS. Data was collected from over 216 HIV/AIDS patients to develop a predictive model. The results from the model showed 81.02% accuracy.

Yang et al. [21] developed a web-based software to predict livestock diseases such as anthrax, swine fever, black quarter, and many others that were included in their study. The software was built to predict the occurrence of a disease cattle would or might have 2 months in advance and then alert the animal husbandry department for preventive measures to be taken. (www.nadres.res.in) is a website that provides a disease forecasting or prediction service. It provides information for farmers about their farm animals and the diseases they could likely have in 2 months and also alerts them for necessary preventive measures to be taken.

Valdes-Donoso et al. [22] predicted the movement of swine animals in the US swine industry. They summarized two networks that produced swine in a state in Minnesota; they utilized a machine learning procedure denoted as random forest an autonomous classification to evaluate the likelihood of pig movements in two countries. Some of the important variables which were used in envisaging animal movements include location remoteness, proprietorship, production type of the farm, and or marketplaces. The weighted kernel approach was used to define spatial discrepancy in the prediction network.

Nusai et al. [23] developed a web-built expert system for swine ailment investigation. It was established mainly for swine agriculturalists and animal husbandmen. The structure covered all swine ailments in the Thai language, and the expert system was alienated into three phases. The three phases included ailment examination, disease analysis employing indications that the handler, i.e., the farmer, noticed, and the last step was the diagnosis of the disease using an animal necropsy lesion. Results later showed that the system could perform ailment examination accurately for 97.50%, identify symptoms accurately for 92.48%, and also establish by lesion accurately for 95.62%.

Schaefer et al. [31] predicted infections in cattle using infrared thermography. The objective of their study was to consider how good the utilization of infrared thermography will be in the early detection of identifying animals that had a systematic infection. They made use of a viral model that contained 15 calves. Some were infected with a particular virus, while some were not. A comparison of both infrared characteristics on both infected and uninfected animals was conducted.

Muriu [32] proposed a mobile-based expert system to predict animal diseases, which promoted a platform between farmers and vet doctors. The results of the research showed that 29 people took part in the exercise; 69% of them were satisfied that the system fulfilled its intended purposes, and 76% of them found the model very easy to use.

Zetian [33] developed a web-built proficient scheme for pig ailment analysis. According to them, once a pig displays indications of a disease, it is very imperative to make an exact analysis to prevent the disease from spreading and to support control strategies. The scheme had beyond 300 rules and 202 images of diverse kinds of disease and indications that affect pigs. It could identify 54 kinds of generally known ailments that affect pigs.

Sinha et al. [34] performed a comparative study of chronic kidney disease prediction using KNN (k-nearest neighbors) and SVM (support vector machine). They defined chronic disease also called chronic renal disease as a medical condition in which an individual has a damaged kidney. Problems, for instance, high blood pressure, anemia which is low blood count, feeble bones, etc. could be cause to damaged kidney.

Table 1 below shows the summary of the previous related works reviewed. The table showed the methods adopted by each previous researcher. The disease they diagnose and predicted was also shown in the table.

3 Materials and Methods

This section discussed the materials used for this study in terms of datasets used for the implementation. The machine learning algorithm adopted for this study was also discussed in this section.

The major objective of this web-based application for detecting swine diseases using the k-nearest neighbor algorithm is to develop an application for diagnosing the disease an infected pig has. It helps the user (i.e., the farmer) to diagnose a disease affecting a pig in real-time by selecting various symptoms through a given list. The symptoms observed are then selected and are then processed to a given conclusion to take out the chances of the disease occurring, and if such diseases are already occurring, an immediate course of medication would be carried out. The use of the web has become a digital fabric in our lives; also, we are living in an age where we cannot do anything without the use of the web. Thus, this is the reason why a web-based application was chosen. Pigs that have suffered from different swine diseases were examined to gather symptoms related to each disease. This was achieved using both primary and secondary data. K-nearest neighbor algorithm, a supervised machine learning algorithm, was then used for training and classifying the dataset.

3.1 Collection of Datasets

Different pig diseases were gathered with their corresponding symptoms. Swine diseases can be categorized generally into five, from which diseases under each category were sampled, and corresponding symptoms were gotten. These five categories were diseases caused by bacteria, viruses, fungi, nutritional deficiencies, and toxicosis. All data obtained and used in the database were taken from the following:

Primary data source: https://www.wattagnet.com/articles/25841-most-common-pig-diseases-worldwide.

Secondary data source: https://www.pig333.com/pig-diseases/.

Table 1 Summary of related works

Authors	Disease	Methods	Conclusion
Adebayo Peter Idowu et al. [26]	Immunizable diseases	Bayesian	The study concluded that various trends and patterns helped a lot in the prediction of these diseases in each location in Osun State
Sohail et al. [27]	Diabetes mellitus	Decision table	They collected data in Nigeria from December 2017 to February 2019, and the rule classifiers used realized a mean accuracy of 98.75%
Sellappan and Rafiah [28]	Heart disease	Decision tree, Naive Bayes, and neural networks	They developed a web-based application and implemented it on a network platform. Chronic diseases were also predicted in hospitals using electronic health records
Brisimi et al. [29]	Heart disease and diabetes	Kernelized and sparse vector machines, logistic regression, and random forests	They made use of two methods to achieve accuracy and interpretability which are K-LRT and a joint clustering and classification
Idowu et al. [30]	HIV/AIDS	Naive Bayes approach	Data was collected from over 216 HIV/AIDS patients to develop a predictive model. The results from the model showed 81.02% accuracy
Yang et al. [21]	Anthrax, swine fever, black quarter		The software was built to predict the occurrence of a disease cattle would or might have 2 months in advance and then alert the animal husbandry department for preventive measures to be taken

(continued)

Table 1 (continued)

Authors	Disease	Methods	Conclusion
Valdes-Donoso et al. [22]	Swine animals	Random forest and weighted kernel approaches	They summarized two networks that produced swine in a state in Minnesota; they utilized a machine learning procedure denoted as random forest an autonomous classification to evaluate the likelihood of pig movements in two countries
Nusai et al. [23]	Swine ailment	Animal necropsy lesion	Results showed that the system could perform ailment examination accurately for 97.50%, identify symptoms accurately for 92.48%, and also establish by lesion accurately for 95.62%
Schaefer et al. [31]	Infections in cattle	Infrared thermography	The objective of their study was to consider how good the utilization of infrared thermography will be in the early detection of identifying animals that had a systematic infection
Muriu [32]	Animal diseases	Not indicated	The results of the research showed that 29 people took part in the exercise; 69% of them were satisfied that the system fulfilled its intended purposes, and 76% of them found the model very easy to use

(continued)

Table 1 (continued)

Authors	Disease	Methods	Conclusion
Zetian [33]	Pig ailment		The scheme had beyond 300 rules and 202 images of diverse kinds of disease and indications that affect pigs. It could identify 54 kinds of generally known ailments that affect pigs
Sinha et al. [34]	Chronic kidney diseases	KNN (k-nearest neighbors) and SVM (support vector machine)	They defined chronic disease also called chronic renal disease as a medical condition in which an individual has a damaged kidney

3.2 *Proposed Algorithm*

When using KNN in prediction or detection, the training set is used to detect the value of interest (which in this case is the right illness) for each member of the target data set. It uses a method called feature similarity. This means a new point is assigned based on how closely it resembles a point in the training set. For each row in the dataset, the k closest members were located out in the collection in instruction. The Euclidean distance measure was then used to determine how near each component of the test data is to the target case being studied. The assessment approaches of the target case of interest for the closest k-nearest neighbor is then found. In other words, weights are the inverses of the distances between objects. The procedure was repeated for the remaining cases until the entire dataset has been completed.

Training set is the initial dataset used to train a system model on performing various actions. This is the data that the development process model would learn with various APLs and algorithms to train the model to work automatically. Figure 1 shows a simple step-by-step procedure of how the algorithm works.

The proposed system requires the user to create an account or register before accessing the system. This would require a name, valid email, or phone number. The system also provides a list of pig diseases, symptoms, and treatments. It also allows the user to input symptoms noticed that are not listed in the system.

The system flowchart in Fig. 2 shows the process flow in which the system goes through in various stages. The flowchart displays the proposed system coupled with the analysis of the system. It shows the step-by-step process of how swine diseases can be detected and also implemented using the KNN algorithm. The system requires the user to log in and input the symptoms observed for the system

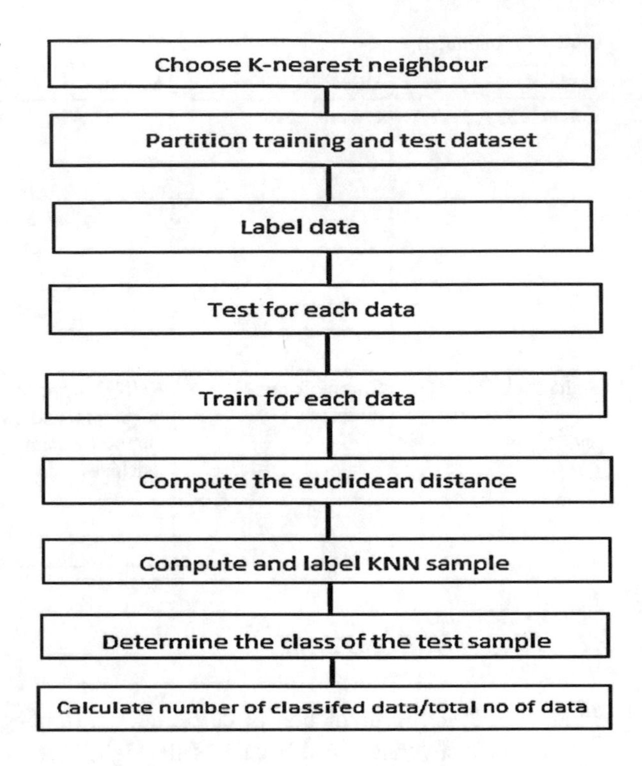

Fig. 1 Proposed system flow

to give a diagnosis. They can also proceed forward to inputting new symptoms that are not found in the system and finding a treatment for the suggested illness. The image below shows the diagrammatic version of the system.

3.3 Cloud Platform

In this study, a database system was created. The cloud-based storage was created utilizing MySQL. The database was established at the backend of the website developed. This cloud-based storage was utilized to store various pig diseases, symptoms, treatments, and new symptoms noticed by the user. In the same database, a table for users was also created which stores the user's information that is to be used in creating an account for the user. This database was named med_diag (a medical diagnosis). The diagram below shows a sample database of how all the databases are interrelated and connected. PHP was used to write the code, in which the algorithm was also embedded for the proposed system; the version used for this system was 7.1.30. Laravel framework under PHP was also used. It is a powerful MVC (model view controller) PHP framework that is used to create simple, elegant full-featured web applications. It was used to create the user interface, making the system appear more user-friendly and easier to operate.

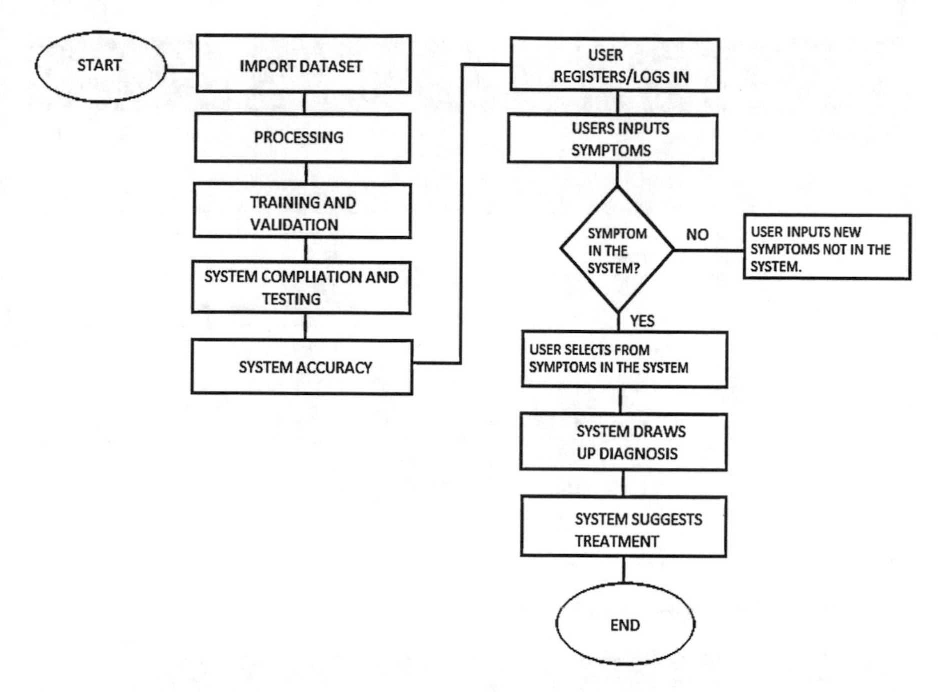

Fig. 2 Flowchart for the proposed system

4 Results and Discussion

This section shows the results gotten from the implementation and the testing of the system using the k-nearest neighbor machine learning algorithm. The interfaces of the implementations are also presented in this section.

The machine learning algorithm was implemented using PHP and MySQL for the database design. The result was obtained, and below are some interfaces from the system implementation and testing.

4.1 Home Page Interface

The home page interface is displayed in Fig. 3. This displays the home page of interactive software created for farmers. It shows what the user has access to from the admin's dashboard. Users need to register and create an account to access the software. They can view diseases, symptoms, and treatments of various swine diseases. From the administrative dashboard, registered users can be monitored; if new diseases come up, the disease list can be updated. The same can be done for symptoms and treatments.

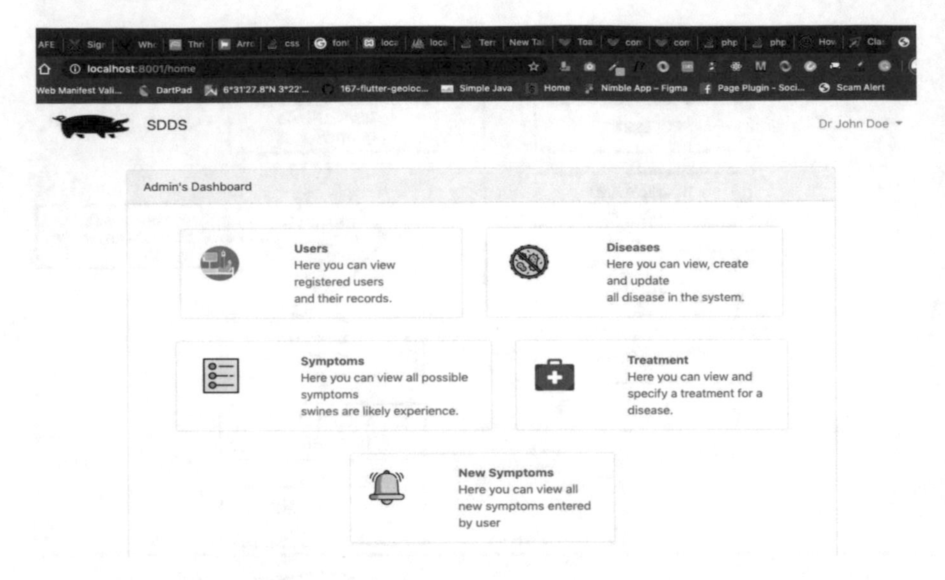

Fig. 3 Proposed system home page interface

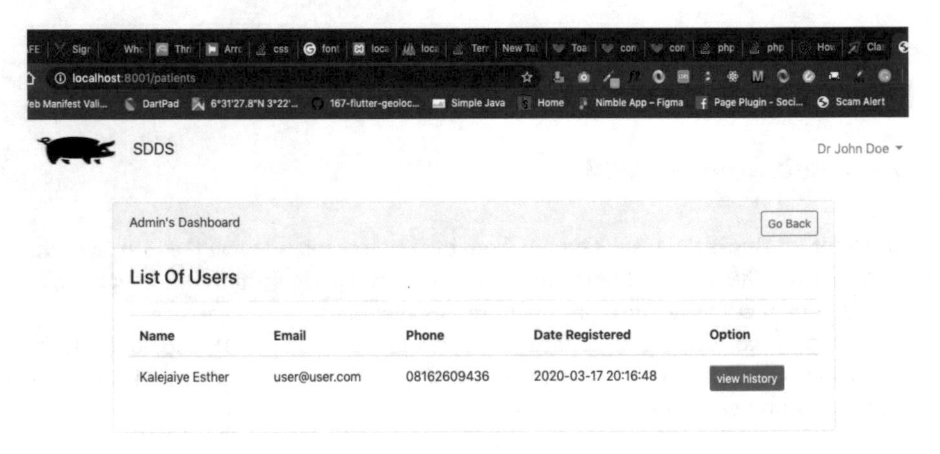

Fig. 4 Administrative back-end dashboard

Figure 4 shows the interface displaying the information of a registered individual or system user with information such as registered user, data registered, email address, and phone number.

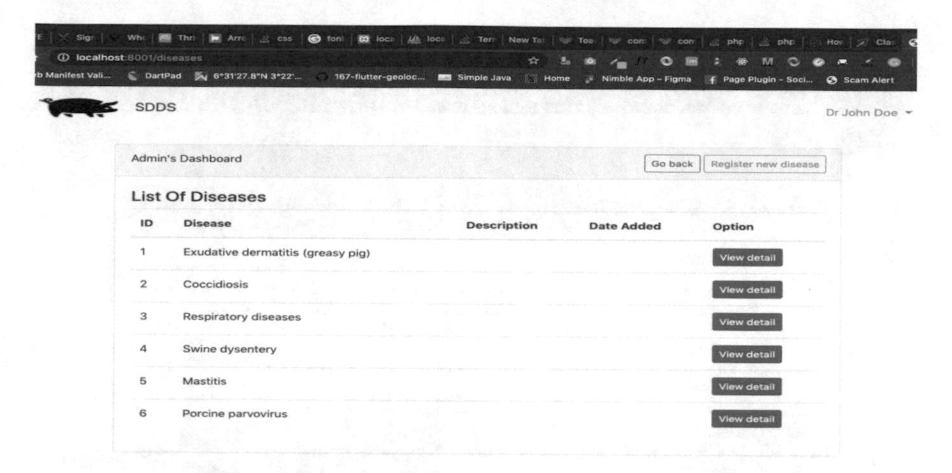

Fig. 5 List of diseases relating to the dataset used

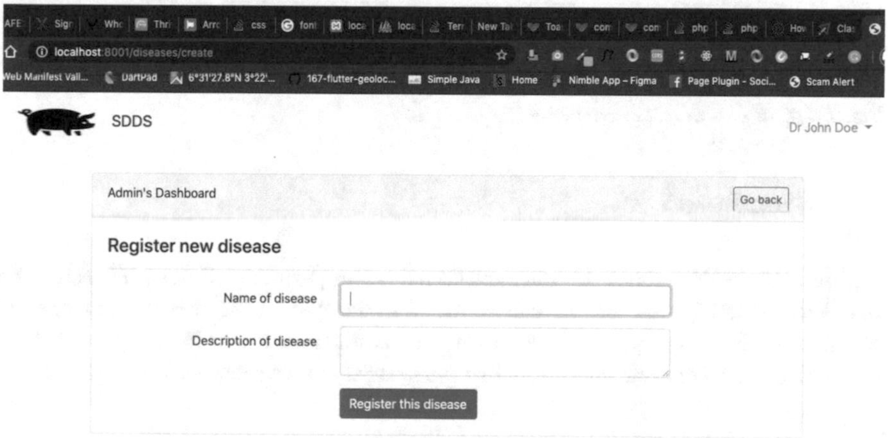

Fig. 6 Registration of new diseases

4.2 *List of Diseases*

Figure 5 shows the list of diverse diseases. This shows the list of diseases that can be found in the system. Five swine diseases were researched extensively; symptoms and treatments for each disease were also drawn from the data source listed above.

Figure 5 shows the list of the different types of disease infection with the pig's datasets used for the implementation of this study.

You can as well register new diseases discovered on the farm into the database system managed by the administrator as seen in Fig. 6.

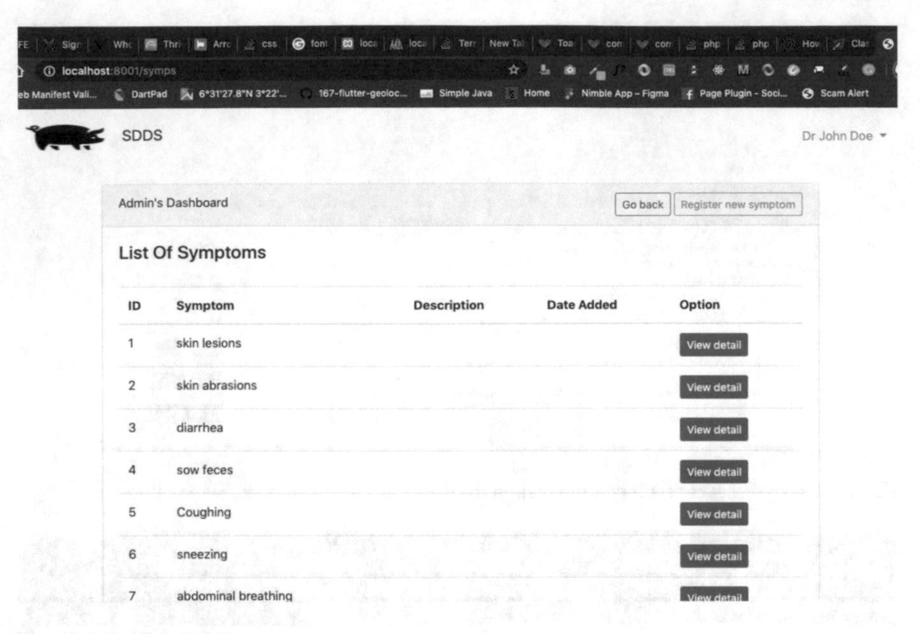

Fig. 7 List of swine disease symptoms

4.3 Symptoms List

Figure 7 shows the list of different symptoms relating to swine disease. This shows a list of registered symptoms that can be found in the system. All these symptoms are a collective list of symptoms drawn from the six diseases found in this system. This is where the farmers select from before the system makes a suggested diagnosis.

4.4 Treatment

Figure 8 shows the treatment for the disease interface. The treatment for the diagnosed disease can be suggested by the system. This offers the farmers an edge to take preventive measures needed to cure their sick pigs without wasting a lot of money to go to the veterinary doctor for a diagnosis. The figures below show that a user can type the disease which its treatment is needed, and it shows the result of the treatment of a disease.

Figure 9 shows the treatment for a particular disease which in this case is exudative dermatitis which is common in a greasy pig. You can also add and manage the system developed. Figures 10 and 11 show the create disease symptoms and create treatment symptoms, respectively.

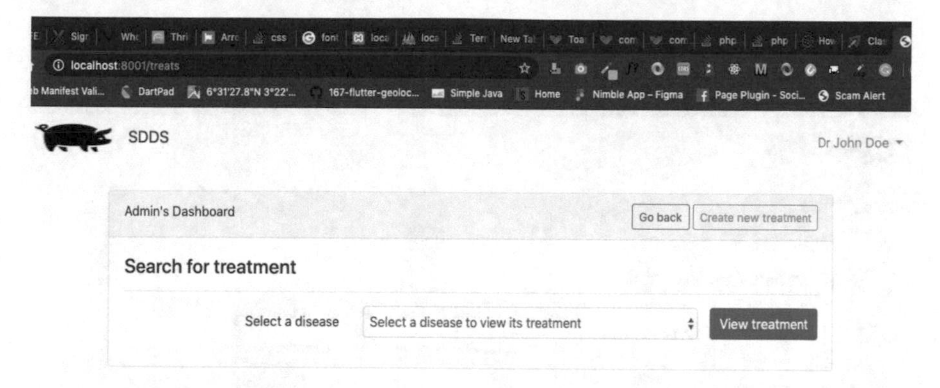

Fig. 8 Disease symptoms interface

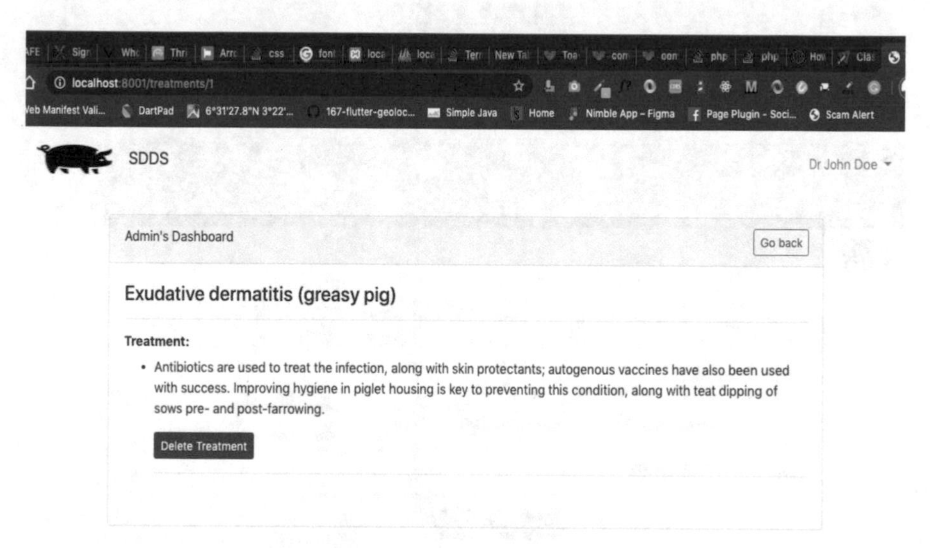

Fig. 9 Treatment for a particular disease

5 Conclusion and Future Works

According to Lasker et al. [35], technology has ushered numerous ways to drive mankind toward a better world, a better life. Mankind would be better off if technology is blended into our lifestyle. People rely on technology to find out solutions for problems they cannot solve by themselves. Farmers of this present age are going digital as a result of the fast growth of technology recently. Animal health-related issues when automated are of great importance to a farmer in need of answers at that particular point in time, especially when those answers come at really easy access. This project focuses on allowing farmers to know what is wrong with an infected pig without the intervention of an animal doctor. This study aims

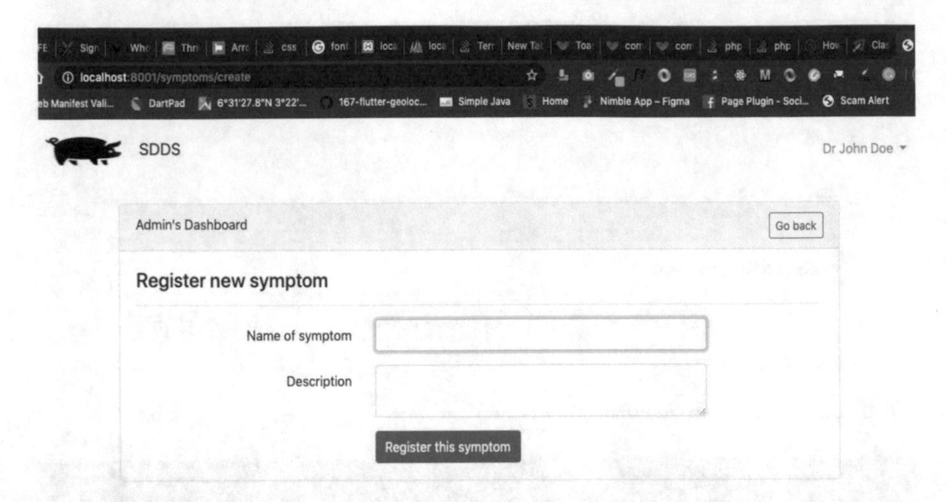

Fig. 10 Adding new symptom detected in the pig into the system database

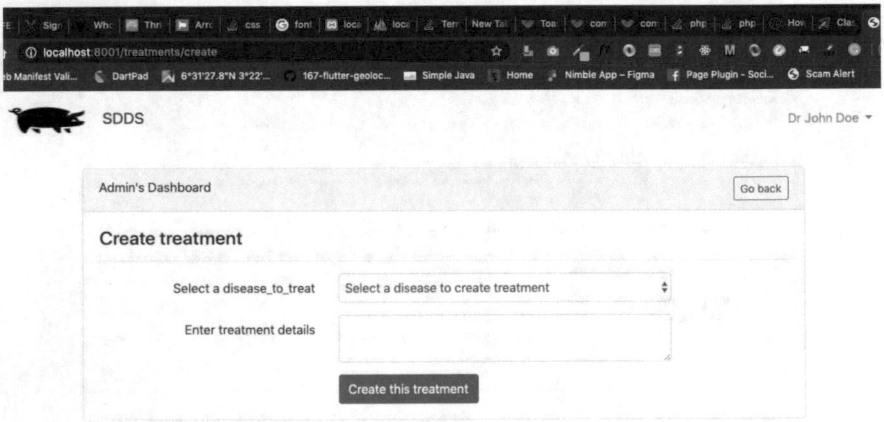

localhost:8001/treatments/create
SDDS
Dr John Doe
Admin's Dashboard
Go back
Create treatment
Select a disease_to_treat
Select a disease to create treatment
Enter treatment details
Create this treatment

Fig. 11 Creating new treatment for the disease detected

to develop a disease detection application that helps farmers to detect the disease a sick pig likely has. The system achieved the following objectives: to design a disease prediction application that detects the likely occurrence of swine disease using predictive algorithms, allow farmers to input the symptoms noticed in the sick pigs, derive correct symptoms from a pig suffering a particular illness, and lastly to prevent swine diseases from affecting other healthy pigs in the farm.

From this research, it is discovered that the symptoms used need to be read for understanding. For future purposes, pictorial symptoms can be added to the symptoms list. This would help farmers who do not have the knowledge of what some of the symptoms mean but recognize the symptoms on the pigs. Also, other detection algorithms should be used besides KNN; this is because as the size of k increases, computing time increases. Therefore, more algorithms should be tried out and tested.

References

1. Adebiyi MO, Ogundokun RO, Abokhai AA (2020) Machine learning-based predictive farmland optimization and crop monitoring system. Scientifica 2020:1–12
2. Jain L, Vardhan H, Nishanth ML, Shylaja SS (2017) Cloud-based system for supervised classification of plant diseases using convolutional neural networks. In: 2017 IEEE international conference on cloud computing in emerging markets (CCEM). IEEE, pp 63–68
3. Dharmasena T, De Silva R, Abhayasingha N, Abeygunawardhana P (2019) Autonomous cloud robotic system for smart agriculture. In: The 2019 Moratuwa engineering research conference (MERCon). IEEE, pp 388–393
4. Dugatkin LA (2020) Principles of animal behavior. University of Chicago Press, Chicago
5. Mulumba-Mfumu LK, Saegerman C, Dixon LK, Madimba KC, Kazadi E, Mukalakata NT et al (2019) African swine fever: update on eastern, central, and southern Africa. Transbound Emerg Dis 66(4):1462–1480
6. Sánchez-Vizcaíno JM, Mur L, Martínez-López B (2012) African swine fever: an epidemiological update. Transbound Emerg Dis 59:27–35
7. Oh SI, Jeon AB, Jung BY, Byun JW, Gottschalk M, Kim A et al (2017) Capsular serotypes, virulence-associated genes, and antimicrobial susceptibility of Streptococcus suis isolated from pigs in Korea. J Vet Med Sci 79(4):780–787. https://doi.org/10.1292/jvms.16-0514
8. Foote AD (2018) Sympatric speciation in the genomic era. Trends Ecol Evol 33(2):85–95
9. Gubbi J, Buyya R, Marusic S, Palaniswami M (2007) Internet of things (IoT): a vision, architectural elements, and future directions. Future Gener Comput Syst 29(7):24
10. Abdmeziem MR, Tandjaoui D, Romdhani I (2016) Architecting the internet of things: state of the art. In: Robots and sensor clouds, Studies in systems, decision and control, vol 36. Springer, Cham, pp 55–76
11. Sun P (2020) Security and privacy protection in cloud computing: discussions and challenges. J Netw Comput Appl 160:102642
12. Aazam M, Huh E-N (2014) Fog computing and smart gateway-based communication for a cloud of things. In: The proceedings of IEEE future internet of things and cloud (FiCloud), Barcelona, Spain, pp 27–29
13. Gubbi J, Buyya R, Marusic S, Palaniswami M (2013) Internet of things (IoT): a vision, architectural elements, and future directions. Future Gener Comput Syst 29(7):1645–1660
14. Aazam M, Huh E-N, St-Hilaire M, Lung C-H, Lambadaris I (2016) Cloud of things: integration of IoT with cloud computing. Architecting the internet of things: state of the art. In: Robots and sensor clouds, Studies in systems, decision, and control, vol 36. Springer, Cham, pp 77–94
15. Yaqoob I, Ahmed E, Gani A, Mokhtar S, Imran M (2017) Heterogeneity-aware task allocation in mobile ad hoc cloud. IEEE Access 5:1779–1795
16. Alamdari N, Tavakolian K, Alhashim M, Fazel-Rezai R (2016) Detection and classification of acne lesions in acne patients: a mobile application. In: 2016 IEEE international conference on electro information technology (EIT). IEEE, pp 0739–0743
17. Oladele TO, Ogundokun RO, Kayode AA, Adegun AA, Adebiyi MO (2019) Application of data mining algorithms for feature selection and prediction of diabetic retinopathy. Lecture notes in computer science (including subseries Lecture notes in artificial intelligence and Lecture notes in bioinformatics), 11623 LNCS, pp 716–730

18. Ikedinachi AP, Misra S, Assibong PA, Olu-Owolabi EF, Maskeliūnas R, Damasevicius R (2019) Artificial intelligence, smart classrooms and online education in the 21st century: implications for human development. J Cases Inf Technol (JCIT) 21(3):66–79
19. Alagbe V, Popoola SI, Atayero AA, Adebisi B, Abolade RO, Misra S (2019) Artificial intelligence techniques for electrical load forecasting in smart and connected communities. In: International conference on computational science and its applications. Springer, Cham, pp 219–230
20. Xu H (2017) Machine learning-based data analytics for IoT devices. Nanyang Technological University. https://doi.org/10.32657/10356/72342
21. Yang K, Sun LP, Huang YX, Yang GJ, Wu F, Hang DR et al (2012) A real-time platform for monitoring schistosomiasis transmission supported by Google Earth and a web-based geographical information system. Geospat Health 6:195–203
22. Valdes-Donoso P, VanderWaal K, Jarvis LS, Wayne SR, Perez AM (2017) Using machine learning to predict swine movements within a regional program to improve control of infectious diseases in the US. Front Vet Sci 4:2
23. Kumar N, Makkar A (2020) Machine learning techniques. CRC Press, pp 67–85. https://doi.org/10.1201/9780429342615-4
24. Gope D, Dasika G, Mattina M (2019) Ternary hybrid neural-tree networks for highly constrained IoT applications. arXiv preprint, arXiv:1903.01531
25. Merenda M, Porcaro C, Iero D (2020) Edge machine learning for AI-enabled IoT devices: a review. Sensors 20(9):2533. https://doi.org/10.3390/s20092533
26. Idowu AP et al (2013) Data mining techniques for predicting immunize-able diseases: Nigeria as a case study. Int J Appl Inf Syst 5(7):5–15
27. Sohail MN, Jiadong R, Uba MM, Irshad M (2019) A comprehensive look at data mining techniques contributing to medical data growth: a survey of researcher reviews. In: Recent developments in intelligent computing, communication, and devices. Springer, Singapore, pp 21–26
28. Sellappan P, Rafiah A (2018) Intelligent heart disease prediction system using data mining techniques. IJCSNS Int J Comput Sci Netw Secur 8(8)
29. Brisimi TS, Xu T, Wang T, Dai W, Adams WG, Paschalidis IC (2018) Predicting chronic disease hospitalizations from electronic health records: an interpretable classification approach. Proc IEEE 106(4):690–707
30. Idowu PA, Aladekomo TA, Agbelusi O, Akanbi CO (2018) Prediction of pediatric HIV/Aids survival in Nigeria using naïve Bayes' approach. J Comput Sci Appl 25(1):1–15
31. Schaefer AL, Cook N, Tessaro SV, Deregt D, Desroches G, Dubeski PL et al (2004) Early detection and prediction of infection using infrared thermography. Can J Anim Sci 84(1):73–80
32. Muriu P (2016) Mobile-based expert system model for animal health monitoring: cows disease monitoring in Kenya. Doctoral dissertation, Strathmore University
33. Zetian F, Feng X, Yun Z, XiaoShuan Z (2005) Pig-vet: a web-based expert system for pig disease diagnosis. Expert Syst Appl 29(1):93–103
34. Sinha P, Sinha P (2015) Comparative study of chronic kidney disease prediction using KNN and SVM. Int J Eng Res Technol 4(12):608–612
35. Schindler LA, Burkholder GJ, Morad OA, Marsh C (2017) Computer-based technology and student engagement: a critical review of the literature. Int J Educ Technol High Educ 14(1):1–28

Privacy and Trust Models for Cloud-Based EHRs Using Multilevel Cryptography and Artificial Intelligence

Orobosade Alabi, Arome Junior Gabriel, Aderonke Thompson, and Boniface Kayode Alese

1 Introduction

There is a rapid transformation of the worldwide network of computers into a sort of single extremely large "virtual" computer, often referred to as the "cloud". The cloud offers extremely high performance computing power, which is accessible via the Internet. In reality, the cloud can be seen as a scalable, effectual and cost-effective solution for handling the increasing storage and computing needs or requirements.

Quite often, cloud computing is seen as an Internet-based computing service (such as servers, networks, storage, applications, and other services). Like other utility services existing in today's electronic society, cloud services are readily accessible on request [1]. As a matter of fact, cloud computing was cited in the year 2006, as the fifth of world's most essential needs (the others being water, electricity, gas, and telephone). Cloud computing is also seen as a blend of technology ideas such as virtualization, distributed computing and high-performance computing, which makes room for enhanced utilization of physical and/or pooled configurable computing resources [2–4]. Besides, the cloud computing paradigm offers new business structures that in turn allow for quite a number of benefits to be enjoyed in the health sector, and even other facets of human existence like education, governance, religion, communication and even sports [5–6].

Stakeholders in the health sectors are rapidly adopting new technologies like the mobile technology, Internet of Health Things (IoHT), wearable devices, and cloud computing. The cloud now has large volumes of health records in digital format that are shared via diverse healthcare institutions or organizations. This increasing adoption of the cloud computing paradigm as a means for rendering

O. Alabi · A. J. Gabriel (✉) · A. Thompson · B. K. Alese
Federal University of Technology, Akure, Nigeria
e-mail: ajgabriel@futa.edu.ng; afthompson@futa.edu.ng; bkalese@futa.edu.ng

S. Misra et al. (eds.), *Artificial Intelligence for Cloud and Edge Computing*,
Internet of Things, https://doi.org/10.1007/978-3-030-80821-1_5

medical services is highly beneficial and brings a lot of ease to the lives of patients and healthcare establishments alike. However, all these come with increasing growth of other concerns, especially those that borders on privacy of patients' information. There is a serious need to make adequate and effectual provisions for enhancing users' trust on cloud service providers and to improve the security/privacy of information. This implies that all hands must be on deck to ensure security from source to destination as health data flies around the Internet highways, whether they be sensors' recorded data, apps, research databases, websites, or even electronic health records (EHR) [7–9].

Indeed, the rapid improvement of technology implies that several smart devices exist, with which access can be made to health records anywhere, anytime. These leaves a lot of data such as consultation reports shared across the Internet, at the mercy of unauthorized individuals who could have malicious intentions [10–12]. Protecting the secrecy of health in the cloud implies that some crucial requirements be carefully considered by electronic health facilities. These crucial requirements include: privacy/confidentiality, integrity/authenticity, accountability, audit, non-repudiation, and anonymity. The security of the data store in an EHR is very crucial to the owners of such medical data. These issues have slowed down the adoption of cloud-based computing.

Several methods exist for hacking patients' data. This underlines the high significance of multiple security/privacy systems. Furthermore, the fact that the exact repository of a user's data is not known to the user implies that the user will require some level of trust from the cloud provider to be sure that his data has not been accessed or controlled by another user. Additionally, privacy risks reside in the location and the multi-tenancy property of the cloud. This creates room for simultaneous servicing of several customers (or clients) from the same instance of software, via secure logical separation of resources.

Indeed, even though privacy issues have been a general concern in computing, it is more hazardous in cloud computing because an information that is to be confidential can be made known to millions of people in a few seconds due to the fact that numerous people in different locations can connect together to get information or share resource that is not even located near them. The integrity of the information shared in a cloud environment is also of great concern, especially as this information can be interrupted, edited or altered. Access control mechanisms could be deployed for determining who can access what information on a system, while a high-level solution for ensuring security and trust in cloud computing could be ones that are based on cryptographic schemes [9].

The rest of the chapter is structured in a way as follows: Section 2 contains the related literature review, while Sect. 3 presents the security/privacy mechanism developed in this work. Then, Sect. 4 contains the trust framework, while Sect. 5 presents the methodology for the design of the proposed or new system. Discussion of system implementation results was done in Sect. 6, while the conclusion is in Sect. 7.

2 Related Works

The Internet offers services that are being enjoyed in almost all facets of human existence today. Some of these facets include education, business, governance, shopping, management of human resources, communication, prediction of climatic conditions, agriculture and electronic voting and even in healthcare for ensuring seamless access to web-based or cloud-based health services [12]. Usually, these services are enjoyed over the public Internet rather than voice or face-to-face interactions. As a result, a major concern becomes that of the privacy of the users and, even more importantly, the trustworthiness of these services. Even though so much efforts have gone into these, there really have been no putative mechanism or tool for specifying, estimating, or even reasoning about the trust. It thus mean that there exist an urgent and serious need for a high-level, theoretical approach (or model) for handling trust issues. Such a model can be easily incorporated into software applications and used on any platform.

Electronic health records are becoming a growing tread in health services and in recent time, with popular uses of portable devices, smartphones and wireless communication. While these facilities have paved way for usage convenience and mobility, the instantaneity of in-patient health care can be significantly improved. Cloud computing has helped tremendously in giving on-demand service to users not minding the location of their data. These cloud services often make life easy for users in terms of saving cost (money, time) and offering huge convenience for users while consuming these services. There have been an urgent need for a cloud-based electronic health record with patient user control, hereby health center can communicate with each other in case of transfer or referral of patient. This will help electronic health record to promote efficient communication of medical information within and outside the health centers and thus reduce costs, eradicate administrative overheads, increase speed and offer a more secure system where environmental disaster, theft or loss of medical information can also be prevented. Cloud computing will help to achieve these and many more by allowing the digital format of the medical record to be shared, among other unified health centers alongside with sharing of resources and medical personnel. Nevertheless, to fully enjoy these budding benefits, certain crucial challenges, especially the privacy of patients' medical information or records, who has the right of access to the medical data and also the trustworthiness of EHR medical data, must be ameliorated. The privacy model will help to address the illegal access of unauthorized user and help to protect the medical data at rest or in transit from eavesdroppers. Lack of proper attention to the privacy situation of an EHR has actually prevented the embracement of this development in the medical world [9]. Patients are worried about the security (confidentiality, integrity, availability, privacy) of their data in the EHR. Furthermore, the fact that the health records of patients become available to be viewed by medical personnel as patients get transferred from one health facility to another implies that trust is a crucial factor [10]. There must be an effective way in determining who is to be trusted with patients' medical information [11]. Indeed, the trust value (or integrity) of EHR medical information/record is subject to the trustworthiness of the medical personnel. Measuring the trustworthiness of an agent could be done using reputation systems. Such systems can provide the compiled confidence measure of an agent over a period of time. Reputation systems are mostly developed to calculate the current (present time, t_0) trust value of an agent. That current trust/confidence measure can then be used to predict the future behavior of the agent or service provider [6, 13].

This current research covered in this chapter was motivated by the numerous advantages of cloud that can promote the medical world. Our research aims to produce an improved health record system that conveniently eliminates the major challenges (privacy and/or security of data) of EHR record system in cloud.

2.1 Review of Selected Related Research Works

A number of researchers have made attempts at scientifically addressing some of these aforementioned issues. A summary of their works is further presented as follows:

In [11], a solution for ensuring the privacy of patients' data by preventing indirect accesses to such data was proposed. The research work also computed the trustworthiness of medical data using beta and Dirichlet reputation system. However, there is a need for better and actual prototyping of their proposed system.

The work in [14] designed a fine-grained encryption solution for protecting individual items within an EMR. Likewise in [15], the authors proposed a secure EMR service system, which they reported as able to offer a secure environment for EMR and safeguard the EMRs especially when they are being transferred between the cloud system and users/clients devices. The security of their EMR system was built on ECC scheme. This offered higher level of security to electronic medical records with reduced or minimal computations on edge devices. However, that single level encryption approach did not consider trust or access control mechanisms towards eliminating the activities of unauthorized users.

The research in [16] highlighted the several barriers that hampers efficiencies of EHR management and healthcare delivery in Nigeria. They highlighted the need for EMRs to be moved to the cloud, in order to improve accessibility to health-care delivery.

In [17], an EMR system was developed to computerize the activities of medical personnel in Nigeria. The study was targeted at eliminating the challenges inherent in traditional paper-based record-keeping systems. However, the proposed system did not consider privacy of patient's records. The system will also suffer from inadequate storage especially due to data explosion.

The work in [18] designed and implemented an EMR system on the cloud for effectively managing medical information dissemination. Their system has a main database for managing patients' records in all collaborating hospitals. However, as a limitation, the security of such information was not considered.

In [19], an ABAC access control mechanism for securing the privacy and security of EHRs was developed. The trustworthiness between the patient and the EHR can be improved by incorporating the "patient content" as an integral part of EHR.

The study in [20] proposed a location-and-biometrics-based user authentication as well as a cryptography-based technique for hiding the existence of EHR data. Their system did not cater for the issue of robust key exchange management between various parties involved.

The research paper in [21] proposed a cloud-based EHR system and access control mechanism. The authors used ElGamal algorithm in ensuring protection of the security of information. The trust aspect of such systems was not considered in this work.

In [22], a privacy-aware ABAC model for EMRs was proposed. Also the use of encryption and digital signatures was reported. However, developing strategies for determining the trustworthiness of such systems was completely ignored. Also in [23], the authors proposed an E-healthcare system that is trust-based. However, this solution is interested only in access control policies and cannot act as a decision support tool for the user.

The work in [24] discussed several access control models, highlighting their strengths, weaknesses and roles towards promoting data privacy in cloud-based applications. Similarly in [25], a survey of cloud privacy risks and challenges was presented. The major existing solutions were presented and examined. The authors also provided guidelines and considerations towards addressing privacy concerns in a better fashion. Now, the study in [26] proposed an ABAC framework for enforcing policies and ensuring that access to the EHR system is granted only to legitimate users. As a limitation, determination of the trustworthiness of such EHR systems was ignored. Even the article in [27] proposed an access control mechanism for securing cloud EHR systems. The study also failed to consider patients' privacy preservation as well as the trustworthiness of such EHR systems.

The research in [28] made attempt at identifying the relevance of trust and privacy to cloud-based education systems. The authors were able to identify a finite set of items, with which they designed questionnaires towards further relevant empirical studies. Likewise in [29], the issue of privacy in cloud computing was discussed extensively. The article focused on the particular components that take part in a cloud privacy system. Also, factors that influence hardware production, as well as customer and provider privacy, were also identified. Their article was more of a positional article than an experiment and result disseminating article.

The article in [30] proposed a system for detecting abnormality in breasts especially at an early and manageable stage. Their system works based on the powers of IoT and even the cloud. Although this work is very good, the secrecy of patients' data was neglected.

The study in [31] presented some major requirements for ensuring security in e-health systems on the cloud. Similarly, the article in [32] discussed the geneses of trust-based computing. However, no experimental findings were reported. Likewise, the article in [33] attempted to highlight the major trust issues with cloud computing. It opined that often cloud customers are not provided with the chance of agreeing or disagreeing with terms and conditions. Indeed, only a few customers ever read their contracts before consenting to its terms. The study in [34] was also able to ascertain probable methods for improving the privacy of cloud-based e-commerce service consumers.

Our chapter contribution therefore has a specific objective of designing a privacy model for patients' medical records using multilevel encryption. The paper also attempts to develop a trust model using reputation estimation technique that is expected to work based on the Josang belief modelling with subjective logic, as presented in [6].

3 The Security/Privacy Mechanism

This research work is partly about the development of EHR privacy model using multilevel encryption. This arrangement combines secret-key (AES) as well as public-key (ECC) cryptography schemes in ensuring that privacy of users' sensitive medical data are preserved. Although public-key crypto-schemes are believed to often provide more robustness, they suffer challenges that pertains to speed and hardware cost. In contrast, private-key crypto-schemes offer cost-effective methods especially in terms of speed and demand for computational resources. Combining both public-key and private-key crypto-schemes could be a better option to using either of them. This chemistry would allow systems to enjoy the strengths of both approaches.

3.1 The Privacy Model

To cater for security/privacy of patients in the proposed system, cryptographic schemes were adopted. The proposed multilevel security framework is made up of two crypto-schemes. The first is a symmetric scheme (i.e. AES), while the second is an asymmetric scheme (i.e. ECC).

The functionality of the proposed privacy-aware EHR system is in two phases. Firstly, all personal information at the local primary hospital where doctors are based and where patients will have to first register requires protection from the eavesdropper. In this scenario, the personal information is first encoded using the AES crypto-scheme on premise, after which, the ECC crypto-scheme is used for encrypting/encoding the key generated from AES algorithm encryption in the cloud. This concept is as presented in Fig. 1.

Furthermore, the privacy model this chapter is proposing for EHRs is as shown in Fig. 2.

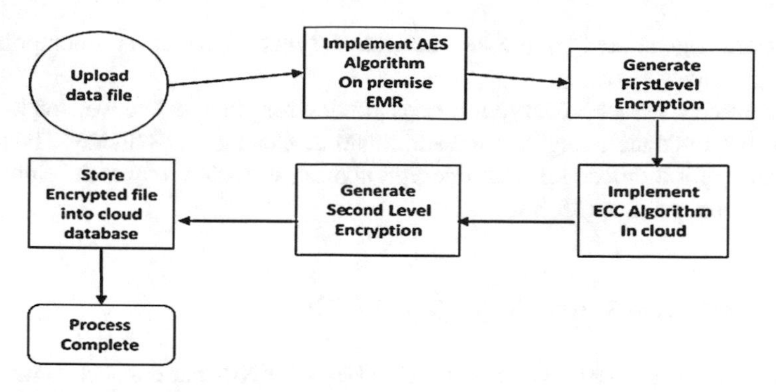

Fig. 1 Block diagram of multilevel encryption

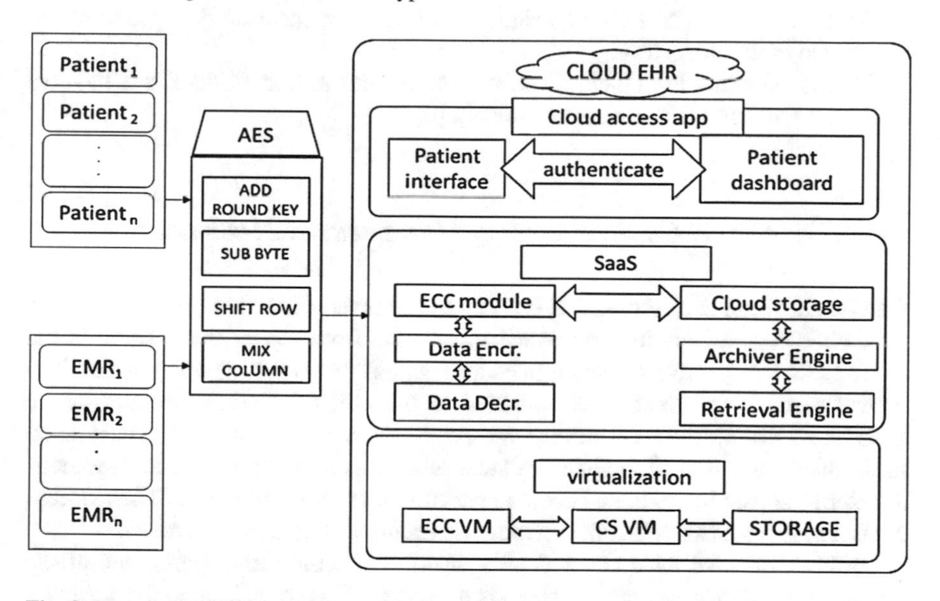

Fig. 2 The proposed EHR privacy model

3.1.1 The Advanced Encryption Standard (AES)

The AES algorithm is a private-key crypto-scheme known to be robustly resistant to attacks and also reasonably fast. It uses permutation as well as substitution network for achieving encoding of plaintext messages, and it is amenable for implementation in hardware and software [9, 35].

The AES algorithm uses a single key for protecting the medical data of the patient on-premise in an EMR. This same key is used by both servers and clients for gaining access to encrypted files. The patients' medical information transmitted across the Internet are quite sensitive and personal data and therefore requires protection via encryption, before transmission. The key required for encrypting these medical data is often generated by the key derivation function which is a component of cryptographic systems.

In summary, the AES encryption steps provide for plain text conversion to cipher text in 10 rounds including the final add round key, using a 128 bit key. The inverse of the encryption procedure is the decryption process, which returns the cipher back to the original plain text form.

3.1.2 The Elliptic Curve Cryptography (ECC)

The second level of encryption is the cloud-based EHR that connects other EHRs (of local primary hospitals) together to facilitate the sharing of resources, or even medical personnel. The ECC algorithm is adopted for ensuring data protection at this (cloud computing) level.

ECC is asymmetric in nature, and its security lies in how difficult it is to solve discrete logarithm mathematics or problem [9].

3.2 The Access Control Aspect of the Security Mechanism

One fundamental part of the EHR is that which connects identities, thereby making it possible for patients to connect with their data from any of the hospital in the cloud EHR irrespective of which primary hospital the patient first registered. The patient medical information is uploaded to the cloud for access by any of the hospital in the network whenever such need arises. Based on some access setup rules, the access control module evaluates all access requests from client systems, to determine whether to grant access to patients' records or otherwise. These access control policies ensures that the privacy and security of data is guaranteed.

In this work, we have adopted what seems to be the most appropriate of the various types of access control schemes for our EHR system. That is the attribute-based access control (ABAC) technique. ABAC seems to be more flexible than other access control techniques. In ABAC, access to patients' records is granted to users according to the policy setup as well as the attribute(s) such users exhibit.

4 The Trust Architecture

The trust architecture of the proposed EHR system is based on the subjective reasoning or logic aspect of the artificial intelligence (AI) field. The theory of subjective probabilistic logic was adopted in order to combine the capacity of probability theory with that of binary logic, to handle likelihood and make inference from argument structures, respectively. This chemistry provides a very powerful formalism.

4.1 Trust Concept

Trust is the proportion in which a given entity meets the performance (or behavior) expected of it by other entities. Trust model helps in the identification of trust concerns affecting a system and how, for instance, reduced levels of trust may negatively affect the reputation (hence, usability) of such systems [36–37].

Conceptually, five aspects of trust were identified in the McKnight's Trust Model. These aspects are *trust activity*, *trust intent*, *trust credibility*, *institution-based trust* as well as *trust nature* [37–39]. Trust activity refers to act(s) causing an increase in the threat to a trustor and thus making the trustor susceptible to the entity being trusted. Trust intent shows that a trustor agrees to participate in trusting behaviors with the trustee. A trust intent denotes a trust decision which usually results in a trust activity. Trust credibility is used to describe the subjective belief that a trustor has on the trustworthiness of a given trustee. Institution-based trust has to do with the conviction that suitable and adequate organizational conditions are available for increasing the prospect of realizing a worthwhile result. Disposition to trust describes the overall tendency of a trustor to rely on others across a wide range of conditions.

Regardless of his trust credibility about them, trust intent as well as trust credibility relies on circumstances and the trustee. Institution-based trust, on the other hand, has more to do with situations/circumstances. Disposition to trust is not reliant on circumstances and trustee.

4.2 Managing Trust

Managing trust is a key parameter in the acceptance and development of cloud computing. The framework for trust management has three different layers, which work together in achieving a trusted and reliable cloud-based system [8]. Specifically, these layers are responsible for sharing trust feedbacks, evaluating trust as well as distributing trust results (Fig. 3).

4.3 The Trust Management Procedure (TMP)

Two major actors (the *trustor* and *trustee*) exist in a system that manages trust. The trustor builds trust, while the trustee manages such trust. Now, the trust management procedure is made up of three major phases: *service discovery, trust management process (TMP) selection and measurement* and *trust assessment* and even the *trust evolution phases.*

Service discovery refers to the automatic detection of devices and services offered over a network. Then, the TMP selection and measurement phase has the

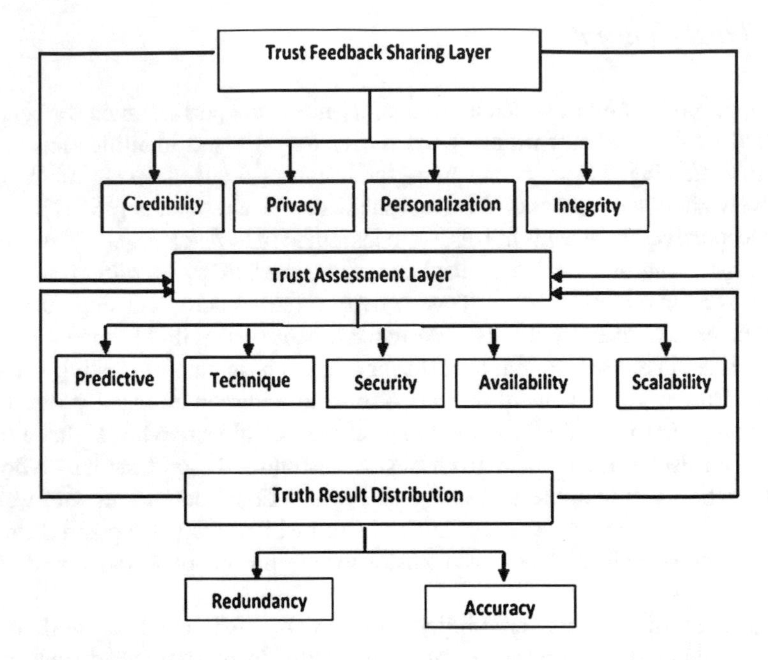

Fig. 3 Framework for managing trust

objective of qualitatively trying to find parameters (such as privacy and security, performance and accountability) for assessing the trustworthiness of cloud services. Now, the trust assessment phase evaluates the trust levels between the trustor and the trustee.

Furthermore, the most common techniques for developing a trust model include the *policy technique*, *reputation technique*, *predictive technique* as well as the *recommendation technique*. In this chapter, the reputation technique is adopted for modelling the trust aspect of the proposed system.

5 The Proposed System Design Methodology

The proposed system adopts a multilevel cryptography approach for achieving security (privacy) in electronic health system. This multilevel cryptography approach comprises two cryptography schemes: First is the AES, which uses a symmetric encryption algorithm for securing data at the local hospitals and also used for generating and sharing keys between the local hospital and the cloud. At this phase, the medical doctors interact with patient records for consultation and possible diagnosis of the patient (subject) through the patient dashboard on the electronic health record and grant permission to medical personnel (object) based on the attribute of the medical personnel. Second, the elliptic curve cryptography, an asymmetric modern cryptography scheme, is adopted for securing the patients' data in the cloud-based electronic health record subsystem to achieve multi-tenancy which permits different hospitals in the cluster to have authorized access to the patients' profiles.

This research work adopted the Software-as-a-Service (SaaS) cloud model, for the handling of the health records in an e-health web portal. Several standalone health center records are linked to a cloud central server that has capacity for admitting virtual machines as tenants. This facilitates synchronization such that it covers the heterogeneity of all the cooperating (or member) local hospital's EMR. The entities that are referred to as tenants here are secure facilities that keep data (or information) in the several (often diverse) healthcare units, given the Internet as the main link of communication between them. The cloud-based EHR houses the interface, a common platform which stands as middleware allowing all the cooperating hospitals' EMR standards. The middleware is open to any type of EMR standard and uses network connections (or the Internet) for facilitating its own communication with each member hospitals (or healthcare unit).

Moreover, to cater for integrity, confidentiality and other common aspects of security in the proposed system, key attributes of targeted users which are the patients attribute, the doctor attributes, the health authorities rules and the environmental condition, are harnessed for use in the access control (based on attributes, i.e. ABAC) mechanism of the system. The ABAC mechanism controls the doctors' request to view the stored patients' profiles or data, at both the local and even the cloud level.

Now, the trust management model is expected to handle the evaluation of the trust value of the medical personnel(s). To cater for this trust feature, the Josang belief model with subjective logic was adopted. This is a framework of probability that represent a specific belief metric called opinion, $\boldsymbol{w}_x^A$, with quadruple parameters:

b denotes the belief
d stands for doubt
u signifies uncertainty
a represents the base rate in the lack of proof or prior trust such that:

$$b + d + u = 1 \tag{1}$$

This implies that $b, d, u \in [0, 1]$ and $a \in [0, 1]$

The extent of uncertainty can be inferred as the lack of proof(s) in support of belief or doubt (disbelief). The values of the quadruple parameters are between 0 and 1 given by the rating of the patient.

The expectation value for the patient's opinion is:

$$E\left[\mathrm{w}_x^A\right] = b + au \tag{2}$$

Taking r and s as positive and negative past observation, respectively, the probability density function is expressed as:

$$\alpha = r + 2a \tag{3}$$

$$\beta = s + 2\,(1 - a) \tag{4}$$

The illustration of trust in subjective logic correctly matches the picture of reputation in Bayesian reputation system.

The Bayesian reputation with rating vector $p = \left[\begin{smallmatrix} r \\ s \end{smallmatrix}\right]$ for patient x of doctor z at rating time t_R given that current time t is P_z^x; therefore, since trustee can change over time λ, longevity factor is introduced as in Eq. (5).

$$p_{z,t_R}^{x,t} = \lambda^{t - t_{R-}}\ P_{z,t_R}^{x} \tag{5}$$

The reputation score at time t is given by the mathematical relation in Eq. (6);

$$Z = \frac{r + 2a}{r + s + 2} \tag{6}$$

Such that, the reputation rating in the light of the score can be captured as in Eq. (7).

$$p^t\,(\mathrm{z}) = E\left[\beta\ \left(p^t\,(\mathrm{z})\right)\right] \tag{7}$$

The estimated value of a cloud provider's reputation is provided by the expected value of the distribution given in Eq. (7).

The score is in the range of [0, 1]. Trust models that work on the principle of reputation are typically used for estimating the trust value at the present time (say, t_0) considering the past behavior(s) of the entity in question. In the proposed model, services are weighed by users using some selected (may be standard) fuzzy-related metrics. Figure 4 shows how the trustworthiness of a cloud provider can be derived for users' feedbacks using fuzzy logic to calculate the reputation value.

Using fuzzy model of trust, a patient (or any cloud user) rates the quality of service (QoS) gotten from the cloud provider in linguistic parameter, using five fuzzy sets to represent the reputation score. Parameters in the range of 0–1 are then mapped using a membership function, while the rules for carrying out reasoning, given such a set of data, are provided by fuzzy logic. The output of the reputation system is the trust recommendation generated based on the experiences of users. Figure 5 shows the process entities in the reputation system, which begins with a cloud/patient's feedback, putting into consideration the cloud provider's expertise, reliability and time decay to calculate the reputation score, which is the trust value. The reputation system then recommend trusts to other users based on the reputation score assigned to the cloud provider.

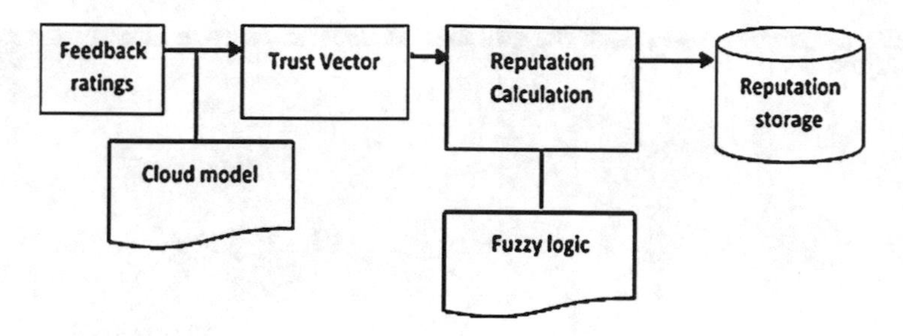

Fig. 4 Reputation measurement approach

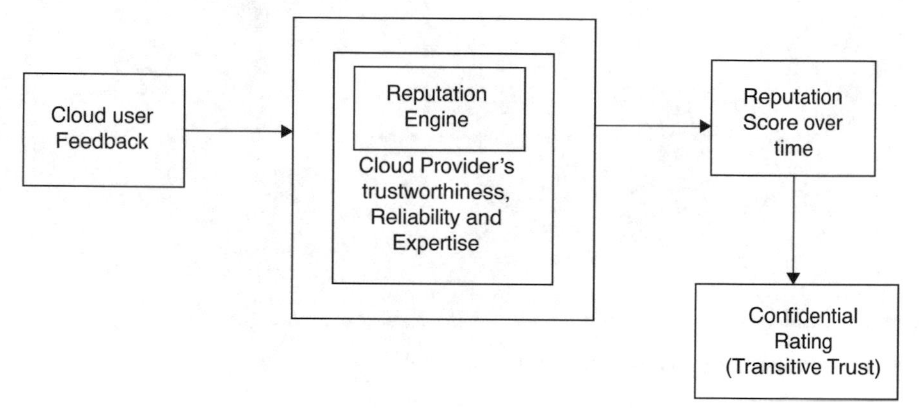

Fig. 5 Architecture of the reputation system

6 Results and Discussions

To evaluate the proposed EHR system, an e-health web portal was developed and configured for end users (doctors, patient, administrators, pharmacist and lab attendant). The tools for development or implementation of the e-health web portal includes the dotnet core 2.0 framework with C Sharp programming language 9.0; Angular JavaScript 6.0 with each module built as a containerized application using docker 2.3, HyperText Markup Language 5 and Cascading Style Sheet 3; MSSQL; and Azure cloud service provider.

The web portal allows for exchange of messages and responses between the cloud server and the local health center. A 2-way access mechanism is provided by the web portal for the end user to receive, view or even modify (where necessary). This access is, however, dependent on the users' privileges.

The information of any patient can be accessed via the web portal, by any authorized person, whether it be on the cloud server or on the local data repositories of healthcare centers. The job of managing the medical personnel as well as the patients on the portal belongs to the individual designated as the administrator. Only

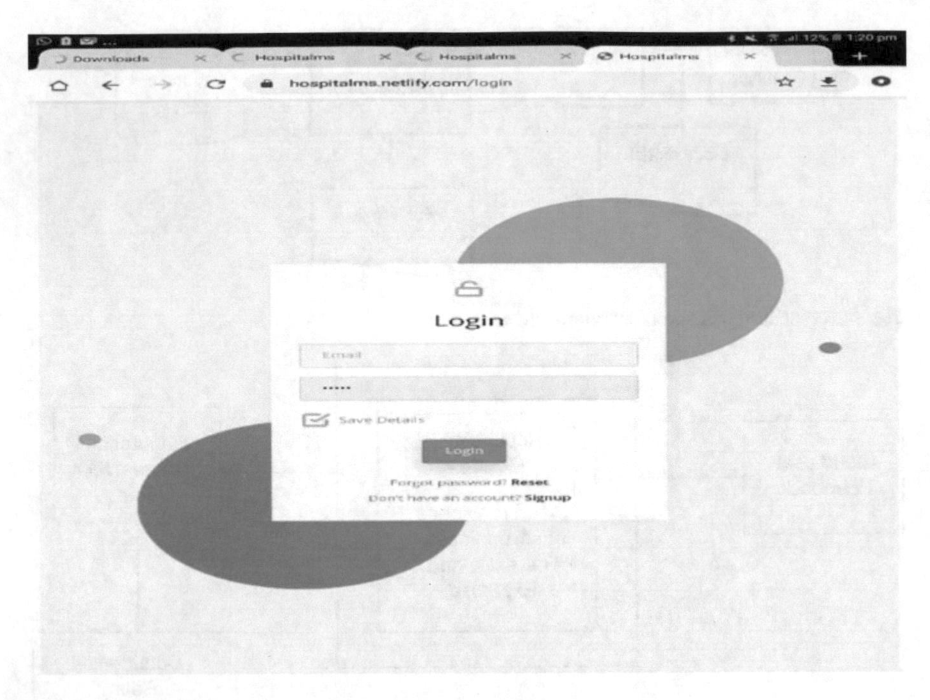

Fig. 6 Patients' login page on the e-health web portal

the medical doctors or consultants are given high-level access privilege to patients' complete medical record, so as to ensure improvement in users' privacy.

As in Fig. 6, the EHR user seeking authorization supplies username and password in his or her respective dashboard in order to gain access into the system.

Figure 7 presents a doctor's dashboard with requests awaiting patients' approval. These requests for more access privileges to patients' records appear on individual patient's dashboard; patients are at liberty to grant/deny access, except during emergencies.

The evaluation of the proposed system was done in two (2) phases. We first evaluated the security/privacy aspect of our proposed system. Then, secondly, we evaluated the trust aspect of the proposed system.

6.1 Performance Evaluation of the Proposed EHR System

Each of the crypto-algorithms used in this multilevel cryptography security approach was carefully selected based on their strengths, which include resistance to timing and other common attacks, faster encryption/decryption speed as well as their suitability for use on resource-constrained devices or environments.

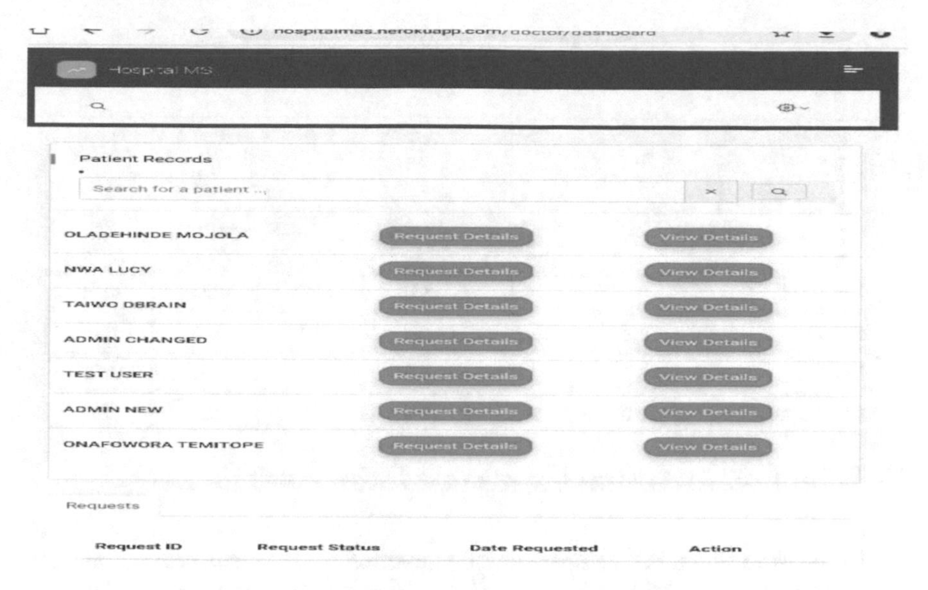

Fig. 7 Doctor's dashboard showing requests awaiting patients' approval

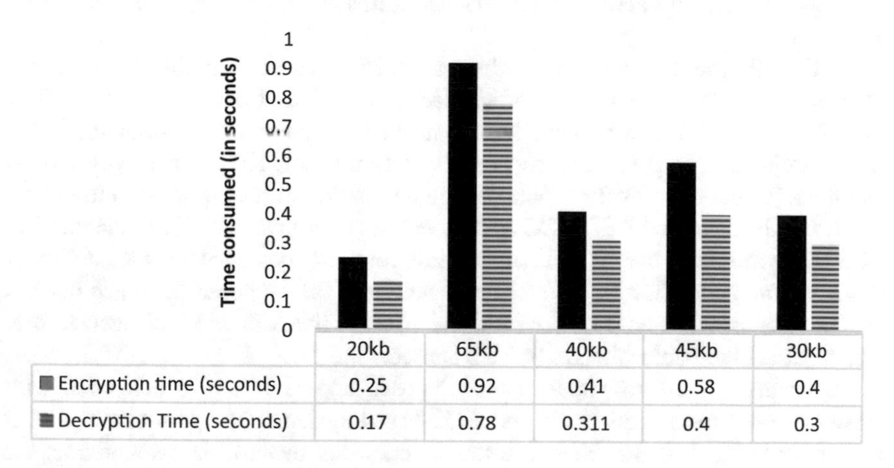

	20kb	65kb	40kb	45kb	30kb
Encryption time (seconds)	0.25	0.92	0.41	0.58	0.4
Decryption Time (seconds)	0.17	0.78	0.311	0.4	0.3

Fig. 8 Chart of the encryption and the decryption time

A performance evaluation of the proposed AES-ECC multilevel cryptography-based privacy model was carried out. The proposed approach was compared to those based on either AES or ECC in terms of overall system's computation time, system throughput as well as output file size. These results are presented in Figs. 5, 6, 7 and 8. Systems' computation time in this case refers to the time taken to encrypt or decrypt a given patient medical record.

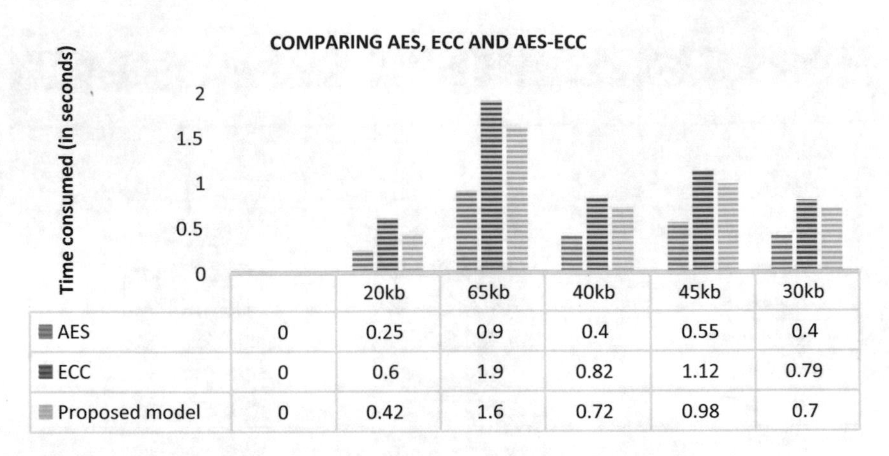

Fig. 9 Overall system computation times across AES, ECC and AES-ECC

The chart in Fig. 8 is a comparison of encryption times versus decryption times recorded for input plaintext medical records of varying sizes. It was noticed that the encryption procedures across the crypto-algorithms were slower than the decryption procedures.

In Fig. 9, the comparison of the proposed AES-ECC multilevel system's computation time to the timings of similar systems based only on either AES or ECC is presented. It can be seen that the new EHR system outperforms the ECC-based equivalent, as it records lesser computation times. However, it was noticed that the AES-based equivalent took lesser computation time compared to the ECC-based and the proposed AEC-ECC approach. This is as expected, due to the fact that AES is a symmetric scheme and requires less computations during the execution of its algorithm. Nevertheless, the multilevel nature of the proposed approach implies that an attacker will require more time and effort to break in or attack successfully than what is obtainable when only AES is used.

Accordingly, the throughput of the proposed approach is also compared with those obtained from AES-based and ECC-based systems. The results are plotted and presented in Fig. 10. Indeed, throughput in this scenario is computed as the amount of data encrypted or decrypted over a given period of time. This is formally captured as in Eq. (8).

$$\text{Throughput} = T_p \ \text{(kilbytes)} / E_t \ \text{(sec)} \tag{8}$$

where T_p is plaintext in kilobytes and E_t refers to the encryption time (in seconds).

Different text file sizes were taken as input and encrypted using these three algorithms. It was observed that AES had the highest throughput, followed by AES-ECC model, while the ECC came last. Worthy of note is that the proposed AES-ECC multilevel approach offers a most robust security/privacy to users of the EHR system than the AES- and ECC-based approaches.

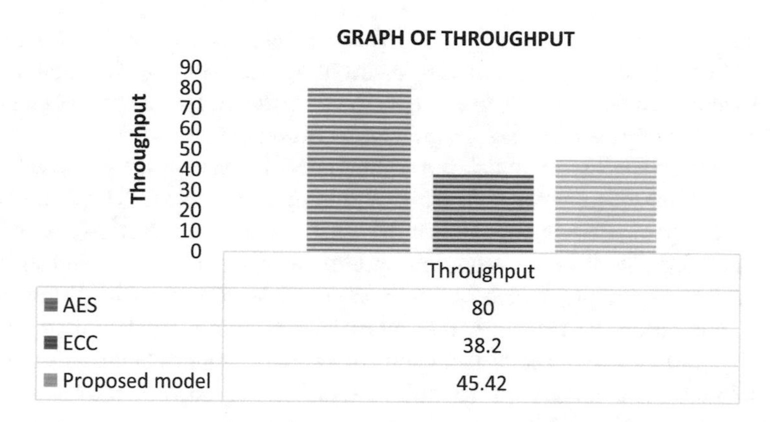

Fig. 10 Comparing the encryption approaches in terms of throughput

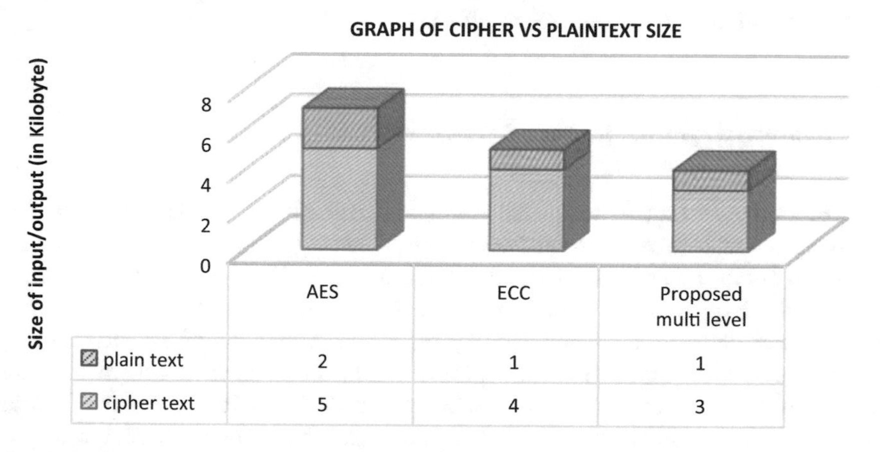

Fig. 11 AES versus ECC versus AES-ECC in terms of ratio of the ciphertext to plaintext

We also considered comparing AES, ECC and AES-ECC approaches in terms of the sizes of both input plaintexts as well as the output ciphertexts. The results are as presented in Fig. 11. It is observed that the size of output files is higher than those of the input plaintexts files across the three crypto-algorithms.

6.2 *Comparison of the Proposed System and Existing Works as per Trust*

In this subsection, the newly proposed approach is compared with two existing most related research works. This is done within the three levels of the framework for managing trust (shown in Fig. 3). These layers are responsible for sharing trust feedback, assessing trust and also distributing trust results. The parameters in Table 1 were used as standards to evaluate the performance of the proposed system compared to other related works, as far as trust is concerned.

In evaluating EHR trust model, the comparative analysis of the proposed model versus two other existing research works is reported in Table 2. The evaluation is based on the parameters mentioned in the trust management framework of Table 1. The technique used in the proposed system (with regard to trust modelling) is the reputation technique, which is the most common and user-friendly technique for trust evaluation. Furthermore, the proposed system provided support for feed credibility, and its strong use of feedback combination implies better integration in contrast to what is obtainable in [13, 40]. In addition, the proposed model used SaaS for its implementation, which is cost-effective compared to the model proposed by Saghar and Stephen in [40], which uses the three service offerings, and Alhaqbani in [13], which did not use any cloud offering.

7 Conclusion and Future Work

A major requirement in cloud computing is trust. Trust assessment is used in deriving a cloud provider's trustworthiness. This approach helps to predict the expected future behavior of a cloud service provider. This paper established and presented privacy and trust management frameworks and/or models for determining the trustworthiness of the health cloud provider and suitability of the services they offer especially in terms of privacy-preserving health or medical record keeping. The multilevel cryptography-based EHR system designed in this paper offers an optimally secure architecture that does not incur high computational cost and is resistant to trivial cryptanalysis, while the belief model guarantees that the trustworthiness of a system can be pre-determined by the clients. Future work will focus on developing quantum attack-resistant privacy solutions.

Table 1 Parameters of the trust management framework in Fig. 3

Layers	Parameters			
Trust feedback sharing layer	Credibility	Privacy	Personalization	Integration
	FC – feed credibility EC – entity credibility None	SP – focus on service provider privacy SR – focus on service requester privacy. None	F – full P – partial None	SFC – strong use of feedback combination NFC – no strong use of feedback combination
The trust assessment layer	Perspective	Technique	Security	Applicability
	SPP – service provider perspective SRP – service requester perspective	PocT – policy techniques RepT – reputation techniques RecT – -recommendation technique PrdT – predictive technique	SAF – support assessment function SCF – Support Communication Function	SaaS – software as a service IaaS – infrastructure as a service PaaS – platform as a service All the forms
The trust result distribution layer	Retort time	Redundancy	Accuracy	Exactness
	SAT – strong emphasis on assessment time NAT – no strong emphasis on assessment time	AR – support assessment redundancy TR – support trust data redundancy None	F – full P – partial N – none	ACL – supports access control level CL – supports communication level None

Table 2 Comparative analysis of other EHR trust model with the proposed model

Model/parameters	Credibility	Privacy	Personalization	Integration	Perspective	Technique	Security	Cloud applica-bility	Retort time	Redundancy	Accuracy	Exactness
Proposed trust model	FC	SR	P	SFC	SRP	Rept	SCF	SAAS	Sat	N	P	ACL
[13]	EC	SR	P	NFC	SRP	Rept	SCF	None	Sat	N	P	SCL
[40]	FC	SR	P	NFC	SRP	Rept	SCF	All	Nat	N	P	ACL

References

1. Buyya R, Ramamohanarao K, Leckie C, Calheiros RN, Dastjerdi AV, Versteeg S (2015) Big Data Analytics-Enhanced Cloud Computing: Challenges, Architectural Elements, and Future Directions. 2015 IEEE 21st International Conference on Parallel and Distributed Systems (ICPADS), pp. 75–84. https://doi.org/10.1109/ICPADS.2015.18
2. Mell P, Grance T (2010) The NIST definition of cloud computing. National Institute of Standards and Technology. http://csrc.nist.gov/groups/SNS/cloud-computing/cloud-def-v15.Doc
3. Thampi SM, Bhargava B, Atrey PK (2013) Managing trust in cyberspace. CRC Press, Taylor and Francis Group, Boston
4. Sultan N (2010) Cloud computing for education: a new dawn. Int J Inf Manag 30:109–116
5. Ajayi P, Omoregbe N, Adeloye D, Misra S (2016) Development of a secured cloud based health information system for antenatal and postnatal clinic in an African Country. In: Advances in digital technologies, Frontiers in artificial intelligence and applications, vol 282, pp 197–210
6. Josang A (2016) Subjective logic: a formalism for reasoning under uncertainty. Springer International Publishing, Cham. https://doi.org/10.1007/978-3-319-42337-1
7. Grandison T, Sloman M (2001) A survey of trust in internet applications. IEEE Commun Surv Tutor, Fourth Quarter 2000. http://www.comsoc.org/pubs/surveys
8. Noor TH, Sheng QZ, Yao L, Dustdar S, Ngu HH (2013) Cloud armor: supporting reputation-based trust management for cloud services. IEEE Trans Parallel Distrib Syst 27(2):367–380., 1 Feb. 2016. https://doi.org/10.1109/TPDS.2015.2408613
9. Alabi O, Thompson AF, Alese BK, Gabriel AJ (2020) Cloud application security using hybrid encryption. Commun Appl Electron 7(33):25–31
10. Olmedilla D, Rana O, Matthews B, Nejdl W (2005) Security and trust issues in semantic grids. In: Proceedings of the Dagsthul seminar, semantic grid: the convergence of technologies, 05271
11. Alhaqbani B, Fidge C (2009) A time variant medical data trustworthiness assessment model. In: Hoang D, Foureur M (eds) Proceedings ofthe 11th international conference on e-health networking, applications and services (HealthCom 2009), 16–18 December, Sydney. IEEE, December 2009. To appear. https://doi.org/10.1109/HEALTH.2009.5406198
12. Gabriel AJ, Adetunmbi AO, Obaila P (2020) A two-layer image-steganography system for covert communication over enterprise network. In: Gervasi O et al (eds) Computational science and its applications – ICCSA 2020, Lecture notes in computer science, vol 12254. Springer Nature, Cham, pp 459–470. https://doi.org/10.1007/978-3-030-58817-5_34
13. Alhaqbani B (2010) Privacy and trust management for electronic health records. Queensland University of Technology, Brisbane
14. Akinyele J, Pagano M, Green M, Rubin A, Lehmann C, Peterson Z (2011) Securing electronic medical records using attribute-based encryption on mobile devices. In: SPSM '11: Proceedings of the 1st ACM workshop on security and privacy in smartphones and mobile device, October 2011, pp 75–86. https://doi.org/10.1145/2046614.2046628
15. Tsai K, Leu F, Tien-Han W, Chiou S, Liu Y, Liu H (2014) Secure ECC based electronic medical record system. J Internet Serv Inf Secur (JISIS) 4(1):47–57
16. Ayeni F, Misra S (2014) Overcoming barriers of effective health care delivery and electronic health records in Nigeria using socialized medicine. In: Proceedings of 11th international conference on electronice, computer and computation, 2014. IEEE CPS Publication, pp 1–4. https://doi.org/10.1109/ICECCO.2014.6997568
17. Ajala F, Awokola J, Emuoyibofarhe O (2015) Development of an electronic medical record (EMR) system for a typical Nigerian hospital. J Multidiscip Eng Sci Technol (JMEST) 2(6). ISSN: 3159-0040
18. Boyinbode O, Toriola G (2015) CloudeMR: a cloud based electronic medical record system. Int J Hybrid Inf Technol 8(4):201–212
19. Joshi M, Joshi KP, Finin T (2018) Attribute based encryption for secure access to cloud based EHR systems. In: Proceedings of the IEEE CLOUD conference, 2018, San Francisco

20. Premarathne U, Abuadbba A, Alabdulatif A, Khalil I, Tari Z, Zomaya A, Buyya R (2015) Hybrid cryptographic access control for cloud-based EHR systems. IEEE Cloud Comput 3(4):58–64. https://doi.org/10.1109/mcc.2016.76
21. Deshmukh P (2017) Design of cloud security in the EHR for Indian healthcare services. J King Saud Univ Comput Inf Sci 29(3):281–287., ISSN 1319-1578. https://doi.org/10.1016/j.jksuci.2016.01.002
22. Seol K, Kim Y-G, Lee E, Seo Y-D, Baik DK (2018) Privacy-preserving attribute-based access control model for XML-based electronic health record system. IEEE Access 6:9114–9128. https://doi.org/10.1109/access.2018.2800288
23. Behrooz S, Marsh S (2016) A trust-based framework for information sharing between mobile health care applications. In: 10th IFIP international conference on trust management (TM), Jul 2016, Darmstadt, Germany, pp 79–95. https://doi.org/10.1007/978-3-319-41354-9_6. hal-01438350
24. Rana ME, Kubbo M, Jayabalan M (2017) Privacy and security challenges towards cloud based access control in electronic health records. Asian J Inf Technol 2(5):274–281
25. Ghorbel A, Ghorbel M, Jmaiel M (2017) Privacy in cloud computing environments: a survey and research challenges. J Supercomput 73:2763–2800. https://doi.org/10.1007/s11227-016-1953-y
26. Afshar M, Samet S, Hu T (2018) An attribute based access control framework for healthcare system. IOP Conf Ser J Phys 933:012020. https://doi.org/10.1088/1742-6596/933/1/012020
27. Rajasekaran P, Mohanasundaram R, Kumar AP (2018) Design of EHR in cloud with security. In: Proceedings of the second international conference on SCI 2018, vol 2. https://doi.org/10.1007/978-981-13-1927-3_45
28. Orehovački T, Babić S, Etinger D (2018) Identifying relevance of security, privacy, trust, and adoption dimensions concerning cloud computing applications employed in educational settings. In: Nicholson D (ed) Advances in human factors in cybersecurity. AHFE 2017, Advances in intelligent systems and computing, vol 593. Springer, Cham. https://doi.org/10.1007/978-3-319-60585-2_29
29. Horn C, Tropmann-Frick M (2018) Privacy issues for cloud systems. In: Elloumi M et al (eds) Database and expert systems applications. DEXA 2018, Communications in computer and information science, vol 903. Springer, Cham. https://doi.org/10.1007/978-3-319-99133-7_3
30. Kavitha M, Venkata Krishna P (2020) IoT-cloud-based health care system framework to detect breast abnormality. In: Venkata Krishna P, Obaidat M (eds) Emerging research in data engineering systems and computer communications, Advances in intelligent systems and computing, vol 1054. Springer, Singapore. https://doi.org/10.1007/978-981-15-0135-7_56
31. Georgiou D, Lambrinoudakis C (2020) Cloud computing framework for e-health security requirements and security policy rules case study: a European cloud-based health system. In: Gritzalis S, Weippl ER, Kotsis G, Tjoa AM, Khalil I (eds) Trust, privacy and security in digital business. TrustBus 2020, Lecture notes in computer science, vol 12395. Springer, Cham. https://doi.org/10.1007/978-3-030-58986-8_2
32. Alofe OM, Fatema K (2021) Trustworthy cloud computing. In: Lynn T, Mooney JG, van der Werff L, Fox G (eds) Data privacy and trust in cloud computing, Palgrave studies in digital business & enabling technologies. Palgrave Macmillan, Cham. https://doi.org/10.1007/978-3-030-54660-1_7
33. Lynn T (2021) Dear cloud, I think we have trust issues: cloud computing contracts and trust. In: Lynn T, Mooney JG, van der Werff L, Fox G (eds) Data privacy and trust in cloud computing, Palgrave studies in digital business & enabling technologies. Palgrave Macmillan, Cham. https://doi.org/10.1007/978-3-030-54660-1_2

34. Fox G (2021) Understanding and enhancing consumer privacy perceptions in the cloud. In: Lynn T, Mooney JG, van der Werff L, Fox G (eds) Data privacy and trust in cloud computing, Palgrave studies in digital business & enabling technologies. Palgrave Macmillan, Cham. https://doi.org/10.1007/978-3-030-54660-1_4
35. Priyadarshini P, Prashant N, Narayan DG, Meena SM (2016) A comprehensive evaluation of cryptographic algorithms: DES, 3DES, AES, RSA and blowfish. Proc Comput Sci 78(1):617–624
36. Corradini F, De Angelis F, Ippoliti F et al (2015) A survey of trust management models for cloud computing, University of Camerino, Via del Bastione 1, 62032, Camerino, Italy
37. McKnight D, Choudhury V, Kacmar C (2002) The impact of initial consumer trust on intentions to transact with a web site: a trust building model. J Strateg Inf Syst 11:297–323
38. Guo J, Chen I A classification of trust computation models for service-oriented internet of things systems. https://ieeexplore.ieee.org/xpl/conhome/7194374/proceedings
39. Boukerche A, Ren Y (2008) A trust-based security system for ubiquitous and pervasive computing environments. Comput Commun 31(18):4343–4351. https://doi.org/10.1016/j.comcom.2008.05.007
40. Saghar B, Stephen M (2016) A trust-based framework for information sharing between mobile health care applications. In: 10th IFIP international conference on trust management (TM), Jul 2016, Darmstadt, Germany, pp 79–95. https://doi.org/10.1007/978-3-319-41354-9_6.hal-01438350

Utilizing an Agent-Based Auction Protocol for Resource Allocation and Load Balancing in Grid Computing

Ali Wided, Bouakkaz Fatima, and Kazar Okba

1 Introduction

Grid computing focuses to enable resource sharing and related problem-solving in multi-institutional, dynamic, virtual organizations [1, 2]. To adjust and control the entities of the grid environment, several researchers used economic approaches. Economic methods give needed tools and standards for the grid. Furthermore, economic models provide sufficient incentives for grid resource providers to pay for their grid resources in the grid environment. Economic-based approaches to resource management in grid computing are efficient and cost-effective. In grid computing, economic-based approaches can include policies, algorithms, and tools for resource management. The field of grid applications has grown to the point that the objective functions and scheduling policies associated with resource provider-controlled applications are difficult to accept from a single model. The survey of existing economic methods agrees on the appropriateness of different approaches in different situations [3].

The English auction (EA), for example, is suitable for maximizing provider profits; however, the model creates a higher cost of communication; on the other hand, the commodity market model (CMM) is sufficient for maintaining the balance between resource offer and demand [4]. Auctions are more suited for resource allocation than other economic models because of dynamic pricing. Nevertheless, some auction models have the disadvantage of high communication requirements when used in large-scale environments [5].

A. Wided (✉) · B. Fatima
Larbi Tebessi University, Tebessa, Algeria
e-mail: wided.ali@univ-tebessa.dz; f_bouakkez@esi.dz

K. Okba
Mohamed Khider University, Biskra, Algeria

S. Misra et al. (eds.), *Artificial Intelligence for Cloud and Edge Computing*,
Internet of Things, https://doi.org/10.1007/978-3-030-80821-1_6

We present a multi-agent-based auction model for grid computing in this paper. This research aims to design an agent-based auction model in grid computing and use a grid simulation toolkit called GridSim to implement the proposed model [6]. We use the terms users to buyers and bidders to resource providers. Based on the studies presented in [7–18], we have designed and realized a descending Dutch auction framework.

The following is the order of the remainder of the paper: The suggested descending Dutch auction protocol is explained in Sect. 2. A description of the current agent-based model follows. In the next section, we discuss the rules of representing descending Dutch auction. Also, we specify subsequent steps in the proposed auction protocol. Section 4 talks about the experimental setup and the simulative study. Section 5 gives the conclusion.

2 Distributed Artificial Intelligence

Artificial intelligence (AI) is an area of science that explores the intellectual behavior of machines. Today, computer scientists have joined the community and have begun executing many intelligent entities that reason separately, interact, discuss, and make choices that upset the whole group. The networked communities have required the appearance of intelligent entities also called agents. Intelligent agents realized as programs are termed as software agents [19]. Grid computing has a lot of similarities to a multi-agent system, and both the multi-agent system and the grid have autonomous behavior. The artificial intelligence-driven load balancing mechanism is a load balancing approach whose main concept is built based on artificial intelligence. Expressly, an artificial intelligence-based framework provides a solution for balancing the load in grid computing by identifying similarities between well-known artificial intelligence algorithms and approaches and grid computing components and notions. We will look at some existing artificial intelligence-based load balancing tools for grid computing environments in this paper.

In the recent application of new technologies, some load balancing approaches in distributed networks based on artificial intelligence have been suggested. Genetic algorithm (GA) [20], artificial neural networks [21], fuzzy logic method [22], and multi-agent system-based technique [23] are some of the techniques used. In the grid environment, there are few AI-based methods for computational intelligence. Among all of the approaches discussed above, the MAS-based solution has shown promise. Multi-agent systems can simulate complex systems involving multiple interactions between autonomous and dynamic agents. Their effectiveness is primarily determined by the organization of the agents. Network maintenance can be problematic for a peer-to-peer organization because all agents must be reformed any time a new component is added. In organizations with central coordination, only the organizer's directory must be updated with new accompaniments. Other difficulties in building MASs arise from their dynamic behavior and the complicated interactions between agents, where aims or allocation of jobs and resources may conflict.

To prevent the drawbacks of both centralized and distributed architectures, we used a centralized model between cluster nodes and a distributed model between grid clusters in the proposed work.

2.1 Benefits of Artificial Intelligence Load Balancing in Grid Computing

The following are some of the benefits of using artificial intelligence to load balance grid computing.

1. Enhanced performance
2. Adding the security
3. Decision-making competency
4. Ensuring the correct use of the resources
5. Higher reliability
6. Decreased response time
7. Higher throughput
8. Efficiency of resource
9. Integrity and flexibility
10. Stability and consistency

3 Related Works

In this paper [4], the author proposed an agent-based framework; it supports auctions happening with multiple providers and users simultaneously. Bidder agent, provider agent, and auctioneer agent are the agents that make up this system, and each of them aims to maximize the goal. The provider agent tries for optimal allocation through maximizing revenue; the auctioneer agent purposes to clear its bundle by selecting the highest bidder among its participants, whereas a bidder agent gets its suitable bundle through budget optimization. This research used an ascending bid auction and focused on provider strategy. The proposed methodology provides competitive performance ever under the constricted budget scenario regarding revenue and utilization. The author likes completing his works by varying resource supply and demand, and he intends to investigate his work in a decentralized environment.

The authors [5] suggested a scheduling algorithm to reduce the cloud resource's response time at a low cost. This study proposed Nimrod-G, a computational economy-driven grid system that provides an economic opportunity for resource owners to share their resources and for resource users to trade off deadlines and budgets.

The approach proposed in this article [9], continuous double auction, is improved by the auctioneer updating bids. A mechanism is also provided for resource providers to decide resource rates based on their workload and for users to determine bids based on job deadlines. This approach also allows for several bids/asks in various auctions, based on the state of the auction's participants. It is suggested that a strategy for determining a more suitable update time be provided to enhance the proposed method, so that users can increase their bids with smaller amounts.

The authors [10] suggested a market method for effectively allocating resources to participants in cloud computing environments. Using this method, users can order a variety of services for workflows and co-allocations, as well as reserve future and existing services in a forward and spot market. Around the same time, the forward and spot markets operate separately to make stable and flexible allocations. In an economic context, the proposed market model seeks only to optimize overall welfare, but this is not always a desirable objective in an ecological sense. The proposed approach would be improved by adding energy consumption optimization using a market method.

In a cloud computing system, the authors [11] developed a novel auction-based scheme for trading free resources. Clients can bid fairly for available processors under the proposed method. Each client imposes a bid for the processors in a second-price auction. The client with the highest bid wins the auction and will continue using the processors to complete its assignments. The winning bid is equal to the second-highest bid. The proposed approach takes into account market demand as well as a cloud provider and client economic interests. The response time for trading processors is not improved in this article.

In the study of [12], in a grid simulator called GridSim, the authors presented a framework for designing auction protocols, and they presented a study of the communication requests of four auctions protocols, namely, English, first-price, Dutch, sealed, and continuous double auctions. The auctioneer, the seller, and the buyers are the main participants in the proposed framework. The user first submits tasks to her broker, who then generates an auction and sets auction parameters including gridlet length, number of auction rounds, auction policy (English or Dutch auction policy), and starting price. In this study, the broker often functions as an auctioneer, posting the auction to itself rather than to an outside auctioneer. Bidders are notified by the auctioneer that a Dutch auction is about to begin. The auctioneer then issues a call for proposals (CFP), determines the starting price, and distributes the CFP to all resource providers. Bids are created by resource providers for the execution of user jobs. When resource providers estimate the CFP for the first time, they are less likely to bid when the available cost is less than what they are willing to pay for the service. As a result, the auctioneer increases the price and creates a new CFP with the new price. Meanwhile, the auctioneer is saving and updating the auction information. Resource providers plan to bid in the second round. The auctioneer concludes the auction by agreeing to the previously stated policy. The user and the bidders are informed of the result after the auction has finished. The experiments reveal that English auctions necessitate further communication; continuous double auctions, on the other hand, have the lowest communication demand. Furthermore, even if the number of rounds needed varies, the final prices in both the English and Dutch auctions are the same.

In the study of [13], an auction algorithm for job scheduling in the grid is proposed. The authors introduced an auction model-based heuristic algorithm for resource management in this paper. The suggested algorithm attempts to assign persons to objects in such a manner that the overall gain is maximized. If an allocation contains in pairs of persons and objects and every object is assigned, it is said to be feasible. The auction algorithm is repeated iteratively until a fair assignment is reached. The bidding phase and the allocation phase are the two stages of the iteration. In this study, the job assignment is solved from the perspective of the resource providers with some restrictions to meet user satisfaction. This study must complete by comparing their results with other existing studies.

The authors [14] proposed an auction-based algorithm for grid computing wireless networks; they developed a reverse online auction technique to assign resources of the grid, where the resource providers arrive in real time and the broker must make several quality decisions, whether or not to sell jobs until the current round ends. A trade-some-with-forecast algorithm is proposed in the proposed method to assist the broker in using his prediction capacity to assign the grid resource in an online environment. Also, two protocols based on reverse online auctions are introduced. The first is the reverse online auction-based (ROAD) protocol, which does not require any forecasting of the degree of satisfaction inputs. The ROOF protocol is another one, where the user broker predicts the potential inputs of satisfaction degree orders based on partial knowledge. In terms of auction stages, accurate estimation, and user satisfaction, experimental results show that the ROOF protocol outperforms the ROAD protocol.

In this paper [15], the authors proposed a genetic auction-based algorithm for resource allocation in grid computing. The proposed algorithm is divided into two parts: an auction module and a genetic module. The auction module is in charge of deciding the price of resource exchange between a resource buyer and a resource provider, and the genetic algorithm is in charge of resource allocation, which takes cost and time constraints into account. The Min-Min algorithm, genetic algorithm, ant colony optimization, and primal algorithms were all compared by the authors. The suggested algorithm is more effective than other conventional algorithms, according to their evaluations.

The Heterogeneous Budget Constrained Scheduling (HBCS) algorithm was proposed by the authors [16] for scheduling budget constrained workflows in a heterogeneous environment. This technique aims to improve scheduling efficiency under QoS constraints, such as reducing time under a budget constraint. The HBCS algorithm, on the other hand, follows the budget constraint. However, it creates an inequitable timetable for the low-priority tasks. Low-priority tasks are most likely to choose the cheapest processor; this will result in a timetable with a longer makespan.

In this paper [17], a new algorithm called OVRAP (Online Virtual Resource Allocation Payment) is proposed for the auction-based cloud resource allocation problem. The following are the defining characteristics of this work: (1) an online auction mechanism is proposed rather than an offline auction mechanism. That is, (1) no time is spent collecting all user requirements before addressing the resource distribution and payment dilemma, (2) the auction process is structured to be fair and honest, (3) each user is allowed to apply several requirements, and (4) social welfare is maximized. Despite the benefits described earlier, the proposed solution's practicality should be increased.

In this paper [18], the combinatorial double auction resource allocation (CDARA) is proposed as a new business paradigm. The suggested technique allows the use of a greedy-based resource allocation algorithm, and the matching service providers' and users' bid price averages are taken into account when calculating the final trade prices. The suggested technique's only flaw is that it is incompatible with incentive systems. Additionally, in terms of resource allocation, it does not support the online and complex aspects of cloud environments.

4 Using Auction Model for Design: A Load Balancing System Based on Artificial Intelligence in Grid Computing

The responsibility of such a load balancer is to maximize the usage of the grid resources. For resource allocation, the authors intended to use artificial intelligence. In this section, they listed the complete auction process. The agents and their available activities on the system will be detailed. We use UML diagrams [24] to designate the proposed model. Also, we use UML use case diagrams (see Fig. 2) to illustrate how agents communicate. Also, we show the workflow by using activity diagrams.

Let us now briefly summarize the functionalities of agents in the system. Grid agent, broker agent, auctioneer agent, and resource provider agent are four artificial intelligence agents that have been considered as shown in Fig. 1.

A grid agent receives user jobs, user budgets, and time deadlines and sends them to the broker agents. Also, the grid agent terminates jobs that are running on a failed resource and restarts them on a new resource.

A broker agent is created by grid agent using GUI with desired details. Each cluster has a broker who schedules and manages jobs in the cluster. The broker

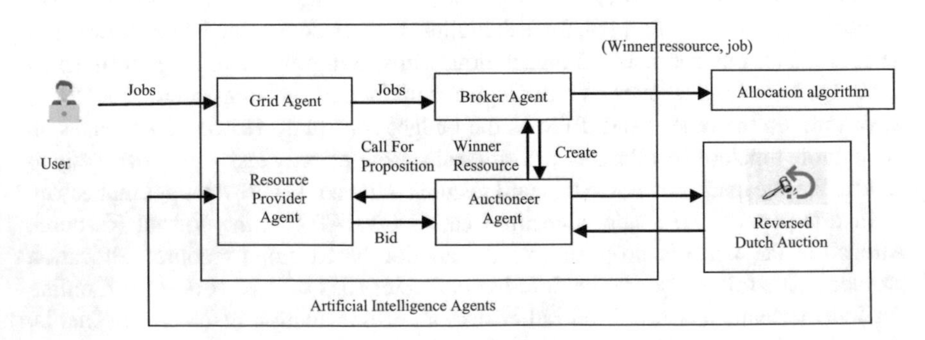

Fig. 1 The proposed system's conceptual framework for artificial intelligence agents

receives the jobs from the grid agent, generates an auctioneer agent for each job, and sets auction parameters including gridlet size, number of auction rounds, first-price, and auction type. (A descending Dutch auction is implemented in our application.) For each job, an auctioneer agent is created. Once created, the auctioneer agent informs the resource provider agent that an auction is about to start. The auctioneer agent then creates a call for proposals (CFP) and distributes it to all resource provider agents, inviting resource providers to participate in the auction. He shall communicate with resource provider agents to know their proposal price.

The auctioneer agent begins with a high price and gradually lowers the price before the auction ends. The bidder with the lowest bid and a balanced load is declared the winner. Also, the auctioneer agent interacts with broker agent to assign the jobs for the winner resource in the auction.

Resource provider agent decides either to accept or reject CFP by checking the cost of the job and its load state. If it rejects, the resource provider agent quits the auction, or else it accepts the proposal of the auctioneer agent and formulates a bid for selling a service to the user to execute her job.

We suggest the following procedure to help the broker in executing the user's jobs, taking into account the characteristics of grid resources.

Bidding phase: Users submit tasks to the grid during the bidding process; the grid agent sends tasks with several QoS requirements such as budget and user time deadline for the broker agent. The broker agent initiates an auction for each job. It also determines the auction's parameters, such as the size of the job, the number of rounds, the deadline, the budget, and the starting price.

Winner determination phase: In this phase, the auctioneer agent gathers information about jobs' deadlines and resource properties, and using the winner determination rules algorithm, the auctioneer agent selects the auction winner. If the auction's round is terminated, the auctioneer agent sends the list of winners for the broker agents. Using the allocation algorithm, the broker agent sends jobs to their winners for execution.

Let us now detail the winner determination rules of a reverse Dutch auction. The proposed protocol prescribes the bidding period (round-time and number of rounds). Resource providers are allowed to submit bids during the bidding period, which ends when the auction closes (Fig. 2).

Rule 1 Bidding period
IF
Nbr-rounds>0
And
Round-time! =0
THEN
Resource providers are allowed to submit bids

The load(R) is determined using the method described in [24, 25].

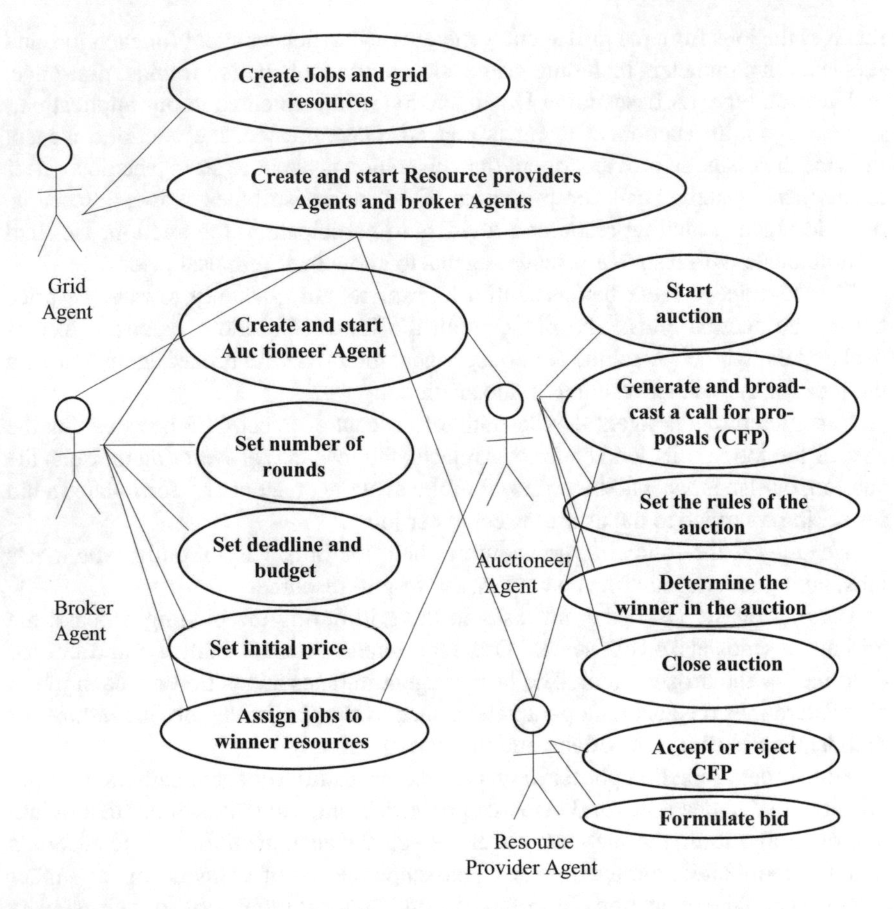

Fig. 2 The proposed agent-based auction model's use case diagram

Rule 2 Auction participation
IF
There is a valid proposal Pr submitted by the resource provider R
And
gridletCost (R)<= budget
And
CpuTime(R)<=deadline
And
Load(R)<Threshold
THEN
Proposal Pr is accepted
ELSE
Proposal Pr is rejected

When the auction's bidding phase expires, the bidder with the lowest bidding price is declared the winner.

Rule 3 Winner determination
IF
GridletCost(R)<Currentprice
THEN
Currentprice =GridletCost
Winner=R

5 Artificial Intelligent Agents Interactions

Communication is the principal concept in distributed artificial intelligent systems. Communication is an interactive process that allows a group of agents to organize their actions in order to solve a problem. The main objective of communication is to resolve agent conflicts. The proposed framework begins with the creation of agents, with the grid agent being the first. Grid jobs and resources are also created. The grid agent also creates and starts resource providers and broker agents, as well as sending jobs to the broker agents. The broker agent in turn initiates an auction for each job. In the proposed protocol, several auctions will run simultaneously under the proposed protocol. The auctioneer agent is in charge of establishing auction rules and managing the auction. Furthermore, the auctioneer agent collects bids from resource providers that are interested in the auction and determines the auction winner (based on the winner determination rules algorithm). Also, it interacts with the broker agent to assign jobs to the winner resources in the auction (see Figs. 3 and 4).

6 Experimental Environment

6.1 System Implementation

The proposed framework is built on JADE 4.5.0, an agent development platform, and the GridSim toolkit, which simulates the grid environment. JADE is a software framework written in Java. It is an open source that is still being developed [26]. JADE makes it easier to create agents that conform to FIPA requirements, allowing multi-agent systems to advance. JADE contains predefined classes for creating agents and for their behaviors. It is a good platform that allows programmers to quickly and conveniently build and interact with agents.

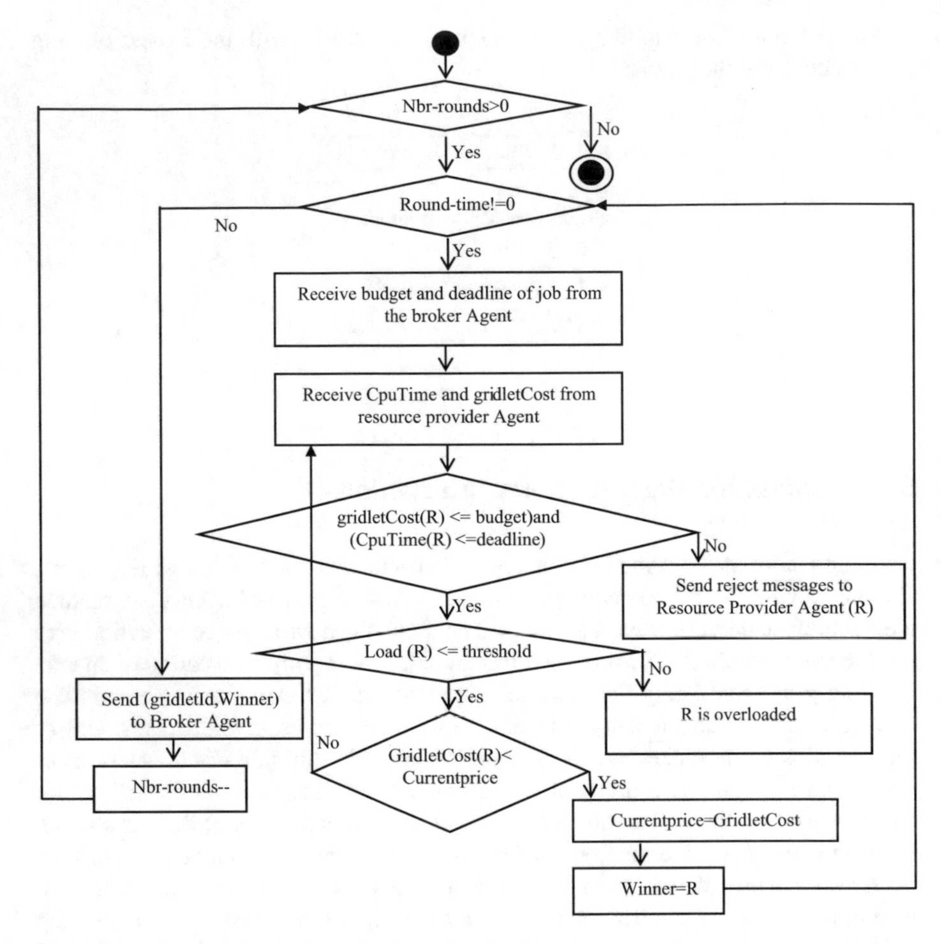

Fig. 3 Flowchart of winner determination rules algorithm

JADE platform is composed of containers, and container contains agents. Everything done by the agent takes place inside the container. The user will build containers based on their requirements.

JADE Agent Classes

In our system, we have implemented four classes, class grid agent, class broker agent, class auctioneer agent, and class resource provider agent, to create agents for the grid agent, broker agent, auctioneer agent, and resource provider agent, respectively. Being a developer or programmer of JADE, we only need to identify the correct class and need to extend our classes from the predefined ones, but they have expanded JADE's agent class and introduced our classes by overriding the setup() and takedown() methods.

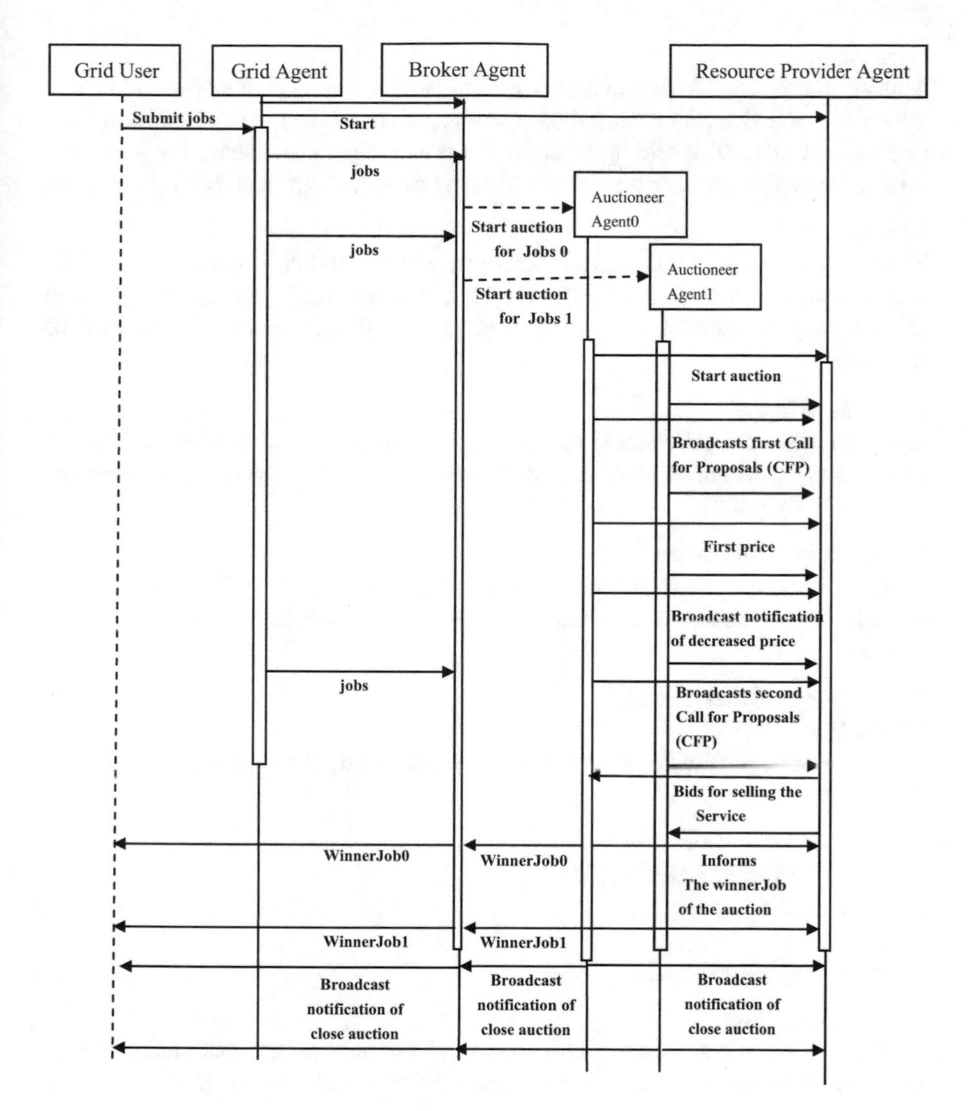

Fig. 4 UML sequence diagram describes agent interactions in the auction model

JADE Agent Behavior Classes

Agent activities, also known as agent activity classes, describe the actual task that an agent will execute after being created. The execution of an instance of a class that extends the class behavior determines the actual function of an agent.

The addbehaviour() method attaches behavior to the agent: the CyclicBehaviour, OneShotBehaviour, and TickerBehaviour classes of the class behavior were all extended in our framework.

Grid Agent

This agent uses OneShotBehaviour. The simulation is set up by the grid agent class, which uses the GridSim standard to create instances of resources, machines, jobs, and other entities. The grid agent starts the broker agents and sends the generated jobs to the broker agents when the simulation time equals the task submission time.

Broker Agent

The broker agent uses a TickerBehaviour; the number of tick is the number of jobs sent to broker agent. In each tick, the broker agent creates an auctioneer agent for assigning one job for a selected resource, and it initiates the auction with its parameters.

Auctioneer Agent

This agent uses a CyclicBehaviour. Each cycle corresponds a round of an auction. In each round, the auctioneer agent starts an auction and invites resource providers for sending their bids.

Resource Provider Agent

This agent uses a CyclicBehaviour. The auctioneer agent sends a CFP to the resource provider agent, who then formulates bids and sends them to the auctioneer agent senders.

The System's Other Classes

User Class

Each user may vary from the other users concerning the following characteristics:

1. Types of job created
2. Scheduling optimization strategy
3. Activity rate
4. Time zone
5. Time deadline and budget

Resource Class

Each resource is associated with a resource provider agent. Each resource may vary from the other resources concerning the following characteristics:

1. Number of processors
2. Cost of the processing
3. Speed of processing
4. Internal process scheduling policy, e.g., time shared or space shared
5. Local load factor
6. Time zone

ComplexGridlet Class

The ComplexGridlet class extends the gridlet class implemented in the GridSim simulator, permitting the simulation of more realistic situations where each job may need other properties like the precise amount of available memory or the specific machine parameters and other constraints in real life.

The FailureLoader Class
Once the time of simulation reaches the time of failure start time, the concerned machine is set to be failed, killing all tasks being at present executed on that machine. When the failure time passes, the machine is restarted.

FIPA Messages for Agent Communication: [27]
The most exciting feature of the agent is their ability to message communication. By sending and receiving messages, the agent connects with other agents. The Foundation for Intelligent Physical Agents (FIPA) suggested a format for these messages, which was defined by FIPA under Agent Communication Language (ACL) (FIPA) [28] and Knowledge Query and Manipulation Language (KQML) [29]. The following messages were used in the proposed system to allow agents to communicate with one another: REQUEST, INFORM, CALL FOR PROPOSAL (CFP), PROPOSE, ACCEPT-PROPOSAL. The auctioneer agent uses the REQUEST message to request that the resource provider agent bid; also the auctioneer agent uses INFORM informative for informing the resource provider agent that the auction is started. The auctioneer agent sends a CALL FOR PROPOSAL (CFP) to all resource provider agents, respectfully inviting them to join in the auction. The resource provider agents respond by sending PROPOSE informative, from which they send their proposal to the auctioneer agent. If the auctioneer agent likes the proposal, it responds with ACCEPT-PROPOSAL to the resource provider agent. The main container and the grid agent are created using the code below (see Fig. 5).

The ACLMessage class implements an ACL message compliant with the FIPA. In the above example is the source code from our program (see Fig. 6).

6.2 *Simulated Parameters*

The implementation is done on grid simulator GridSim. The GridSim toolkit provides a comprehensive set of capabilities for simulating a wide range of heterogeneous systems (users, resources, resource brokers, applications, and schedulers). It can be used to simulate application schedulers for distributed computing networks such as grids and clusters of different or single administrative domains. We utilized the following parameters:

Resource Parameters These parameters give information about available resources during the load balancing period such as:

1. Number of resources in each cluster
2. Size of memory (RAM)
3. Number of clusters
4. Date to send load information from resources

```
program create main container {
AgentGrid agentgrid = new AgentGrid() ;
Runtime rt= Runtime.instance();
Properties p=new ExtendedProperties();
p.setProperty("gui","true");
ProfileImpl pc=new ProfileImpl(p);
jade.wrapper.AgentContainer container
=rt.createMainContainer(pc);
try {
container.start();
} catch (ControllerException e) {
// TODO Auto-generated catch block
e.printStackTrace();
}
AgentController agentcontroller;
Profile pp;
pp=new ProfileImpl();
pp.setParameter(Profile.CONTAINER_NAME,"container -1");
jade.wrapper.AgentContainer
agentcontainer=rt.createAgentContainer(pp);
agentcontro l-
ler=agentcontainer.createNewAgent("agentgrid",
"AgentBasedLoadBalancing.AgentGrid", new Object[]{});
agentcontroller.start(); }
```

Fig. 5 Source code for creating the main container

5. Tolerance factor

Job Parameters These parameters are:

1. Number of jobs queued at every resource
2. Arrival time, submission time, processing time, finish time, waiting time, and start time
3. Job length
4. Job priority

Network Parameter LAN and various WAN bandwidth sizes

Performance Parameters We were focused on two objectives to evaluate the performance of the proposed model during our experiments: resource usage and load balancing.

Creating a Grid Resource A grid resource simulated in GridSim contains one or more machines. Similarly, a machine contains one or more PEs (processing elements or CPUs) [30]. Figure 7 is the output of our test program.

Creating Gridlets (Jobs) A gridlet is a job that can run on a grid resource. Every gridlet has a unique ID that distinguishes it from the others. The number of gridlets to be generated, as well as the length, file size, and output file size for each gridlet, is all determined by the user. For simulation, we need to know the gridlet length, input file size, output file size, and unique gridlet ID. Gridlets can be manually generated,

```
program winner determination {
if (msg.getContent().startsWith("GridletCost")) {

gui.showMessage("sender:"+msg.getSender().getLocalName(),
true);
gui.showMessage("time:"+System.currentTimeMillis(),true);
System.out.printf( " Start auction " +
this.getAgent().getLocalName());
gui.showMessage(" Start auction " +
This.getAgent().getLocalName(),true);
sender=msg.getSender().getLocalName();
winneride=ResourceProviderAgent.resources.get(sender);
bid-
ers.put(winneride,Double.valueOf(msg.getContent().substri
ng(11)));
System.out.println(bidders.get(winneride));
GridletCost= Dou-
ble.valueOf(msg.getContent().substring(11));
System.out.println( "gridlet     " +
this.getAgent().getLocalName()+ sender
 +"cost"+GridletCost + "
time"+System.currentTimeMillis() );
 Sys-
tem.out.println(sender+"***************"+Scheduler.Load.g
et(winneride) +
"time" + System.currentTimeMillis());
 if(Scheduler.Load.get(winneride)>algo2.thresholdH2){
   System.out.println("*******the resource is overloaded"
   +sender +Scheduler.Load.get(winneride)) ;
                    }
                   else
                    {
   if ( GridletCost<=price)
     {
        counter=counter--;
        if(counter!=0){
         price=GridletCost;
         winnerid=msg.getSender().getLocalName();
 winnerident=ResourceProviderAgent.resources.get(sender);
  System.out.println("the resource is underloaded"+
Auction"+          this.getAgent().getLocalName()
+sender+"Load" + Scheduler.Load.get(winneride) +
"price"+price );
 System.out.println("WinnerJob for "
+this.getAgent().getLocalName()+winnerid +
"currentprice"+price);
 Min=GridletCost;
           }}
  }}}}
gui.showMessage("WinnerJob for auction" +
this.myAgent.getLocalName() +"=  " +winnerid +
"cost"+price  ,true);
System.out.println("WinnerJob for " +
this.getAgent().getLocalName() +"=  "   +winnerid +
"cost"+price);
winner.put(this.getAgent().getLocalName(),winnerident);
```

Fig. 6 Source code for winner determination

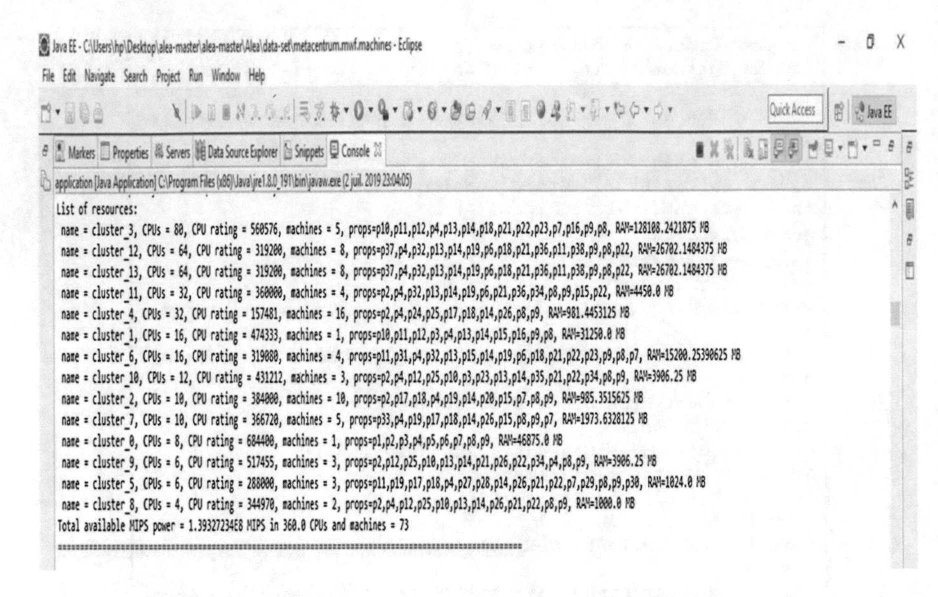

Fig. 7 Grid Resource Creation in GridSim (the output of the test program)

Properties Servers Snippets Console

<terminated> Application [Java Application] D:\programmation agents\eclipse\jre\bin\javaw.exe (26 oct. 2018 19:06:31)

============= OUTPUT for User_0 ==========

Gridlet ID	Gridlet Length	Gridlet input file size	Gridlet out put file size
0	900.0	900	900
1	600.0	600	600
2	200.0	200	200
3	300.0	300	300
4	400.0	400	400
5	500.0	500	500
6	600.0	600	600
7	900.0	900	900
8	600.0	600	600
9	200.0	200	200

Fig. 8 Jobs creation in GridSim (the output of the test program)

or they can be generated at random. The steps to create gridlets are presented in Fig. 8.

The broker receives jobs from the grid agent and generates an auctioneer agent for each one. It also determines auction parameters such as gridlet size, number of auction rounds, first bid, and auction type. Figure 9 shows the broker agent interface.

The resource provider agent decides to either accept or reject CFP by checking the cost of the gridlet and its load state. If it rejects, the resource provider agent quits

Fig. 9 Broker agent interface

the auction, or else it accepts the proposal of the auctioneer agent and formulates bids for executing jobs of the user as shown in Fig. 10.

The auctioneer agent communicates with resource provider agents to know their proposal cost. The auctioneer agent begins with a high price and gradually lowers the price before the auction ends. The bidder with the lowest bid and a balanced load is the winner. Also, the auctioneer agent interacts with the broker agent to assign the jobs for the winner resource of the auction. Figure 11 depicts the auctioneer agent's interface.

7 Performance Evaluation

In the proposed system, a user submits jobs to the grid agent, which in turn sends them to the broker agents; each cluster in the grid has a broker agent attached to it. Following that, the broker agent runs an auction for each job. We used Dutch auctions here. Therefore, the resource provider agents are the bidders, and they bid for job execution. Table 1 shows the 14 resource settings that we simulated. The limit of auction rounds is set to 5, and each round with a time of 2 minutes. Jobs are represented as gridlet objects, which contain all job data as well as information about job execution management.

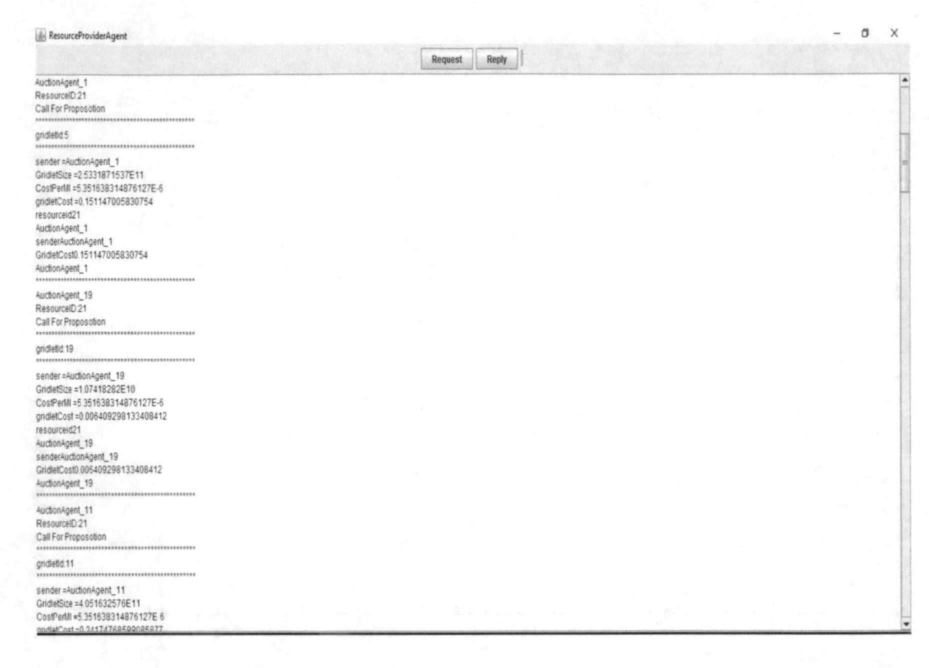

Fig. 10 Resource provider agent interface

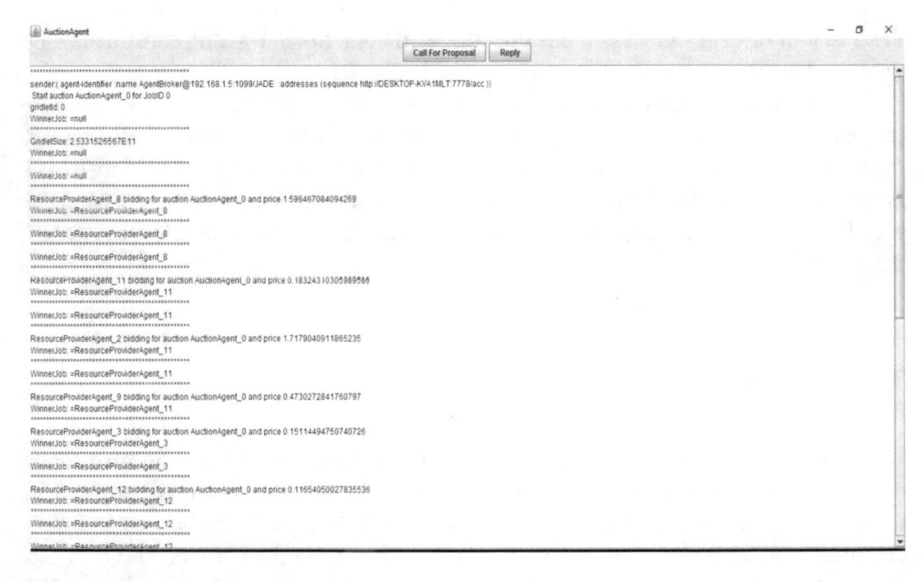

Fig. 11 Auctioneer agent interface

The processing capacity of each resource per second is represented in Column 2 of Table 1 (million instructions per second). Column 3 shows the cost of processing a million instructions for each resource.

Table 1 Details of resources processing capacity

Resources	Resource capability Mips Rating	Processing cost (Cost/MI)
R0	5,475,200	4.38
R1	7,589,328	6.32
R2	1,152,000	7.81
R3	8,969,216	5.35
R4	1,889,772	1.90
R5	1,728,000	1.04
R6	5,105,280	9.40
R7	3,667,200	8.18
R8	1,379,880	8.69
R9	3,104,730	5.79
R10	5,174,544	6.95
R11	11,520,000	8.33
R12	20,428,800	9.39
R13	1,118,400	1.07

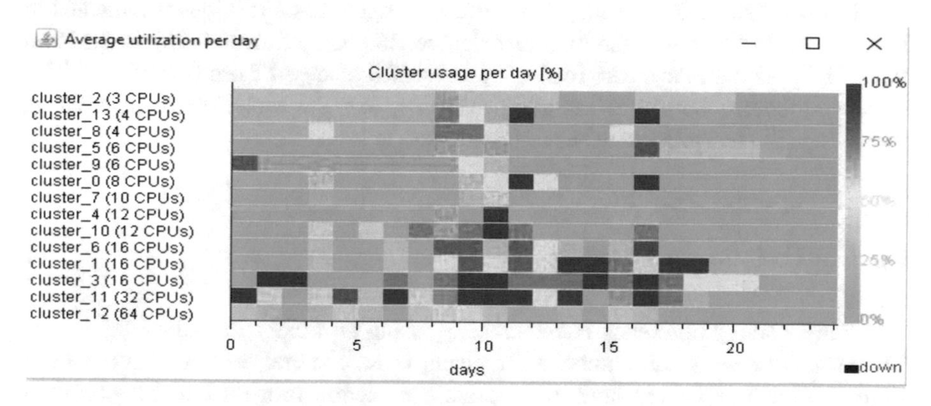

Fig. 12 Average resources utilization per day

All simulations were conducted on an Intel I3 Duo 2.00 GHz PC with 4 GB of RAM running Windows 2010, with JADE (Java Agent Development Framework) for agent deployment, for analyzing experimental results.

Experiment 1: Resource Utilization

Figure 12 shows that the resource R12 which is having the highest processing capacity (20,428,800) is prevented (cluster_12 usage = 21.51%), while there is an increase of utilization of cluster cluster_13 (processing capacity = 8,969,216) to 52.35%. This is because the proposed algorithm allows the job to be spread around the most available resources, whereas overloaded resources are prohibited.

Experiment 2: Load Balancing

The suggested algorithm distributes the load and avoids overloading of resources by excluding overloaded resources from the list of winners in the auction.

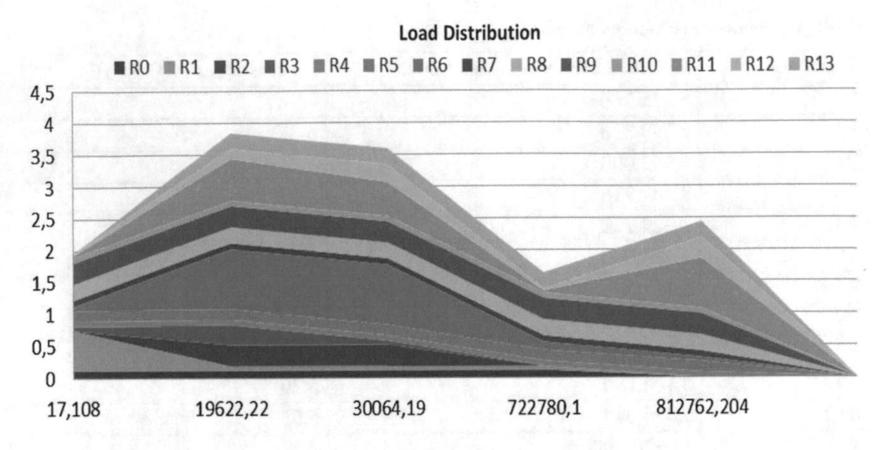

Fig. 13 Load distribution with 100 jobs and 14 resources

The load distribution among 14 resources with a total of 100 jobs is indicated in Fig. 13. We indicate that the load of resource R6 changed from 0.937 to 0.125 in time 722780.1 sec. Also, the load of resource R6 changed from 0.0625 to 0.937. The overloaded resource is reducing overtime, and the underloaded resource is increasing till the system reaches a stable state.

8 Conclusion and Future Works

The grid system's efficiency is enhanced by using artificial load balancing agents. The artificial agents communicate for giving users cheaper resources in a shorter time. We have explained how to integrate the auction process into the proposed agent-based model for solving the grid computing resource management problem. We specified the system's main functions and design details, as well as the interaction protocol. The objectives were to improve resource utilization and grid load balancing. The results revealed that the proposed model can be an effective resource allocation solution in grid computing. As a result, the proposed model can effectively balance the load, reduce the number of underloaded or idle resources, and improve resource utilization. We think that it would be interesting to other auction protocols such as (English auction, first-price, and a sealed-bid auction), and how implementing, and integrating them in our Agent-based model. To develop the proposed algorithm, we will use another artificial intelligence approach such as artificial neural networks and genetic algorithms. Furthermore, we will intend to use fuzzy logic rules to know cluster and node states.

References

1. Czajkowski K, Fitzgerald S, Foster I, Kesselman C (2001) Grid information services for distributed resource sharing. In: 10th IEEE international symposium on high performance distributed computing. IEEE Press, New York, pp 181–184
2. Foster I, Kesselman C, Nick J, Tuecke S (2002) The physiology of the grid: an open grid services architecture for distributed systems integration. Technical report, Global Grid Forum
3. Haque A, Alhashmi SM, Parthiban R (2011) A survey of economic models in grid computing. Futur Gener Comput Syst 27:1056–1069
4. Haque A (2018) Iterative combinatorial auction for two-sided grid markets: multiple users and multiple providers. Glob J Comput Sci Technol Cloud Distrib 18(1). ISSN: 0975-4350
5. Buyya R, Abramson D, Giddy J, Stockinger H (2002) Economic models for resource management and scheduling in grid computing. Concurr Comput Pract Exp 14(13–15):1507–1542
6. Buyya R, Murshed M (2002) GridSim: a toolkit for the modeling and simulation of distributed resource management and scheduling for grid computing. J Concurr Comput Pract Exp (CCPE) 14:13–15
7. Veit DJ, Gentzsch W (2008) Grid economics and business models. J Grid Comput 6(3):215–217
8. Altmann J, Routzounis S (2006) Economic modeling for grid services. In: e-Challenges, Barcelona
9. Teymouri S, Rahmani AM (2012) A continues double auction method for resource allocation in economic grids. Int J Comput Appl 43:7–12
10. Fujiwara I (2010) Applying double-sided combinational auctions to resource allocation in cloud computing. In: IEEE 10th annual international symposium on applications and the internet
11. Alsarhan A, Al-Khasawneh A (2016) Resource trading in cloud environments for utility maximisation using game theoretic modelling approach. Int J Parallel Emerg Distrib Syst 31:319–333
12. Assunção D, Buyya R (2006) An evaluation of communication demand of auction protocols in grid environments. In: 3rd international workshop on grid economics and business models. World Scientific Publications, Singapore
13. Gahrouei AR, Ghatee M (2018) Auction-based approximate algorithm for grid system scheduling under resource provider strategies. arXiv preprint. arXiv: 1803.04385
14. Ding L, Chang L, Wang L (2016) Online auction-based resource scheduling in grid computing networks. Int J Distrib Sens Netw 12(10):1–12
15. Satish K, Reddy AR (2018) Resource allocation in grid computing environment using genetic–auction based algorithm. Int J Grid High Perfor Comput 10(1):1–15
16. Arabnejad H, Barbosa JG (2014) Budget constrained scheduling strategies for on-line workflow applications. In: International conference on computational science and its applications. Springer, Cham
17. Zhang J, Xie N, Zhang X, Li W (2018) An online auction mechanism for cloud computing resource allocation and pricing based on user evaluation and cost. Future Gener Comput Syst 89:286–299
18. Samimi P, Teimouri Y, Mukhtar M (2016) A combinatorial double auction resource allocation model in cloud computing. Inf Sci 357:201–216. Elsevier
19. Moulin B, Draa BC (1996) An overview of distributed artificial intelligence. In: O'Hare GMP, Jennings NR (eds) . Fundamentals of distributed artificial intelligence, Wiley, pp 3–56
20. Santhosh J, Aruna R, Balamurugan P, Arulkumaran G (2020) An efficient job scheduling and load balancing methods using enhanced genetic algorithm. Eur J Mol Clin Med 7(08). ISSN 2515–8260

21. Negi S, Panwar N, Vaisla KS, Rauthan MMS (2020) Artificial neural network based load balancing in cloud environment. In: Advances in data and information sciences, Lecture notes in networks and systems, vol 9. Springer, Singapore
22. Hajoui Y, Bouattane O, Youssfi M (2018) New hybrid task scheduling algorithm with fuzzy logic controller in grid computing. Int J Adv Comput Sci Appl 9(8):547–554
23. Wooldridge M (2002) An introduction to multi agent systems. Wiley, Chichester
24. Wided A, Okba K, Fatima B (2019) Load balancing with job migration algorithm for improving performance on grid computing: experimental results. Adv Distrib Comput Artif Intell J 8(4):5–18
25. Wided A, Okba K (2019) A novel agent based load balancing model for maximizing resource utilization in grid computing. Informatica 43(3):355–361
26. Bellifemine FL, Caire G, Greenwood D (2007) Developing multi-agent systems with JADE. Wiley, Hoboken. ISBN: 978-0-470-05747-6
27. Bellifemine F, Poggi A, Rimassa G (1999) JADE–A FIPA-compliant agent framework. In Conference on the practical application of intelligent agents and multi-agent technology, pp 97–108
28. Poslad S (2007) Specifying protocols for multi-agent systems interaction. ACM Trans Auton Adapt Syst 2(4):15
29. Tim F, Fritzson R, McKay D, McEntire R (1994) KQML as an agent communication language. In: Proceedings of the third international conference on information and knowledge management, CIKM, pp 456–463
30. The Gridbus project, GRIDS Lab. http://www.cloudbus.org/gridsim/ 15.02.21

Optimization Model of Smartphone and Smart Watch Based on Multi Level of Elitism (OMSPW-MLE)

Samaher Al-Janabi and Ali Hamza Salman

1 Introduction

Edge computing "doesn't replace the cloud; it simply puts the parts of the applications that need to be closer to the endpoints where they belong. It's a type of hybrid cloud, in which all data doesn't have to shuttle back and forth between far-away servers and user devices." This study used this idea to solve the problem of recognition; the activities of human take from smartphone and smart watch.

Internet of Things (IoTs) is a term which refers to interconnected hardware and software technologies that allow sharing of resources among the devices or use of the service of some devices by others; also it is used to produce the data from the connected device and sensor to the Internet. In the simplest concept of the Internet of Things, it is any device that is able to connect to the Internet, and the ability to convey data through the network does not need interaction between the user and computer [1].

A smartphone is a phone that includes a compact computer and other software that is not originally connected to phones, such as the operating system and the ability to work to run other applications. Smartphones are nowadays an integral part of people's daily lives. Smartphones have been used for communication between people and for browsing on Internet platforms and social media [2].

Smart watches, in light of the recent progress taking place in the world, spread, and competition began to produce design and superior features. It is a wrist watch that performs multiple tasks such as accounts, games, and answering calls and works on playing audio files, including those equipped with a camera. It may contain several applications such as the measurement of the number of pulses and the

S. Al-Janabi (✉) · A. H. Salman
Department of Computer Science, Faculty of Science for Women (SCIW), University of Babylon, Babylon, Iraq
e-mail: samaher@itnet.uobabylon.edu.iq

S. Misra et al. (eds.), *Artificial Intelligence for Cloud and Edge Computing*, Internet of Things, https://doi.org/10.1007/978-3-030-80821-1_7

measurement of blood pressure [3]. The ten best smart watches of 2019 have the following features: best overall, best for Samsung owners, best fitness tracking, best health features, best rugged, best battery, best value, best for minimalists, best for kids and best for music.

Biometric is the result of the combination of two words bio (life) and metric (for measurement). It can be defined as the science of statistically analyzing biological data and the technology of measuring. This is used to measure the characteristics of a person (physiological and behavioral) that can be used to identify his or her identity. Biometric is a science for measuring behavioral and/or physical characteristics that are unique to each person and they belay that a person is who he or she pretends [4].

"Biometric" means "life measurement." The unique physiological characteristics of individual identification are related to this term. A number of biometric aspects have evolved and are used to verify an individual's identity [23].

Computational Intelligence (CI) is the theory, design, application and development of biologically and linguistically motivated computational paradigms. Traditionally the three main pillars of CI have been Neural Networks, Fuzzy Systems and Evolutionary Computation. Also, CI study the design system based on intelligent agents [5]. Agent acts in an environment—it does something. Agents consist of society, dogs, and humans. An intelligent agent is a system that plays intelligently. The goal of smart computing is to understand smart behavior, and then it can be applied in industrial and natural systems [22].

Optimization is part of machine learning under the supervised learning. It means change for the better, and it is a way to achieve continuous improvement, and it can be applied in all aspects of life. The optimization aims to continuously develop processes and activities related to individuals and the production path [13, 24].

The problem of this study is how can develop the technology to recognize the Biometric Activities related to smartphones and Smartwatch as one of the main activities performed by any person. Such as [Walking, Jogging, Stairs, Sitting, Standing, (Eating – Soup, Pasta, Chips, Sandwich), Brushing Teeth, Drinking, kicking (Soccer Ball), Clapping, Writing, (Playing – Tennis, Basketball), Typing, and Folding Clothes). On other side, time is considered one of the main important challenges facing us today this is due to speed development technologies and increase number of activities required to achieves from any person in specific time. Therefore, time management is one of the secrets of success in everyday life. The challenge of this study is how to accomplish eighteen different smartwatch and smartphone activities that fifty-one people can perform in less than one hour (54 minute). Therefore, this study will be described new a model.

While the main objectives of this paper are Find the main split point (Pivot or start point) to divided the complex network into multi subgraphs. Design optimization model for hug dataset related to activities, based on cooperative between two of deep learning techniques (i.e., DSA & ALO).

In general, we can summarization the aim of this study is design multi-level optimization system, in each level find the best decision than pass it into next level through applied the elitism principle. In general; the system can deal with the hug dataset by combination between two deep learning techniques (i.e., Deterministic Selection Algorithm and Ant lion optimization algorithm).

The system divided the complex network into multi subgraph then find the optimal subgraph for each complex network and find the correlation among the activities in each level based on the relationships among Time, coordinations and activities.

The remined of this chapter is organization as follow: Sect. 2 show the complex network with Characterizing Networks, Sect. 3 explain main stage of design suggest model "OMSPW-MLE", Sect. 4 appears the results of the OMSPW-MLE and give justification of that results through analysis it. Section 5 discussion while, finally Sect. 6 shown the Conclusions and Recommendation of Future Works.

2 Complex Network

Networks consist of set of elements known as nodes or vertices and edges that connected between them. The system taking the structure of networks also called graphs. Complex networks are networks that have important topological features, non-trivial and patterns that not exit in simple networks.

Complex network is characterizing that is expansive, consisting of huge number element; associations patterns are not regular or simply random. Developments in technology cause the increasing computational power and the size of storage area, so that can be gathering of large volumes of information from it; this can be useful for the analysis complex networks much more detail. Complex networks are mostly dynamic in their nature and with time its advance their topology [6, 7].

2.1 Characterizing Networks

Recently, the analysis of topological properties of complex networks is considering one of the important directions for scientific researchers. It can be representing as any topological changes to graph [18]. To understand the behavior of network must use the measurement that capable of illustrating the most relevant topological features of that network. And by using these measures we can synthesis, analysis and discrimination of complex networks in useful way [18].

Chosen specific set of topological measures is depending on the task we perform and networks. There are large set of topological measures that can be used, each one express a specific feature in the networks such as connectivity, clustering coefficient, degree distribution, diameter, centrality, connected component, betweenness centrality and etc. [8].

Connectivity

Consider have graph $G\ (V, E)$, it vertex is V and edges is E. if path present between all vertices pair then the graph is called connected, if not is disconnected. Connected components (components) is consider sub graphs of G if have largest connected (i.e. Networks consider highly connected if there are number of vertices in one components) [9].

If there is path from v1 to v2 and there is also path from v2 to v1 then this graph is called strongly connected.

There are straightforward ways to compute the connected components for the graph G (i.e. the numbers of edges and node in one component) such as breadth-first search or depth-first search. In order to discover whole connected component in the graph, loop over its nodes, whenever the search reaches a node that existing in connected components earlier found, starting a new loop using depth first or breadth first search.

Clustering Coefficient

Clustering coefficient CC for node $v \in V$ represent how whole neighbors connected to node v. Can also be defining as relationship between the present connections of its neighbors and possible connections among them as follows:

$$\boldsymbol{CC\ (v) = \frac{2mv}{nv\ (nv - 1)}} \tag{1}$$

Where nv is represent the total number of vertex neighbors, and mv is represent the number of edges between them [10]. Sometimes CC also called local clustering coefficient. To compute average $\overline{CC}$ for graph G is by compute CC values for all networks nodes as follows:

$$\boldsymbol{\overline{CC}\ (G) = \frac{1}{n} \sum_{v \in V} cc\ (v)} \tag{2}$$

Where the number of nodes is representing by n. The low values of $\boldsymbol{\overline{CC}}$ is represent that the pairs of nodes have low connectivity [11, 12].

Diameter

Average diameter is one of networks property, mean that any two nodes are connecting by minimum number of edges, average over the possible set of pairs in the networks. Naturally, the diameter has important impact on network dynamics. For example, over the large diameter networks the data takes long time to flow [13].

The diameter of G is the length of the greatest distance in G [14, 15]. The eccentricity for networks vertexes is represent as greatest distance from any node to vertex v can be illustrated as follow:

$$\sigma(V) = \max_{u \in V} d\ (u, v) \tag{3}$$

Therefore, the diameter for connected graph is representing as the maximum distance between any two vertices [16, 17].

However, there are many disadvantages to diameter. First, it's more difficult to compute it. Second, its number instead of as an alternative of a distribution, therefore contains far less information that the distribution of large distances between vertices. Finally, the diameter impact of any changes to the graph so it more sensitive [18].

Vertex Degree
For the graph *G*, the number of edges that connected to vertex *V* is representing its degree [19]. It considers important parameter to analysis the complex networks. The whole degrees of graph vertices are double number of its edges [20] it represents as follows:

$$\sum_{v \in V} \delta(v) = 2m \tag{4}$$

Where, *m* is number of edges and $\delta(v)$ is representing vertex *v* degree.

Modularity and Community
Measure that uses in structure of graph or networks. It was used to measure the power of dividing the network into units that can be called communities, cluster or modules [21]. Community is group of nodes that sharing common property or feature such as in topic, color and other features. The network has a high value of modularity indicates that it has density between the nodes the communities but sparse connection between the nodes in the other communities. It considers optimization methods for community detection algorithms. The modularity value is lies between [−1, 1]. Can be defining as following [22]:

$$Q = \sum_{i=1}^{k} \left(e_{ii} - a_i^2 \right) \tag{5}$$

Where, *e_ii* is edges percentage bettween subgraphsmodule *i* and is the percentage of edges with at least one end in module *i*.

Density
Graph with number of edges is near to maximal number of edges is called dense graph different to graph with small number of edges is called sparse graph. If graph *G* is unweight the density can be defined as ratio between the actual numbers of edges to number of possible edges as follow [22]:

$$\text{Density}(G) = \frac{2m}{n\,(n-1)} \tag{6}$$

Where, the number of exiting edged is representing by m and possible number of edges in the graph with node n is represent by $n\,(n{-}1)$.

3 Design OMSPW-MLE

Time is considered one of the main important challenges facing us today this is due to speed development technologies and increase number of activities required to achieves from any person in specific time. Therefore, time management is one of the secrets of success in everyday life. The challenge of this thesis is how to accomplish eighteen different smartwatch and smartphone activities that fifty-one people can perform in less than one hour (54 minute). Therefore, this chapter will be described new a model called optimization model of smart phone and smart watch base on multi level of elitism (OMSPW-MLE).

OMSPW-MLI consist of four stages: The first stage related to collection and preprocessing data sets, in general preprocessing include two steps: split data into two groups of activities (activities achieve through using smartphone or smartwatch) then draw graph for each group (i.e., the number of nodes in that graph is constant while the labels of link are different therefore this graph is considered complex graph).

The second stage split each complex graph into multi subgraph through apply the Deterministic selection algorithm (Dselect) to find the set of roots help in split that complex graph into multi-subgraph.

The third stages find the optimal subgraph through Ant lion optimization algorithm (ALO).

The final stage evaluation the patterns through apply three measures, the connected of the nodes bettween subgraphsand without the graph, in addition to modularity.

Figure 1 explain block diagram of OMSPW-MLI. While the main stages of that model show in algorithm (1).

In general, we can summarize the main points of this thesis as explain below:

- It deals with a large database collected by sensors (accelerometer and gyroscope) implanted on the smartphone and Smartwatch.
- Preprocessing the data through convert the complex graph into multi subgraph by Dselect.
- Use the Ant Lion Optimization algorithm to find the optimal subgraph for each group of datasets.
- Evaluations the optimal sub graph based on three measures; the connected of the nodes bettween subgraphsand without the graph, in additions to modularity.
- Find the relationships among the time, coordination's and activities represent by multi-levels.

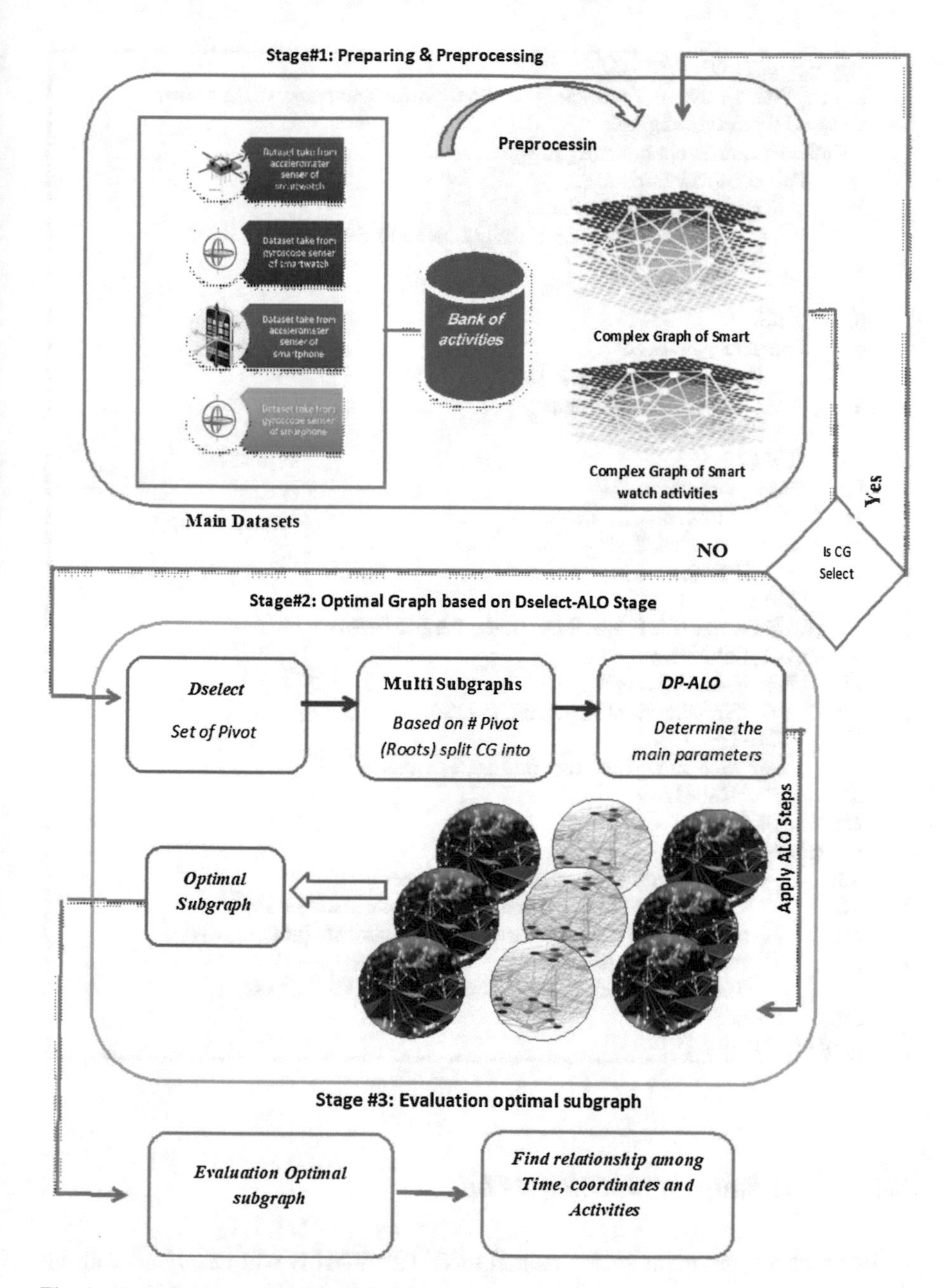

Fig. 1 Block diagram of *OMSPW-MLE*

```
Algorithm#1: OMSPW-MLE
Input: Dataset take in real-time from smartphone and smartwatch senses
Output: Optimal subgraph
// Collection & Preprocessing Stage
1:      For each row in datasets
2:          For each column in datasets
3:                  Call split base on activity related to SP and SW
4:           End for

5:       End
6:      For each row in SP
7:            For each column in SP
8:                    Call draw graph
9:               End for
10:      End
11:      For each row in SW
12:              For each column in SW
13:                    Call draw graph
14:             End for
15:      End
// Find Optimal Graph based on Dselect-ALO Stage
// Determine the Pivot
16:      For each id_dataset
17:             Find the Pivot according to DSA
18:       End for
19:       For each sub graph from multi subgraphs
20:             Call ALO
21:       End
// Evaluation Stage
 23:         For all patterns generated from optimal graph
24:               Compute Degree of connected node within graph
25:            compute degree of connected node without graph
26:            compute Modularity
27:            find relationships (time, corredinaties, Activities)
28:          End;
 End OMSPW-MLI
```

3.1 Main Stage of OMSPW-MLE

In this section, the main stages related to OMSPW-MLE will described with all details for each stage.

3.1.1 Data Collection and Preprocessing Stage

The data sets used in this thesis is an open-source database called Wireless Sensor Data Mining (WISDM).[1] This data was collected in a global laboratory in the department of computer and information sciences of for dham university through the use of four sensors (two of each type), which are the accelerometer and the gyroscope, these sensors were implanted into smart devices (smart watch and smartphone). The dataset contains 15,630,426 samples. It was recorded by relying on 51 people who can carry out 18 daily activities for the average person per day, and each activity does not exceed 3 minutes to implement it (54 minutes). The preprocessing is one of the important processes for analyzing the data, in this study the preprocessing includes split datasets based on the source that collection from it then convert each dataset into a graph. In general, there are several types of graph, which are as follows:

The random graph, which is either direct or indirect, and has no fixed shape or behavior, which means that the number of nodes in the graph is subject to increase or decrease. The undirected graph in which all the edges are bidirectional means that the edge is two different directions. The directed graph all edges in this type of graph in one direction. Algorithm 2 show the process convert dataset to graph.

After we have collected the data, at this stage we need to do the pre-processing process, which is one of the important processes for analyzing the data, here we convert the database into a graph as a pre-processing of the collected data. There are several types of graph, which are as follows:

After convert database to graph, we encountered difficulty in dealing with the graph, so the deterministic selection algorithm was proposed that divides the graph into multiple subgraphs. This algorithm chooses the optimal root from among the multiple roots of the sub-graphs. After applying the DSA algorithm, several optimal sub-graphs were found.

3.1.2 Apply Deterministic Selection Algorithm (Dselect)

The proposed system is implemented on a raw dataset to take results. DSA used to determine the best seed for each cluster from dataset as in Algorithm 3.

3.1.3 Apply Ant Lion Optimization Algorithm (ALO)

ALO is one of the main optimization algorithms described in Algorithm 4.

[1] http://archive.ics.uci.edu/ml/datasets/WISDM+Smartphone+and+Smartwatch+Activity+and+Biometrics+Dataset+

```
Algorithm # 2: Covert Data into Graph(CDG)
Input:    Dataset_id take from smartphone &smartwatch sensors
Output:   Complex Graph_id
// Random graph
1:  For each sample in Dataset_id Do
2:        G=nx.random_geometric_graph (len(Dataset_id.count()))
3:  End for
// Undirected Graph
4:  For each sample in Dataset_id Do
5:        G=G1.to_directed()
6:  End for
// Directed Graph
7:  For each sample in Dataset_id Do
8:        G=G1.to_directed()
9:  End for
End CDG
```

4 Results and Analysis

We will present the results of implementing the main stages of OMSPW-MLE that described with details in section three. Where, the model includes five main stages; the first stage related to collection and preprocessing data sets from smartphone and smartwatch by two types of sensor (Gyroscope and Accelerometer) Then draw graph for each dataset. The second stage finds the optimal root "seed" for each subgraph. The third stage, Find the optimal subgraph from multi subgraph using the ALO. The Fourth stage related to verification and Evaluation the optimal subgraph generated from each dataset based on combination Deep learning techniques (DSA & ALO) through compute three measures degree of connected nodes inside and outside the graph add to the Modularity of the optimal sub graph.

4.1 Implementation OMSPW-MLI Stages

In this part; we will show the results for each stage in OMSPW-MLI model and justification for all results will be given.

```
Algorithm #3 Dselect
Input:      Complex Graph
Output:     Set of Pivot to Split the Complex Graph
// Split Dataset_id into multi groups and sort each group
1:  Split Dataset_id into groups of 5, sort each group
2:    create empty matrix called C number of element n\5  // n total number of samples of
Dataset_id
3:  P = Dselect (C, N\5, N\10)                    // recursively computer median of C
4:  Partition Dataset_id around P
5:  For I in range 1 to N-J
6:      For j in range N-I to N
7:          IF I=J
8:              Return P
9:          End IF
10:         IF J<I
11:             Return Dselect (1st part of Dataset_id ,J-1,I)
12:           Else
13:                 Return Dselect (2nd part of Dataset_id, N-J,I-J)
14:         End IF
15:     End for
16: End for
End Dselect
```

4.1.1 Description of Database

In this section, we will describe the raw data used in this work. The database used in this thesis is an open-source database called Wireless Sensor Data Mining (WISDM). This data was collected in a global laboratory in the Department of Computer and Information Sciences of Fordham University through the use of four sensors (two of each type), which are the accelerometer and the gyroscope, these sensors were implanted into smart devices (smart watch and smartphone). The database contains 15,630,426 lines (4,804,403 lines for phone accel data/ 3,604,635 line for phone gyro data/ 3,777,046 line for watch accel data/ 3,440,342 line for watch gyro data). It was recorded by relying on 51 people who can carry out 18 daily activities for the average person per day, and each activity does not exceed 3 minutes to implement it (54 minutes). The person coding in range (1600–1650) and the activity in formula character range from (A to S) without using (N) as shown in Table 1.

Algorithm #4. ALO

Input: multi subgraphs

Output: optimal subgraph

1: Initialize the first population of ants and antlions randomly

2: Calculate the fitness of ants and antlions

// Find the best antlions and assume it as the elite (determined optimum)

3: While the end criterion is not satisfied

4: **For** every ant

5: Select an antlion using Roulette wheel

6: Update c and d using equations Eqs

7: $c^t = \frac{c^t}{I}$; $d^t = \frac{d^t}{I}$

8: Create a random walk and normalize it using

9: $X(t) = [0, cumsum(2r(t_1) - 1), cumsum(2r(t_2) - 1), \ldots, cumsum(2r(t_n) - 1)]$

10: $x_i^t = \frac{(x_i^t - a_i)*(d_i - c_i^t)}{(d_i^t - a_i)} + c_i$

11: Update the position of ant

12: $Ant_i^t = \frac{R_A^t + R_E^t}{2}$

13: ***End for***

Calculate the fitness of all ants Replace an antlion with its corresponding ant it if becomes fitter //

14: sum(x.^ 2 - 10 * cos (2 * pi. * x)) + 10 * dim

15: **-20*exp(-.2*sqrt(sum(x.^2)/dim))-exp(sum(cos(2*pi.*x))/dim)+20+exp(1)**

16: Update elite if an antlion becomes fitter than the elite

17: **End while**

Return elite

4.1.2 Draw Network

In this stage, we will be drawing the network for each dataset after split database into two datasets come from smartphone and smartwatch. as shown in Fig. 2.

4.1.3 The Results of Dselect Algorithm

We found that the Smartphone Accelerometer sensor network was divided into 65 subgraphs, while it was found that the Smartphone Gyroscope Sensor network was divided into 50 subgraphs, while the Smartwatch Accelerometer sensor network was divided into 47 subgraph and the other type, the Smartwatch gyroscope sensor,

Table 1 Show the name of each activity with the code of it

Seq.	Name of activity	Code
1	Walking	A
2	Jogging	B
3	Stairs	C
4	Sitting	D
5	Standing	E
6	Typing	F
7	Brushing teeth	G
8	Eating soup	H
9	Eating chips	I
10	Eating pasta	J
11	Drinking from cup	K
12	Eating sandwich	L
13	Kicking (Soccer Ball)	M
14	Playing catch w/tennis	O
15	Dribbling (Basketball)	P
16	Writing	Q
17	Clapping	R
18	Folding clothes	S

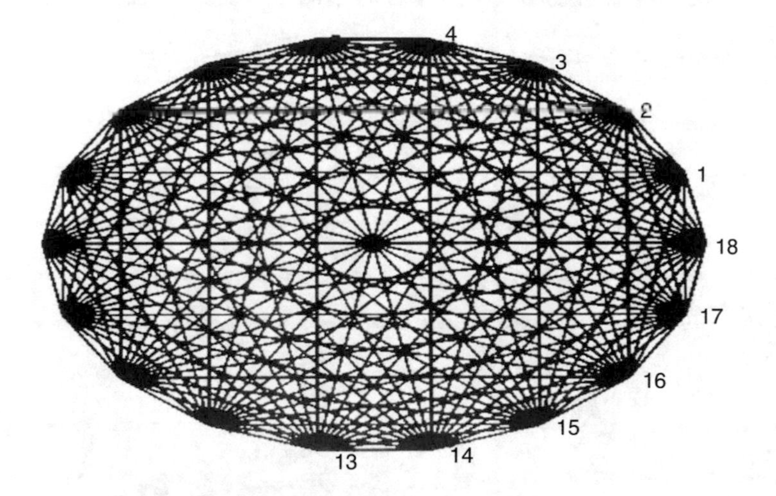

Fig. 2 Complex network represents the activity

is divided into subgraph 59, depending on the Dselect. As a result, Table 2 shown the main result get by Dselect into Four sensors. While Figs. 3, 4, 5 and 6 shown distribution subgraphs in each network and number of nodes related to it.

Table 2 The main result gets by Dselect into four sensors

	Result of DSA			
	Phone		Watch	
	Gyroscope	Accelerometer	Gyroscope	Accelerometer
#SG	50	65	59	47
#Nodes	3,608,635	4,804,403	3,440,342	3,777,046

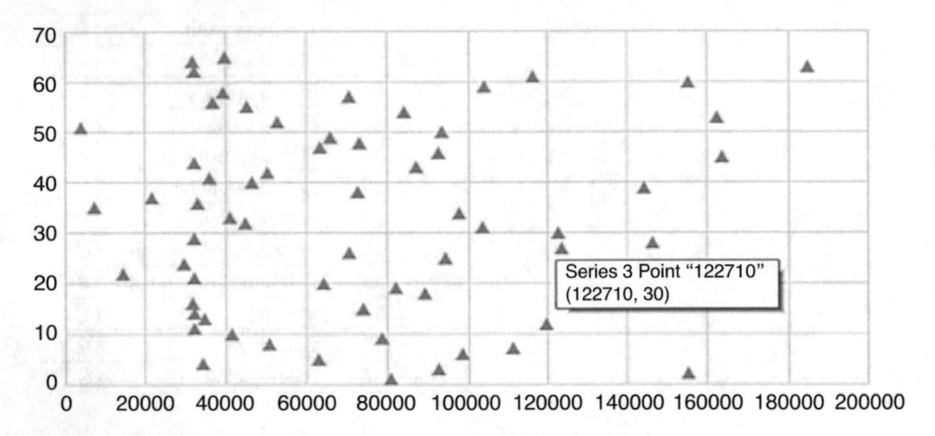

Fig. 3 Sixty-five pivot generation by Dselect related to Smartphone Accelerometer

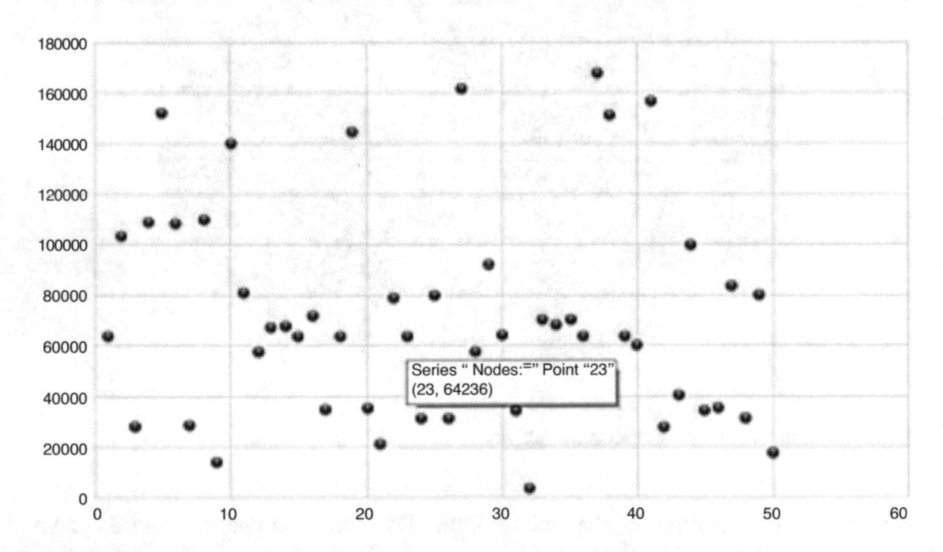

Fig. 4 Fifty pivot generation by Dselect related to Smartphone Gyroscope

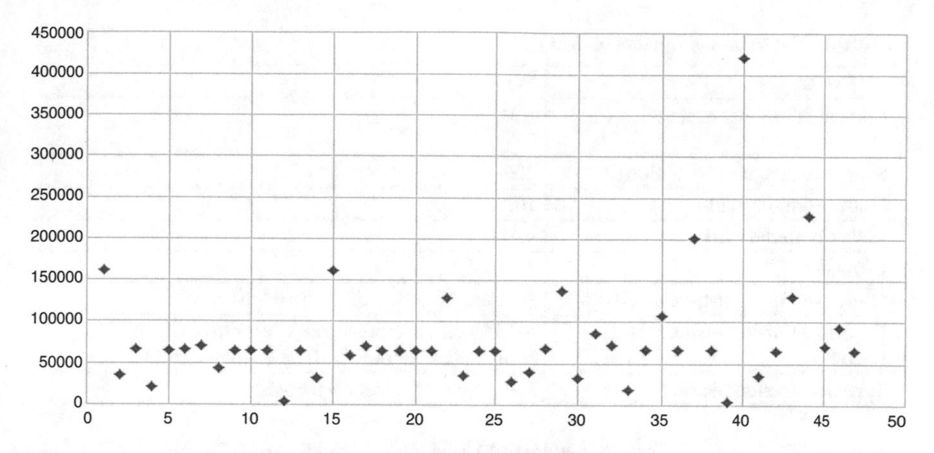

Fig. 5 Forty-seven pivots generation by Dselect related to Smartwatch Accelerometer

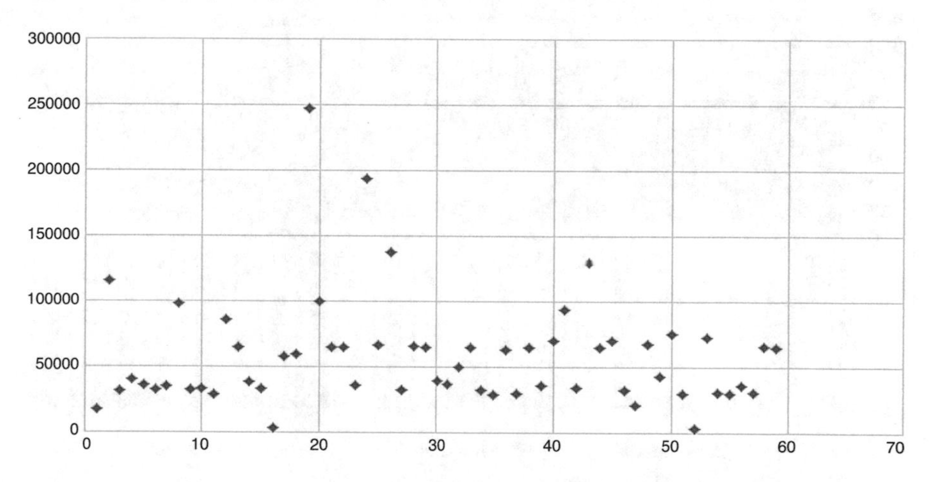

Fig. 6 Fifty-nine pivots generation by Dselect related to Smartwatch Gyroscope

4.1.4 Results of ALO Algorithm

In this section, we will explain the main results get by ALO, where in that algorithm using the parameters shown in Table 3. the input of that algorithm is output of Dselect (i.e., number of pivots and number of nodes related to that pivots) while the output is optimal subgraph for each dataset (i.e., smartphone Accumulator, smartphone Gyroscope, smartwatch Accumulator and smartwatch Gyroscope).

The best sub-graph is No. (51), where the value of the first objective function for it during iterations No. (500) is (80.849) and the value of the second objective

Table 3 The Main Parameters of ALO

Parameter	Value
Maximum number of iterations (Max-it)	500
SearchAgents_no(Population)	200
Upper Boundary (ub)	100
Lower Boundary(lb)	−100
Dimintions(dim)	4
Objective Function-Linear (OBJ#1)	=sum(x.^2–10*cos(2*pi.*x))+10*dim
Objective Function-SIGMID (OBJ#2)	−20*exp.(−0.2*sqrt(sum(x.^2)/dim))-exp(sum(cos(2*pi.*x))/dim) + 20 + exp(1)
Type of selection method	Deterministic selection method

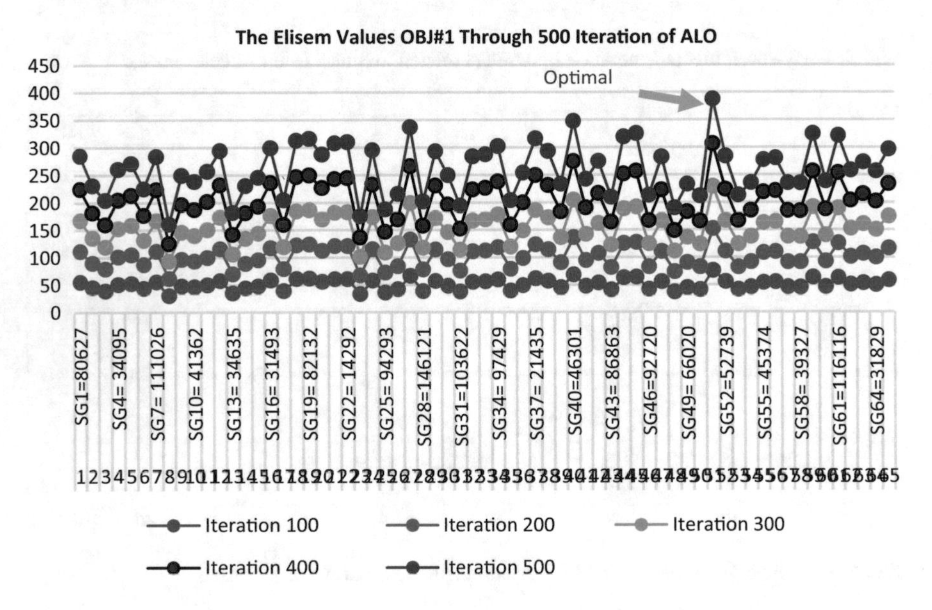

Fig. 7 The value of objective function number one (OBJ#1) of sixty-five subgraph related to Smartphone Accelerometer

function for the same iteration is (6.136). As explain in Figs. 7 and 8. Therefore, that sub-graph is passed to Evaluation stage.

The best sub-graph is No. (9), where the value of the first objective function for it during iterations No. (500) is (96.089) and the value of the second objective function for the same iteration is (1.778). As explain in Figs. 9 and 10. Therefore, that sub-graph is passed to Evaluation stage.

The best sub-graph is No. (39), where the value of the first objective function for it during iterations No. (500) is (72.219) and the value of the second objective function for the same iteration is (5.427). As explain in Figs. 11 and 12. Therefore, that sub-graph is passed Evaluation stage.

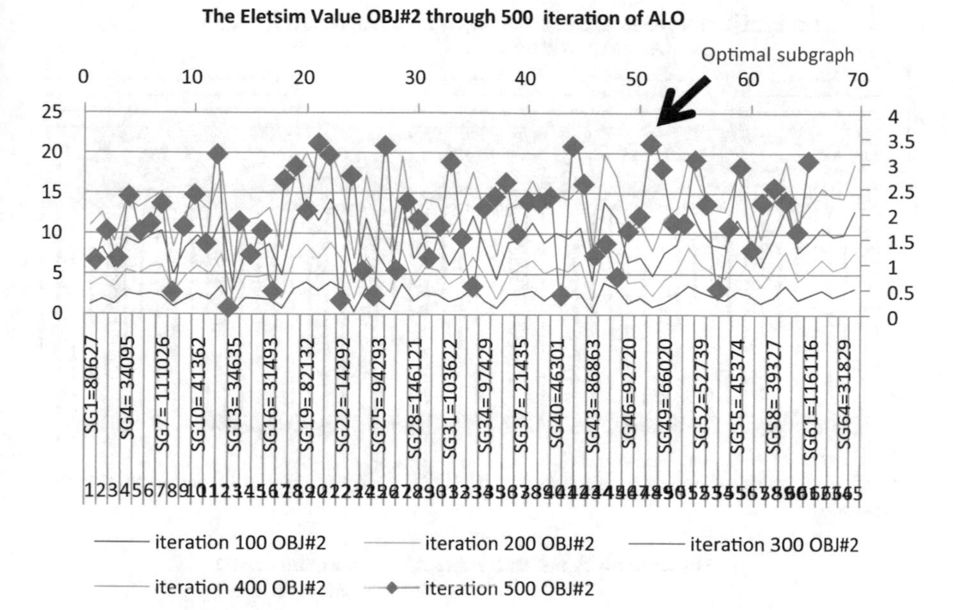

Fig. 8 The values of objective function number two (OBJ#2) of sixty-five subgraph related to Smartphone Accelerometer

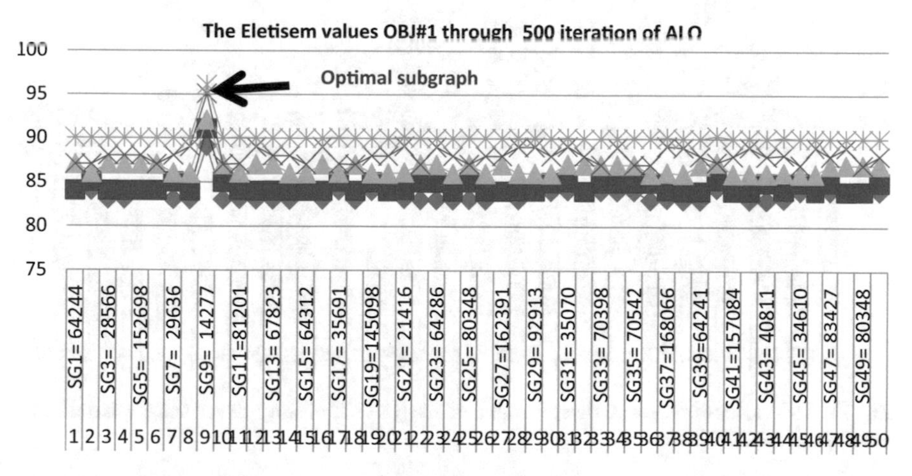

Fig. 9 The values of objective function number one (OBJ#1) of fifty subgraphs related to Smartphone Gyroscope

The best sub-graph is No. (1), where the value of the first objective function for it during iterations No. (500) is (96.307) and the value of the second objective function for the same iteration is (1.807). As explain in Figs. 13 and 14. Therefore, that sub-graph is passed Evaluation stage.

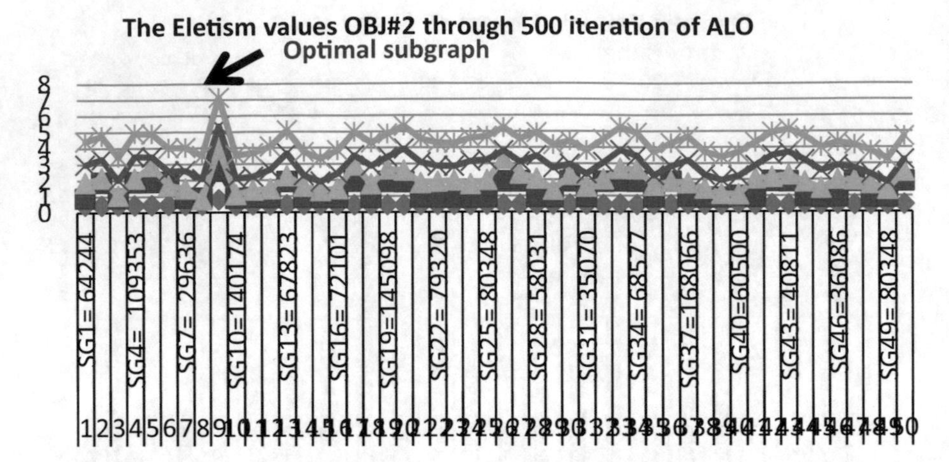

Fig. 10 The values of objective function two (OBJ#2) of fifty subgraphs related to Smartphone Gyroscope

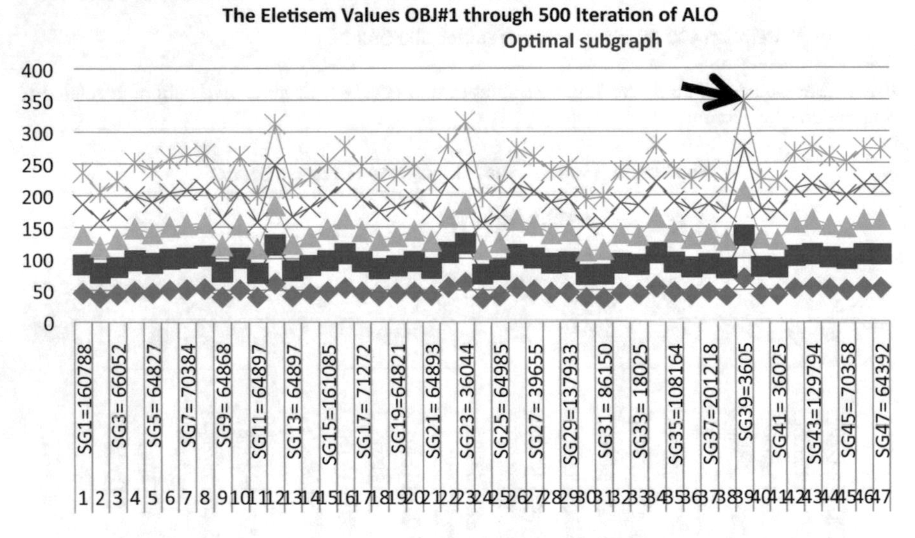

Fig. 11 The values of objective function number one (OBJ#1) of forty-Seven subgraphs related to Smartwatch Accelerometer

4.1.5 Evaluation

The verification from the results of OMSPW-MLE we will use three evaluations measures, the first two measures determined the degree of connected nodes between subgraphs for each dataset while the other measure represent the connected of nodes within the optimal subgraph. Finally using the modularity measures the result of all that measure shown in Table 4.

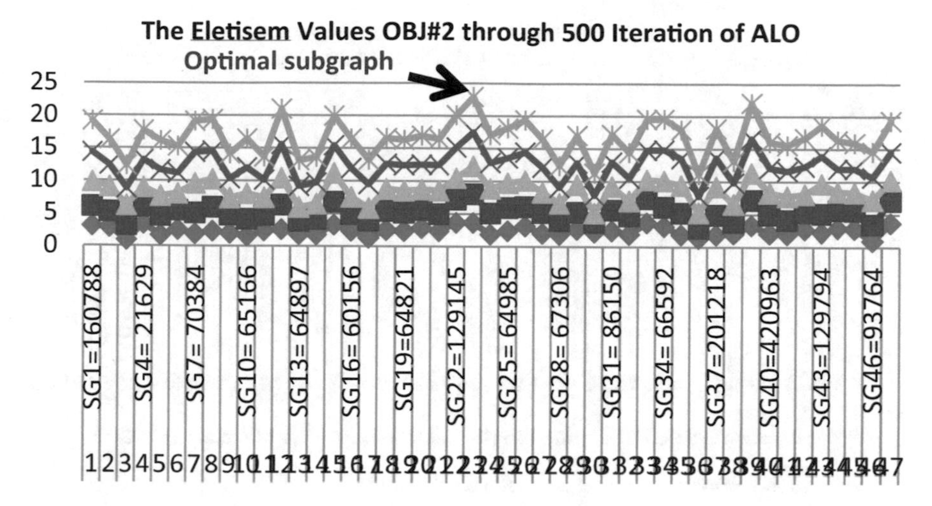

Fig. 12 The values of objective function number two (OBJ#2) of forty-Seven subgraphs related to Smartwatch Accelerometer

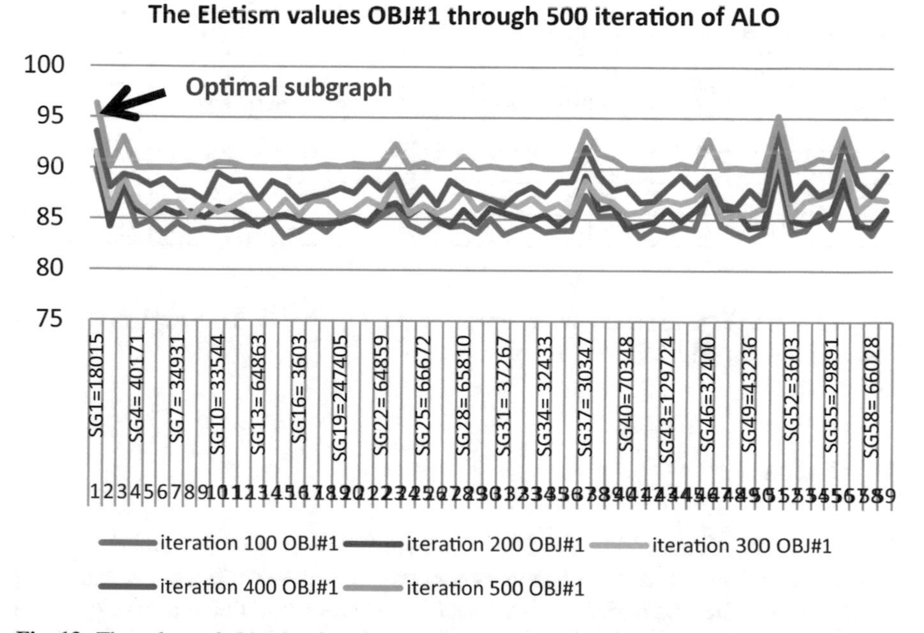

Fig. 13 The values of objective function number one (OBJ#1) of fifty-nine subgraphs related to Smartwatch Gyroscope

Finally, we extraction three levels of connection among the activity as explain in Fig. 15.

As a result, we found correlation among activities (A, B, C, D, E, F) as explain in Level one of figures (Fig. 15), among activities (G, H, I, J, K, L) as explain in Level

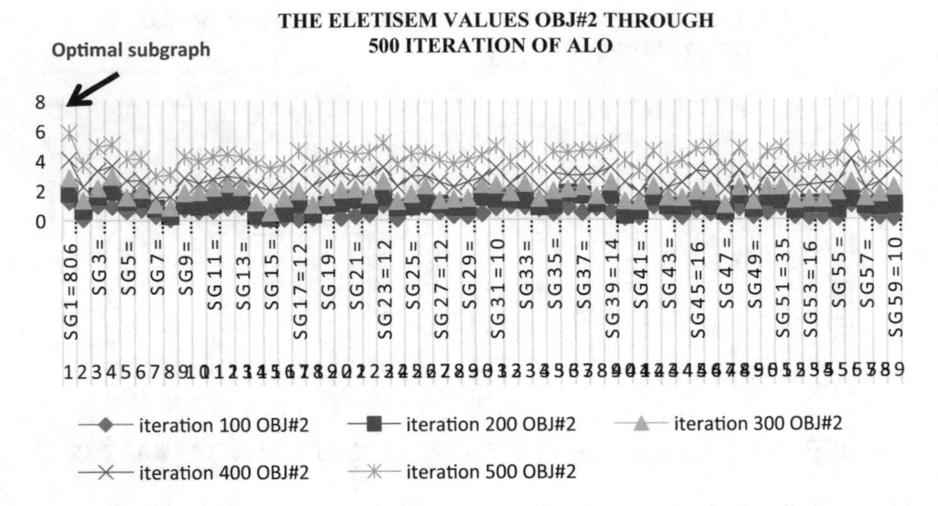

Fig. 14 The values of objective function number one (OBJ#2) of fifty-nine subgraphs related to Smartwatch Gyroscope

Table 4 The value of Evaluation measures

		connected		
#Seq	Optimal subgraph	Within subgraphs	Between different subgraphs	Modularity
1	SP-A	87.12	12.88	0.27
2	SP-G	87.11	12.89	0.21
3	SW-A	87.10	12.90	0.23
4	SW-G	98.81	1.19	0.52

two of figures (Fig. 15) and among activities (M, O, P, Q, R, S) as explain in Level Three of figures (Fig. 15).

5 Discussion

The world has witnessed the rapid development of the internet and smart environments and devices, which have become widely applicable to various areas, including medicine and other industries. In this research, we attempt to find a way to generate optimal graph to distinguish human activities by using a deep learning technique (Dselect and ALO). Data from four sensors were collected in real time. The data obtained were preprocessed and converted into a graph. The optimal pivots for each subgraph were then identified using the DSA algorithm. Then, the ALO was applied to find the optimal subgraph among the dynamic number of subgraphs. From the apply ALO, on the sixty-five subgraphs related to smartphone Accelerometer find the optimal subgraph number (51) and the value of object function one is (80.849)

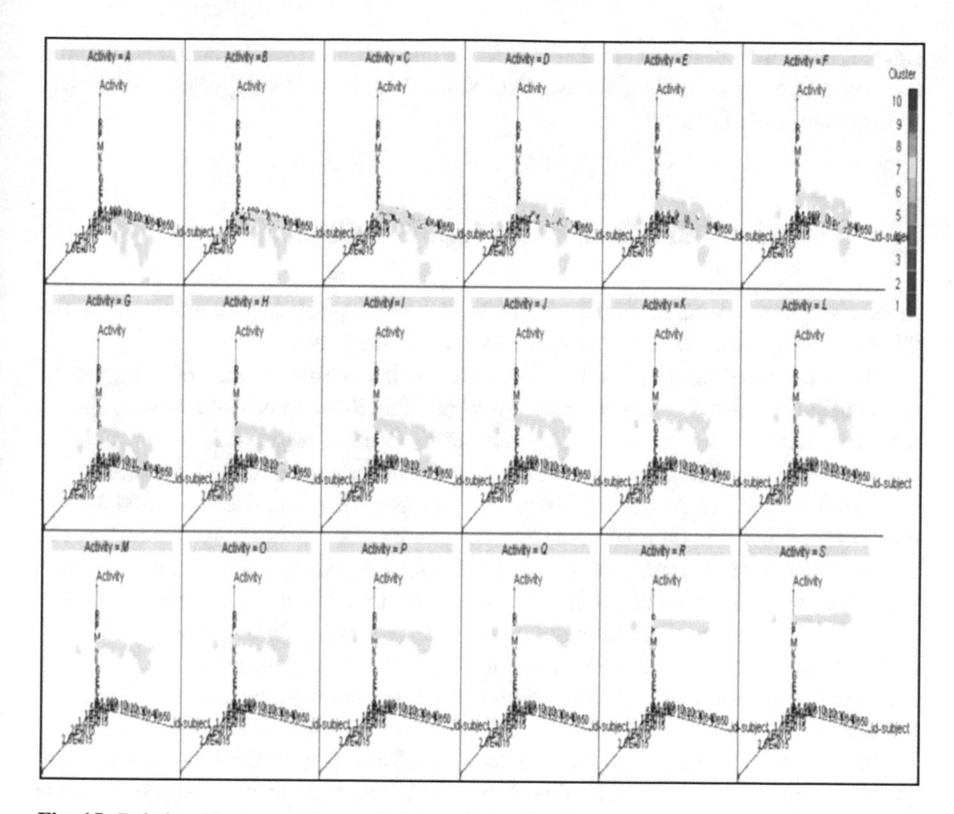

Fig. 15 Relationship among time, activities and coordination

and the value of object function two is (6.136), the value of within subgraph is (87.12), the value of between different is (12.88) and the value of modularity is (0.27).

On the Fifty subgraphs related to smartphone Gyroscope find the optimal subgraph number (9) and the value of object function one is (96.089) and the value of object function two is (1.778), the value of within subgraph is (87.11), the value of between different subgraph is (12.89) and the value of modularity is(0.21).

On the Forty-seven subgraphs related to smartwatch Accelerometer find the optimal subgraph number (39) and the value of object function one is (72.219) and the value of object function two is (5.427), the value of within subgraph is (87.10), the value of between different is (12.90) and the value of modularity is (0.23).

On the fifty-nine subgraphs related to smartwatch Gyroscope find the optimal subgraph number (1) and the value of object function one is (96.307) and the value of object function two is (1.807), the value of within subgraph is (98.81), the value of between different is (1.19) and the value of modularity is (0.52). As a result, three levels of interdependence were reached between the activities through the relationship between time, activity and the ID of the person, where the first level linked the six activities represented (A, B, C, D, E, F), while the second level

Linking activities between (G, H, I, J, K, L) and the last (third) level, also linking six other types of activities, namely (M, N, O, P, Q, S) as we explained previously in Figure number (15).

6 Conclusions and Recommendation of Future Works

In this section, we can summarize the main benefit point to explain them in the future also the entire wall to attempt to avoid it in the future:

The sensors of smartphones and smart watch are one of the most important devices that can provide us with a torrent of big data that can be used to characterize human activities as well as to diagnose some disease cases. It is a rich field of knowledge that is characterized by the accuracy of the data obtained through it.

Decision making process of huge data is very important matter due to the continuous growth in data volumes and advancement of technology. Therefore, the combination of two techniques of deep learning proposed in this thesis in its two strategies, was the stage of finding the best point to divided the complex networks into set of subgraphs, then finding the best subgraph for each network on the bases of which the interconnectedness between the activity is determined as a constructive solution and its effectiveness was followed by the results we have reached, with its three levels.

The Dselect is prove highly and effective performance in dividing huge dataset, as it divided the four-complex network into a different number of important points (pivots) and each pivot belongs to it a different number of points. Therefore, it is recommended to use it in many applications due to its dependence on the principles of division, arrangement and selection continuous to reach the best division points.

Using the elitism principle inside the ALO give a pragmatic result based on the both activation functions used inside it. Add to that, the linear and sigmoid functions prove ability to deal with different number of subgraphs to choose the optimal from them.

The three-evaluation measure used give new trend to analyze the characterized of graph in addition, it gives idea of complex network and how can analysis the activities based on the multilevel of elitism.

The following points may be good ideas for future work;

It is possible to perform another objective function for other Deep Learning (DL) algorithm, such as the Long Short-Term Memory (LSTM) algorithm or the CapsNet algorithm.

It is possible to build a recommendations system based on one of the mining methods such as the gSpan algorithm for optimal subgraph.

Other types of sensors can be used to distinguish other types of activities.

References

1. Gonçalves S, Dias P, Correia A (2020) Computers in human behavior reports nomophobia and lifestyle : smartphone use and its relationship to psychopathologies. Comput Hum Beh Rep 2:100025. https://doi.org/10.1016/j.chbr.2020.100025
2. Maglogiannis I, Ioannou C, Spyroglou G, Tsanakas P (2014) Fall detection using commodity smart watch and smart phone. In: IFIP advances in information and communication technology, vol 436, pp 70–78. https://doi.org/10.1007/978-3-662-44654-6_7
3. Duch W (2007) What is computational intelligence and where is it going? Stud Comput Intell 63:1–13. https://doi.org/10.1007/978-3-540-71984-7_1
4. Harkeerat Kaur G (2016) A comparative study of different biometric features. Int J Adv Res Comput Sci 7(6):2776–2784
5. Al-Janabi, S., Alkaim, A., Al-Janabi, E. et al. Intelligent forecaster of concentrations (PM2.5, PM10, NO2, CO, O3, SO2) caused air pollution (IFCsAP). Neural Comput & Applic (2021). https://doi.org/https://doi.org/10.1007/s00521-021-06067-7
6. Boothalingam R (2018) Optimization using lion algorithm: a biological inspiration from lion's social behavior. Evol Intell 11:31–52. https://doi.org/10.1007/s12065-018-0168-y
7. Weiss GM, Yoneda K, Hayajneh T (2019) Smartphone and smartwatch-based biometrics using activities of daily living. IEEE Access 7:133190–133202. https://doi.org/10.1109/access.2019.2940729
8. Alsamhi SH, Ma O, Ansari MS, Meng Q (2019) Greening internet of things for greener and smarter cities: a survey and future prospects. Telecommun Syst 72:609–632. https://doi.org/10.1007/s11235-019-00597-1
9. Ray A, Holder LB, Bifet A (2019) Efficient frequent subgraph mining on large streaming graphs. Intell Data Anal 23:103–132. https://doi.org/10.3233/IDA-173705
10. Al-Janabi S, Al-Shourbaji I (2016) A hybrid image steganography method based on genetic algorithm. In: 2016 7th International conference on sciences of electronics, technologies of information and telecommunications (SETIT), Hammamet, 2016, pp 398–404. https://doi.org/10.1109/SETIT.2016.7939903
11. Zanin M, Papo D, Sousa PA, Menasalvas E, Nicchi A, Kubik E (2016) Combining complex networks and data mining: why and how. Phys Rep 635:1–44
12. El Khaddar MA, Boulmalf M (2017) Smartphone: the ultimate IoT and IoE device. In: Smartphones from an applied research perspective. INTECH
13. Alkaim AF, Al-Janabi S (2020) Multi objectives optimization to gas flaring reduction from oil production. In: Farhaoui Y (ed) Big data and networks technologies. BDNT 2019, Lecture notes in networks and systems, vol 81. Springer, Cham. https://doi.org/10.1007/978-3-030-23672-4_10
14. Masoud M, Jaradat Y, Manasrah A, Jannoud I (2019) Sensors of smart devices in the internet of everything (IOE) era: big opportunities and massive doubts. J Sens 2019. https://doi.org/10.1155/2019/6514520
15. Sareen P (2014) Biometrics – introduction, characteristics, basic technique, its types and various performance measures. Int J Emerg Res Manage Technol 9359(34):2278–9359
16. Chen K, Dumitrescu A (2020) Selection algorithms with small groups. Int J Found Comput Sci 31(3):355–369. https://doi.org/10.1142/S0129054120500136
17. Chen K, Dumitrescu A (2020) Selection algorithms with small groups. Int J Found Comput Sci 31(3):355–369. https://doi.org/10.1142/S0129054120500136
18. Al-Janabi S, Al-Shourbaji I, Shojafar M, Abdelhag M (2017) Mobile cloud computing: challenges and future research directions. In: 2017 10th international conference on developments in eSystems engineering (DeSE), Paris. IEEE, pp 62–67. https://doi.org/10.1109/DeSE.2017.21
19. Grzimek B, Schlager N, Olendorf D, McDade MC (2004) Grzimek's animal life encyclopedia. Gale Farmington Hills, Detroit

20. Kaveh A, Mahjoubi S (2018) Lion pride optimization algorithm: a meta-heuristic method for global optimization problems. Sci Iran 25(6B):3113–3132. https://doi.org/10.24200/sci.2018.20833
21. Lu L, Ren X, Qi L, Cui C, Jiao Y (2019. Springer) Target Gene Mining algorithm based on gSpan. LNICST 268:518–528. https://doi.org/10.1007/978-3-030-12981-1_36
22. Al-Janabi S, Alkaim AF, Adel Z (2020) An innovative synthesis of deep learning techniques (DCapsNet & DCOM) for generation electrical renewable energy from wind energy. Soft Comput 24:10943–10962. https://doi.org/10.1007/s00500-020-04905-9
23. Al-Janabi S, Alkaim AF (2021) A comparative analysis of DNA protein synthesis for solving optimization problems: a novel nature-inspired algorithm. In: Abraham A, Sasaki H, Rios R, Gandhi N, Singh U, Ma K (eds) Innovations in bio-inspired computing and applications. IBICA 2020, Advances in intelligent systems and computing, vol 1372. Springer, Cham. https://doi.org/10.1007/978-3-030-73603-3_1
24. Alfa AA, Misra S, Ahmed KB, Arogundade O, Ahuja R (2020) Metaheuristic-based intelligent solutions searching algorithms of ant colony optimization and backpropagation in neural networks. In: Proceedings of first international conference on computing, communications, and cyber-security, Lecture notes in networks and systems, vol 121. Springer, Singapore, pp 95–106

K-Nearest Neighbour Algorithm for Classification of IoT-Based Edge Computing Device

Micheal Olaolu Arowolo, Roseline Oluwaseun Ogundokun, Sanjay Misra, Jonathan Oluranti, and Akeem Femi Kadri

1 Introduction

The Internet of Things (IoT) is an advanced model intended to offer enormous applications that are presently a portion of our everyday existence. Under dynamic networks, millions of smart devices are installed to provide lively functionality, including connectivity, tracking, and control of vital infrastructures [1, 3]. However, due to the shortage of bandwidth and money, this enormous creation of IoT gadgets and the resulting big data traffic caused at the edge of the network imposed further loads on the advanced centralized cloud storage model. Edge computing (EC) is now evolving as a revolutionary approach that puts data conversion and depository close to ultimate consumers, contributing to what is known as EC-aided IoT. While this model offers new functionality and quality of service (QoS), it likewise raises tremendous dangers in the areas of data protection and confidentiality [2].

Presently, the IoT scenario has attained a great deal of popularity. It encompasses a software and hardware infrastructure that ties the corporeal world to the Internet.

M. O. Arowolo · R. O. Ogundokun (✉)
Department of Computer Science, Landmark University, Omu Aran, Nigeria
e-mail: arowolo.micheal@lmu.edu.ng; ogundokun.roseline@lmu.edu.ng

S. Misra
Department of Computer Science and Communication, Ostfold University College, Halden, Norway
e-mail: sanjay.misra@hiof.no

J. Oluranti
Covenant University, Ota, Nigeria
e-mail: oluranti@covenantuniversity.edu.ng

A. F. Kadri
Department of Computer Science, Kwara State University, Malete, Nigeria
e-mail: akeem.kadri@kwasu.edu.ng

S. Misra et al. (eds.), *Artificial Intelligence for Cloud and Edge Computing*, Internet of Things, https://doi.org/10.1007/978-3-030-80821-1_8

The number of IoT devices has risen significantly owing to the exponential advancement of concentration in this archetype. More than 75 billion gadgets are predicted to be coupled to the Internet by 2025, contributing to a financial impact on the worldwide economy. Usually, IoT gadgets have restricted processing capacity and minor reminiscences and can produce huge quantities of data. In our homes, towns, cars, and industries, reduced-power and coupled systems, primarily sensors, would be utilized. Cloud storage can be sufficient for the growth of the IoT market, but together with potential frequency range capacity, the interruption instigated by data transmission is inappropriate for certain assignments (e.g., well-being keep tracking) [2].

Limitations in resource-scarce computer computational resources limit the application of multifaceted machine learning algorithms; multiple software agent frameworks provide stable and efficient resolutions for optimizing the implementation of various edge computing. Low-data fusion is concerned with tasks for delivering edge components, while calculation needs to be delegated to more powerful systems for a profound understanding of the data and administrative intentions. Transferring raw data to cloud servers upsurges the cost of transmission, causes device responses to be delayed, and exposes private data. A real-world approach is to examine computing data nearer to its origin and convey it to the inaccessible servers that solitary, the information required for additional cloud processing to resolve the problems. Instead of distant areas, edge computing pertains to the data transformation accomplished as close as possible to the information sources [3].

Effective artificial intelligence (AI) algorithms can be implemented on IoT gadgets, such as AI techniques include support vector machine (SVM), deep learning (DL), and, among other things, neural networks (NNs) to make this scenario possible. It is noteworthy that NNs in the application process need less computational resources than in the training phase. In terms of RAM, the AI technique can be run on a computer with low power (random access memory). AI has been used for the prediction and detection of diseases [4–6]. For various subjects, for instance, smart cities, computer vision, medical care, self-propelled, and machine learning algorithms are used [7, 8]. There are numerous instances of what way machine learning may be applied on edge gadgets in these areas [9–11].

Edge computing is increasingly linked with AI, advancing one another in the equivalent of efficiency: edge intelligence and smart edge. The intellect of the edge and the smart edge are not distinct from each other. Edge intelligence is the target, and intelligent edge DL techniques are as well a component of edge intelligence. Consecutively, the smart edge could deliver edge intelligence with advanced service material and asset utilization [12].

This research, therefore, uses machine learning systems to deploy an edge computing device and evaluate the algorithms of machine learning that can be applied in edge computing using the edge analysis approach to machine learning.

The article's remaining part is structured as thus: Unit 2 discussed the literature review on AI, ML, edge computing, and IoT devices as well as the interrelated studies. Unit 3 discussed the materials and methods. The ML procedure used for the execution of the system in the study was conferred in the unit as well. Unit 4 discussed the execution and testing of the system. Here, the findings discovered are also conferred in this unit. The study came to an end in Unit 5.

2 Background and Literature Review

In this section, we provide the background of the work which includes an architecture for IoT edge-centric, AI and edge computing, machine learning, and other related works.

Upon the accelerated spread of IoT technologies by 2025, it is estimated that there will be 77.44 billion IoT devices [13]. Different IoT architectures [14–16] were introduced by diverse organizations from diverse backgrounds for the vast number of IoT applications, and edge computing has been documented as an imperative abutment for IoT structures [15].

There are four key parts of the edge-centric IoT architecture: the cloud, the end machines of the IoT, the edge, and its apps. The sketch of the architecture carves over together the resources obtainable and every group's unique functionality. To make their lives more comfortable, consumers use intelligent IoT software, while more commonly they connect with IoT end devices via cloud or edge interactive interfaces somewhat than unswervingly communicating with IoT finishing gadgets. Deeply rooted in the real universe are the IoT end units. They feel things as they are and assume action to regulate things as they are, but in computer-heavy activities, they are not advanced. There are almost unlimited resources in the cloud; nevertheless, they are generally physically situated distant from the final gadgets. As a result, an IoT cloud-centric structure is generally unable to perform effectively [17], particularly once the structure possesses the requirements for the present. This could together coordinate the further three groups to concur, agree, and counterpart the cloud and IoT client gadgets for enhanced accomplishment, with the edge being a dominant element of the entire architecture.

IoT users submit IoT data access queries or IoT device control commands to the edge-centric IoT architecture. Via a net or mobile application-enabled interface supported by the cloud or the edge, such requests and commands can ultimately enter the edge layer. The edge layer would then manage them, either routing them to IoT users' gadgets or managing them on behalf of IoT gadgets on the edge layer. The edge layer, not solitary, bridges them with consumers and the cloud by communicating with IoT gadgets but can also stock the gathered records submitted from IoT gadgets and unload major computational requirements, for instance, big data scrutiny and robust protection procedures from IoT users' gadgets. It is also possible to move several current IoT user's system facilities from the cloud to the edge and to configure them depending on the requirements of IoT users' gadgets. The edge can operate independently of the cloud in relation to the interaction between the cloud and the edge, or collaboratively the edge can work with the cloud. The edge is efficient enough to manage IoT application needs in the first model. It will, for example, provide storage and processing resources to satisfy all IoT device demands. Furthermore, the edge receives cloud resources to control the edge layer or to better manage the needs of the IoT framework. Besides, it is possible to move several current IoT end system facilities from the cloud to the edge and to configure them depending on the requirements of IoT users' gadgets. The edge can operate independently of the cloud in terms of the interaction between the edge and the cloud or the edge can collaboratively work with the cloud. The edge can sufficiently manage IoT implementation requirements in the first model. It will, for example, provide storage and processing resources to satisfy all IoT device demands. In model 2, the edge receives cloud resources to control the edge layer or to better manage the needs of the IoT framework. Second, mechanically, the layer of the edge is similar to the IoT users' units. It could meet the instantaneous necessities required in the plan for security [18]. Third, data from several IoT end devices is obtained and stored by the edge layers. Therefore, the edge is a safer area to decide protection verdicts relative to end devices, so a top protection verdict relies on the algorithm's performance and the availability of appropriate information simultaneously. The edge layer, for example, will detect intrusion more effectively with more data [19–21]. Several security procedures would be transformed to steering strategies with the recognition of software-defined networks and virtualization; however, they can clash with each other. We will overcome these disputes at the edge with an analysis of the entire system coupled by the edge. Fourth, it is mostly not feasible to install and maintain firewalls on any IoT end system, given resource limitations, maintenance expense, and incredibly large size of end devices. Instead, installing firewalls on the edge layer allows approaching threats to be more easily cleared and obstructed. Fifth, the edge layer should keep track of the movement of these devices, given the versatility of end devices, and provide them with a continuous secure relation. Moreover, the very secure link between the users' gadgets and the edge layer helps to create a good trust between them. It alleviates the issues of confidence building between these devices. Sixth, with the cloud, the edge typically has a huge-velocity communication. The edge can communicate with the cloud layer for protection support whenever it is required. For instance, the cloud can offer position and role authentication for the edge, as shown in Sha, Alatrash, and Wang [22], and the cloud can create powerful protection measures to secure the edge. First, we discuss edge-centered IoT security resolutions.

The advent of edge computing in IoT networks has brought many research challenges, such as the design of dynamic computing offloading schemes, allocation of resources (e.g., computing resources, spectrum resources), and power management transmission. Over the last year, IoT generated enormous interest in science. The concept of IoTs is seen as part of the future Internet and will consist of billions of smart things communicating with each other. In the nearest future, International Networks will constitute hierarchically organized embedded devices that expand the world's borders with physical entities and artificial resources. IoT would bring new capabilities to connected things. IoT can communicate with one another without human interference. Some early IoT implementations in the healthcare, transportation, and automotive industries have already been developed [23, 24]. IoT systems are currently in their infancy; however, there have been several recent advances throughout the integration of sensor objects into the cloud-based Internet [23, 25, 26]. IoT creation includes several issues, including communication networks, applications, protocols, and standards [23, 27]. The term Internet and Things refers to a worldwide interconnected network with an emphasis on visual technologies, networking, information-processing, and connectivity, the latest version of interaction which could be ICT [28]. Smart sensing and wireless communication technologies, in particular, became some of the IoT's components and the newest opportunities and forensic horizons appeared [29].

IoT is triggered by using Radio Frequency Identification technology, which is becoming much more popular in various industries such as manufacturing and retail [30, 31]. Disruptive technologies are the term used for technological advances that have changed the way ordinary products are done in the world. The IoT has been implemented in many aspects of daily life as part of the Disruptive Technology community. It is projected that the demand for IoT gadgets will upsurge to 20–30 billion by 2020 [32].

There have been numerous innovative pieces of machinery intended at operating at the edge of the net in the advancement of edge computing, with similar perceptions but distinct emphases, for instance, cloudlet, micro data centers (MDCs), fog computing, and mobile edge computing (now multi-access edge computing). The edge computing group, however, is yet to attain a compromise on the homogenous concepts, architectures, and edge computing protocols. For the suite of new technologies, the common word "edge computing" is used [33]. Because of its benefits of lowering information transfer, enhancing facility potential, and simplification of cloud computing strain, edge computing has become an important solution for breaking the bottleneck of emerging technologies. The architecture of edge computing will transform to being a significant accompaniment to the cloud, even substituting the position of the cloud in few instances [34]. Edge computing is an enabling resource that facilitates network IoT applications, lower data on behalf of cloud facilities, and upstream information on behalf of IoT services. They are interchangeable as opposed to fog computing, but edge computing focuses further on the stuffed side, whereas fog computing concentrates further on the infrastructure side. Both computing and network resources along the path between data sources and cloud data centers can be an "edge." Edge computing's rationale is that computing can take place near data sources. At the edge, not only do stuff request cloud service and content but also perform cloud computing tasks, as well as transmitting requests and distribution services from the cloud side to the user side. Edge can perform computing offloading, data storage, caching, and processing [35].

Edge computing consists of important metrics, such as the delay in the definition. Besides, high bandwidth, which is accomplished by restricting the flow of data. Edge computing servers, and with a lengthy latency rate for a centralized server. Energy usage is also a big concern, apart from this. External resource systems-related computational work contributes to an improved battery life of user devices. Edge computing is fast, stable, scalable, and reliable in terms of speed [36].

Edge computing may deliver several benefits. The edge computing model, for example, can be flexibly applied from a single house to society or even the size of the city. Edge computing for services requiring consistent and low latency, such as health emergencies or public safety, is an effective paradigm since it can save time for data processing as well as simplify the layout of the network. Decision and predictions are possible to diagnose and transmit from the edge of the network, which is more effective compared to data collection and central cloud decision-making. Edge computing exceeds geographic-based applications such as transport and utility management. Due to the understanding of place, that is, cloud computing data could be obtained and processed based on geographical position in edge computing, without being loaded into the cloud [37].

The machine learning procedures that could be utilized in resource-constrained edge computing sceneries are explored in this report. The ML procedures implemented in the subsequent paragraphs are the utmost commonly utilized and present the issue of taking artificial intelligence to hardware-resource-constrained devices. Machine learning models currently require sufficient memory to store training data and computing power to train large models. New computer models have been designed to work effectively on edge equipment by using shallow models that require sufficiently low processing power to be used on IoT devices. Alternatively, reducing the size of the inputs of the model for classification applications will increase the learning speed [38].

The algorithm k-nearest neighbors (KNN) is a procedure utilized in the identification of patterns, built on the features of the artifacts similar to the one deliberated. This approach is utilized for both problems with classification and regression. There are numerous updated variants of KNN that help execute the procedure on hardware-constrained gadgets, with ProtoNN being the most creative. It is an algorithm based on KNN. The key KNN problems for edge computation are the size of the training data (the algorithm uses the entire dataset to produce predictions), the time of prediction, and the choice of distance metrics [39]. For classification and regression problems, tree-based ML procedures are utilized, which is a very known routine in the IoT area. Nevertheless, the normal tree algorithms could not be implemented on them owing to the inadequate resources of the gadgets. An algorithm that is evolving is Bonsai. The tree technique is specifically intended for IoT gadgets that are strictly resource-constrained, and it retains forecasting accuracy whereas lessening the model size and cost of estimation. It initially discovers that the size model is decreased by a single parse tree and then makes nonlinear estimation through the internal nodes and leaf ones. Bonsai finally studies the sparse matrix, jutting all the knowledge into a low-dimensional space in which the tree is taught. This makes it possible to carry the algorithm to tiny devices such as IoT ones [40]. At the embedded stage, one of the most commonly used ML algorithms is the SVM. SVM is an algorithm for supervised learning that can be utilized for problems with both classification and regression. By outlining an optimum hyperplane that split up every group, the algorithm discriminates between two or more classes of data. Help vectors are the data nearest to the hyperplane, which will appear in the hyperplane itself being redefined if removed. The critical elements of the dataset are considered for these purposes. The loss function utilized by the procedure is usually the loss of the hinge and the optimization function is the descending gradient form [2].

We assume AI and edge computing's confluence is natural and imminent. In effect, there is an engaging interaction between them. On the one hand, AI provides technology and approaches for edge computing, and edge computing can unleash its power and scalability with AI. On the other hand, edge computing provides scenarios and platforms for AI, and with edge computing, AI can extend its applicability.

AI Provides Edge Computing with Technologies and Methods Edge computing is a disseminated computing model in general, where software-defined networks are developed to distribute data and offer sturdiness and elasticity of services. Edge computing faces resource utilization challenges in multiple layers, for instance, frequency of the CPU loop, authority of entry, radio frequency, bandwidth, and so on. As a consequence, several effective optimization methods are highly challenging to increase machine performance. AI technologies are capable of fulfilling this mission. Essentially, AI models extract from actual scenarios unrestricted optimization issues and then iteratively find the asymptotically optimal solutions using stochastic gradient descent (SGD) methods. Either mathematical methods of learning or methods of deep learning may provide the edge with support and guidance. Also, reinforcement learning, like multi-armed bandit theory, multi-agent learning, and deep Q-network (DQN), plays an increasing and significant role in edge resource distribution issues [4].

Edge Computing Provides AI with Scenarios and Platforms The explosion of IoT devices renders the Internet of Everything (IoE) a fact [41]. Rather than mega-scale cloud data centers, ubiquitous and globally dispersed smartphone and IoT devices are generating more and more data. Many further implementation examples will significantly accelerate the realization of AI from theory to reality, such as in intelligent networked cars, autonomous driving, smart house, smart city, and real-time public security data collection. Also, AI systems can be moved from the cloud to the edge with high connectivity efficiency and low computing power requirements. Edge computing, in a phrase, offers a heterogeneous framework full of rich capabilities for AI. AI chips with computational acceleration such as Field Programmable Gate Arrays (FPGAs), Graphics Processing Units (GPUs), Tensor Processing Units (TPUs), and Neural Processing Units (NPUs) are now increasingly becoming possible to combine with smart mobile devices. To support the edge computing model and to promote DNN acceleration on resource-limited IoT chips, more companies are interested in the design of chip architectures. Edge hardware updates also inject vigor and energy into AI.

2.1 Related Works

The key problems related to QoS specifications for applications operating on traditional paradigms such as fog and cloud environments were explored by Guevara et al. [42]. A standard machine learning classification strategy was then presented. The paper's key gain is to concentrate professionally on facets of machine learning. However, the paper usually suffers from inadequate numbers of articles presented, and especially incomplete and insufficient records of the recently written manuscript. Furthermore, the investigation agonizes from the deficiency of the high content of the papers studied. The organization of the study is still not very pleased.

Wang et al. [43] studied several offloading procedures that are discussed in three major groups constrained by two edges, mobile gadgets, and edge clouds, namely, contact connections called intermediate carriers. The key drawback of their investigation was that the topic entirely is considered and the complete relevant studies are checked collectively. The investigation, however, agonizes from the absence of newly issued manuscripts in the relevant research area (i.e., about 7% issued in 2018, 4 out of 54). Additionally, this survey has sacrificed specific and comprehensive facets of each individual in the two-end method referred to above for the sake of honesty. Some essential factors, for instance, fault tolerance and protection matter in the scheme, which consecutively unswervingly influence the complete competency of the system, are also neglected. This degradation, which is QoS, QoE, could deceive the operator's viewpoint. There is no systemic format for the paper to select papers, too.

Aazam et al. [44] reviewed few offloading structures in the study interest of computational models, for instance, edge, fog, cloud, and also IoT. After these paradigms, they introduced an arrangement. It also discusses permissive expertise for offloading as middleware and associated features. The evaluation was well classified as a strength in the predefined standard, with adequate instances for separately described standards. A judicious proportion of recently issued manuscripts (approximately 24% printed in 2017 – 12 out of 51) is also included in the topic, nevertheless, with certain crucial features, for instance, coarseness and flexibility in previous researches, which does not evaluate those studies. The paper does not possess a systematic setup for choosing manuscripts.

Mach et al. [45] also pronounced certain use cases, range of capabilities, standardization, and offloading of computation in MEC settings. The investigation was made as an advantage by examining many MEC-connected studies and papers, of which 23% were recently issued at the period of acceptance (e.g., available studies in 2016 – 29 out of 124). The manuscript, however, in some respects, such as the proposed granularity, is not well organized. Some important topics and methods are also not enclosed in the previous studies. By way of additional disadvantage, some aspects are simplified that lead to loss of generality. The manuscript also does not provide a methodical setup for selecting papers. It is also worth noting that this paper does not cover future paths well.

Boukerche et al. [46] examined energy sensitivity in Mobile Cloud Computing (MCC) in the area of MCC and Green Cloud Computing techniques. They studied procedures, design, setting up, and balance procedures. Next, the benefits and drawbacks of those experiments were contrasted and classified in terms of offloading mechanism and resource manager forms. The quality of their manuscript is its proficient analysis of the topic that has been sufficiently structured to address energy-conscious concerns. Nevertheless, it is a difficult task to make a firm indeterminate state amid evolving pieces of machinery which include MCC and MEC. They provide a precise sense and description for all of them as a result of the lack of principles in the discussion, and this is exactly one of the feeble goals of this study. As another example, the analysis does not fully cover similar topics. This study also hurts from a deficiency of newly issued manuscripts in the re-examined sector (approximately 5% available in 2018, 7 out of 141, and approximately 8% in print in 2017, 11 out of 141) despite the consideration of significant numbers of checked articles.

Peng et al. [47] examined several papers relevant to MEC in terms of design, amenity acceptance, and delivery. Computing offloading and data offloading are categorized as two important facets of the MEC model for amenity acceptance. Edge server (ES), amenity delivery and its practical predictor, ES implementation, and resource utilization are checked for service provision. This survey also explores several other problems, such as MEC implementations. The analysis is supplied with reasonable recently published articles as a strength (about 30% published in 2017 – 37 out of 123). The investigation is not, however, theoretically well organized. The investigation likewise hurts from a fluent summary of the literature with inadequate technical descriptions and predefined fields in each division.

Shakarami, Shahidinejad, and Ghobaei-Arani [48] proposed a systematic analysis of game-theoretical offloading methods in the MEC setting. In the form of tables and maps, they have contrasted different critical aspects of literature. Their manuscript is the solitary comprehensive study evaluation that covers the topic in an efficient technique, as a power. The investigation likewise encompassed similar newly in print manuscripts about the relevant field. By way of a downside, the article is not linked to approaches of ML, which is not, certainly, its focus.

Cao et al. [49] suggested an investigation on basic principles in MEC with an emphasis on ML and the foremost implementations. As a strong point, their investigation emphases competently on a single issue that has no literature context. This investigation agonizes a deficiency of ample convincing papers to accurately include the field of study. As an additional downside, there is no systematic format for the paper to pick papers, too. In this article, potential instructions are still not adequately powerful and well protected.

In another work, Taleb et al. [50] presented MEC (mobile edge computing) survey that explores the major technical enablers in this domain. It discusses MEC implementation, taking into account both the experiences of individual providers and a network of mobility enabling MEC platforms. Also discussed here are various potential MEC implementation solutions. It also discusses the study of the MEC reference architecture and its core implementation scenarios that can help application developers, service providers, and third parties with multi-tenancy support. The latest standardization efforts and future open testing issues are also detailed in this work.

In the paper by Madakam et al. [51], the key aim of this study is to describe the Internet of Things, architectures, and critical innovations in our everyday lives and their applications. In this paper, the key observations are made. (a) The traditional definition of IoT is not present. (b) Universal standardization at the architectural level is needed. (c) Technologies differ from vendor to vendor and thus interoperability is essential. (d) There is a need to develop uniform protocols for better global management.

The study by Dolui et al. [52] examines the effectiveness and contrasts the feature sets of various forms of edge computing models, including fog computing, cloudlet, and mobile edge computing. Edge computing has been a field of focus for researchers, with a lot of exposure to IoT and smartphones that require real-time answers.

Mao et al. [53] presented a study of state-of-the-art mobile edge computing technology with an emphasis on radio (network) optimization and computational tools.

3 Materials and Methods

This section discussed the materials used for this study in terms of datasets used for the implementation. The machine learning algorithm adopted for this study was also discussed in this section.

3.1 Datasets

The datasets used for this study is presented as an addition to the work published in the MDPI Sensors Journal entitled "Process Management in IoT Operating Systems: Cross-Influence between Processing and Communication Tasks in End-Devices." From the perspective of Edge Computing, this thesis focuses on process management in IoT operating systems. Specifically, in sensor end devices, we carried out an empirical analysis of the cross-influence between processing and communication tasks [54].

3.2 K-Nearest Neighbor (KNN) Algorithm

The ML procedure used for this system's implementation is KNN. In several aspects, for instance, expert networks and intellectual structures, artificial learning has been widely used. An intelligent machine can hardly be considered a real intelligent system without learning capacity. To resolve the deficiencies of expert systems and smart structures, several pieces of research began with ML. The KNN procedure is a popular method of ML built on computer codes to give a run down and creäte human-acquired knowledge. It is straightforward, spontaneous, and quick to execute and possesses an extensive variety of applications. It is practically ideal for multiple kinds of data structures. Consequently, the KNN method has been commonly used as a non-parametric technique appropriate for non-inactive and nonlinear active structures for forecasting analysis in different research areas: Davis and Nihan [55] utilized the KNN technique in interim forecasting of highway traffic. Mehrotra and Sharma [56] proposed a version of the k-nearest neighbor resampled optimized for utilization of manifold prognosticator variables. By incorporating two separate methodologies for the k-nearest neighbor approach, Bannayan and Hoogenboom [57] developed a method to forecast the comprehension of instantaneous regular climate results.

Mangalova and Agafonov [58] considered a modeling technique aimed at forecasting the power production of wind ranch energy generators, and the KNN process is the fundamental modeling solution. An improved KNN model was suggested by Cai et al. [59] to increase prediction exactness built on spatiotemporal similarity and to accomplish multiphase prognostication. In the present centuries, in the researches for the time series prognostication utilizing KNN technique, comprehensive and elegant methods have been published, and few researchers have enhanced the KNN technique and extended it to various research areas. For short-term traffic forecasting, Cheng et al. [60] presented an adaptive spatiotemporal k-nearest neighbor technique. Martínez et al. [61] predict using a dissimilar specialized KNN learner every different season. Fan et al. [62] suggest an innovative immediate load prognostication method to evaluate the energy characteristics and regulations built on the weighted k-nearest neighbor algorithm.

Xu et al. [63] presented a procedure to simulate road traffic situations in time series dependent on kernel k-nearest neighbors (kernel-KNN). Multidimensional k-nearest neighbor (MKNN), autoregressive moving average model (ARMA), and quadratic regression model are included (2020) to develop the EEMD-FFH method. In a detailed analytical analysis using a broad data collection of M3-competition industrial time series, Kück and Freitag [64] analyze the predicted success of local-nearest neighbor methods, which are based on the concept of dynamical structures. Furthermore, the KNN procedure can also be utilized to classify (Hattori and Takahashi [65], Hattori and Takahashi [66], Jiang et al. [67], Jiang et al. [68], Miao et al. [69], Tan [70], Tan [71], Wan et al. [72], Yoon and Friel [73], Zhang et al. [74], Zhang et al. [75]).

The performance of the KNN procedure relies on the demonstrativeness and comprehensiveness of the data, comparable to further data-driven procedures. The KNN procedure has a very basic operating process as a classical regression model: k signifies the sum of closest neighbors. K training models can be initiated in the training set, built on the distance extent technique. Determine the k value that minimizes the error and the sample length l. Meaning, the basic postulation of the KNN procedure is that a condition can be found in the previous neighborhood extra or a reduced amount comparable to that projected in the upcoming. The KNN technique does not level ancient data, particularly once there are special events; the performance of the forecast is improved than the parametric process.

4 Result and Discussion

This section shows the results gotten from the implementation and the testing of the system using the k-nearest neighbor machine learning algorithm. The interfaces of the implementations are also presented in this section.

The machine-learning algorithm was implemented using MATLAB 2015 package. The datasets were inputted into the MATLAB and were then passed into the KNN classifier algorithm, and the result was obtained. The performance metrics used for the evaluation were confusion matrix, and below are some interfaces from the system implementation and testing.

Figure 1 shows the datasets used for the implementation. This particular interface in Fig. 1 is the datasets inputted into MATLAB, and the KNN classification algorithm was implemented on it.

Figure 2 shows the k-nearest neighbor scattered plot showing the classification of the datasets.

Figure 3 shows the confusion matrix for the k-nearest neighbor implemented on the system. The KNN of the system true-positive (TP) value is 35, true-negative (TN) value is 16, false-positive (FP) value is 5, and false-negative (FN) value is 4.

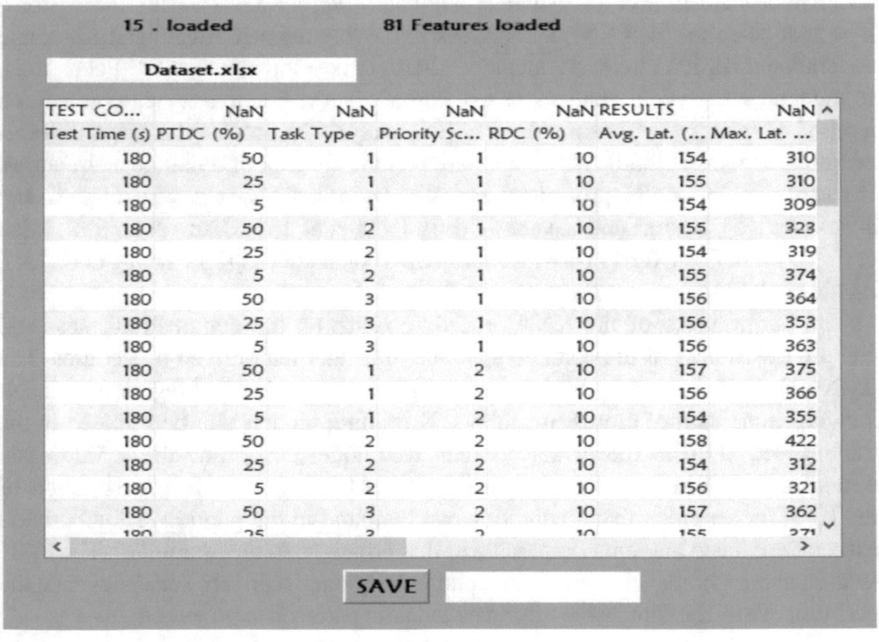

15 . loaded
81 Features loaded
Dataset.xlsx

TEST CO...	NaN	NaN	NaN	NaN	RESULTS	NaN
Test Time (s)	PTDC (%)	Task Type	Priority Sc...	RDC (%)	Avg. Lat. (...	Max. Lat. ...
180	50	1	1	10	154	310
180	25	1	1	10	155	310
180	5	1	1	10	154	309
180	50	2	1	10	155	323
180	25	2	1	10	154	319
180	5	2	1	10	155	374
180	50	3	1	10	156	364
180	25	3	1	10	156	353
180	5	3	1	10	156	363
180	50	1	2	10	157	375
180	25	1	2	10	156	366
180	5	1	2	10	154	355
180	50	2	2	10	158	422
180	25	2	2	10	154	312
180	5	2	2	10	156	321
180	50	3	2	10	157	362
180	25	[illegible]	[illegible]	10	155	371

SAVE

Fig. 1 Sample datasets

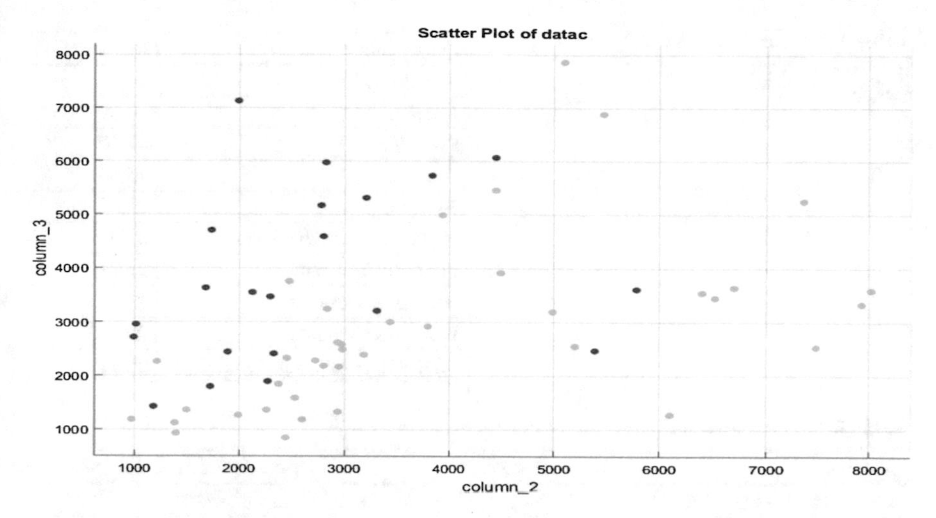

Fig. 2 Scatter plot of the dataset

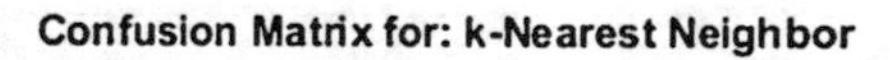

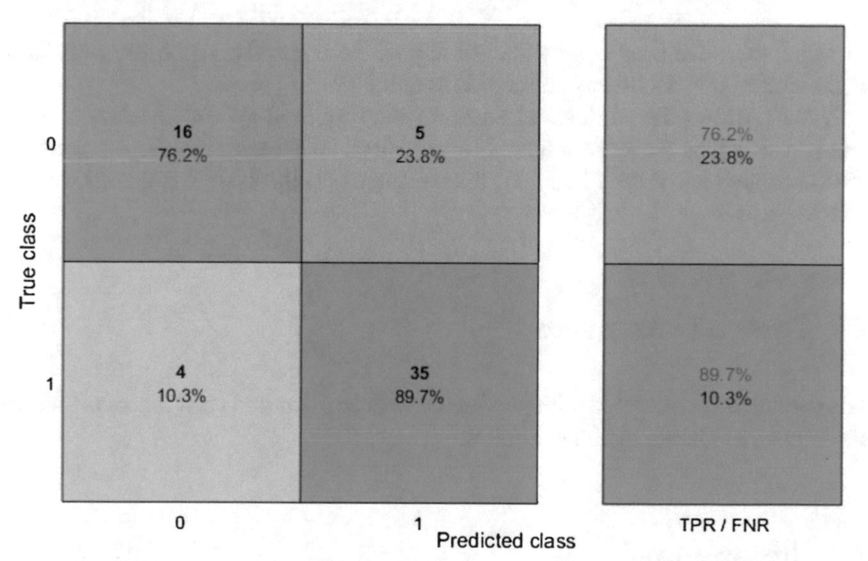

Fig. 3 K-nearest neighbor confusion matrix. Confusion matrix for KNN: TP = 35, TN = 16, FP = 5, FN = 4

4.1 System Evaluation

The developed system was appraised employing the confusion matrix metrics. The measures used in this study were sensitivity, specificity, precision, accuracy,

Table 1 Confusion matrix for proposed system evaluation

Performance metrics	Results (%)
Sensitivity	89.74
Specificity	76.19
Precision	87.50
Accuracy	85.00
F1 score	88.61
Negative predictive value	0.8000
False-positive rate	23.81
False discovery rate	12.50
False-negative rate	0.1026
Mathews correlation coefficient	66.71

Table 2 Comparative analysis with previous researches

Authors	ML methods used	Disease predicted	Accuracy
Khateeb and Usman [76]	NB, KNN, and bagging classifiers	Breast cancer	79.20%
Hashi et al. [77]	Decision tree and KNN	Diabetes	76.96%
Proposed system	KNN	COVID-19	85%

F1 score, negative predictive value, false-positive rate, false discovery rate, false-negative rate, and Matthews correlation coefficient.

Table 1 shows the confusion matrix for the proposed system evaluation, and it was revealed that the system had 88% precision, 90% sensitivity, 76% specificity, 85% accuracy, and 89% F1 score. This indicated that the system is an efficient and effective one.

4.2 System Comparison

The system was compared with previous researches that had used machine learning algorithms, as shown in Table 2 below.

4.3 Discussion

The proposed system was compared with two previous studies, and it was realized that the proposed study performance supersedes those previous works with an accuracy of 85%. Khateeb and Usman [86] employed three ML algorithms in their study and those algorithms' performance was compared, but the KNN accuracy was given as 79.20% which showed that our proposed work performs better than theirs. Hashi et al. [77] used two algorithms on the diabetes dataset, and the two accuracies were compared as well, but the KNN accuracy for the research was 76.96% which also signifies that our posed ML algorithm performs better than the two works compared with, with an accuracy of 85%. It is finally concluded that the proposed system performed better, effectively, and efficiently in terms of accuracy being used as the performance metrics.

5 Conclusion

The implementation of ML on IoT gadgets decreases overcrowding issues in the network by permitting data to be computed near data sources, protecting data upload privacy, and decreasing energy consumption for unremitting wireless broadcasting to gateways or cloud servers. The goal of this investigation was to deliver an efficient approach to ensure the implementation of low-performance ML techniques on hardware in the IoT paradigm, allowing the means for IoT awareness. In this study, a machine learning approach using the KNN classification algorithm was developed using a process management in IoT operating systems: cross-influence between processing and communication tasks in end devices data. The result of the experiment shows an 85% accuracy which outperformed other methods that have been suggested in the literature. However, this study proves to be relevant and can be adopted for better efficiency. Future work suggests a robust dataset and combines efficient machine learning approaches such as KNN with other optimizer algorithms that can help better the technology world.

References

1. Adeniyi EA, Ogundokun RO, Awotunde JB (2021) IoMT-based wearable body sensors network healthcare monitoring system. In: IoT in healthcare and ambient assisted living. Springer, Singapore, pp 103–121
2. Merenda M, Porcaro C, Iero D (2020) Edge machine learning for AI-enabled IoT devices: a review. Sensors 20(9):2533. https://doi.org/10.3390/s20092533
3. Liu Y, Yang C, Jiang L, Xie S, Zhang Y (2019) Intelligent edge computing for IoT-based energy management in smart cities. IEEE Netw 33(2):111–117. https://doi.org/10.1109/MNET.2019.1800254
4. Odusami M, Abayomi-Alli O, Misra S, Shobayo O, Damasevicius R, Maskeliunas R (2018) Android malware detection: a survey. In: International conference on applied informatics. Springer, Cham, pp 255–266
5. Adeyinka AA, Adebiyi MO, Akande NO, Ogundokun RO, Kayode AA, Oladele TO (2019) A deep convolutional encoder-decoder architecture for retinal blood vessels segmentation. In: International conference on computational science and it applications, Lecture notes in computer science (including subseries Lecture notes in artificial intelligence and Lecture notes in bioinformatics). Springer, Cham. 11623 LNCS, pp 180–189
6. Oladele TO, Ogundokun RO, Kayode AA, Adegun AA, Adebiyi MO (2019) Application of data mining algorithms for feature selection and prediction of diabetic retinopathy. Lecture notes in computer science (including subseries Lecture notes in artificial intelligence and Lecture notes in bioinformatics), International conference on computational science and it applications. Springer, Cham. 11623 LNCS, pp. 716–730

7. Ikedinachi AP, Misra S, Assibong PA, Olu-Owolabi EF, Maskeliūnas R, Damasevicius R (2019) Artificial intelligence, smart classrooms and online education in the 21st century: implications for human development. J Cases Inf Technol (JCIT) 21(3):66–79
8. Alagbe V, Popoola SI, Atayero AA, Adebisi B, Abolade RO, Misra S (2019) Artificial intelligence techniques for electrical load forecasting in smart and connected communities. In: International conference on computational science and its applications. Springer, Cham, pp 219–230
9. Xu H (2017) Machine learning based data analytics for IoT devices. Nanyang Technological University. https://doi.org/10.32657/10356/72342
10. Ieracitano C, Mammone N, Hussain A, Morabito FC (2020) A novel multi-modal machine learning based approach for automatic classification of EEG recordings in dementia. Neural Netw 123:176–190. https://doi.org/10.1016/j.neunet.2019.12.006
11. Panesar A (2021) Machine learning algorithms. Apress, Berkeley, pp 85–144. https://doi.org/10.1007/978-1-4842-6537-6_4
12. Yazici M, Basurra S, Gaber M (2018) Edge machine learning: enabling smart internet of things applications. Big Data Cogn Comput 2(3):26. https://doi.org/10.3390/bdcc2030026
13. Portal S (2018) Internet of things (IoT) connected devices installed base worldwide from 2015 to 2025. https://www.statista.com/statistics/471264/iot-number-of-connected-devices-worldwide/
14. Gubbi J et al (2013) Internet of things (IoT): a vision, architectural elements, and future directions. Futur Gener Comput Syst 29(7):1645–1660
15. Lin J, Yu W, Zhang N, Yang X, Ge L (2017) On data integrity attacks against route guidance in transportation-based cyber-physical systems. In: Proceedings of the 14th IEEE annual conference in consumer communications and networking conference (CCNC 2017)
16. Singh D, Tripathi G, Jara AJ (2014) A survey of internet-of-things: future vision, architecture, challenges and services. In: Proceedings of 2014 IEEE world forum on internet of things (WF-IoT)
17. Chen X, Jiao L, Li W, Fu X (2016) Efficient multi-user computation offloading for mobile-edge cloud computing. IEEE/ACM Trans Netw 24(5):2795–2808
18. Lee I, Lee K (2015) The internet of things (IoT): applications, investments, and challenges for enterprises. Bus Horiz 58(4):431–440
19. Sha K et al (2018) On security challenges and open issues in internet of things. Futur Gener Comput Syst 83:326–337
20. Brewster T (2016) How hacked cameras are helping launch the biggest attacks the internet has ever seen. https://www.forbes.com/sites/thomasbrewster/2016/09/25/brian-krebs-overwatch-ovh-smashed-by-largest-ddos-attacks-ever/#705007235899. Sept 2016
21. Russon M-A (2016) Hackers turning millions of smart CCTV cameras into botnets for DDoS attacks. http://www.ibtimes.co.uk/hackers-turning-millions-smart-cctv-cameras-into-botnets-ddos-attacks-1525736. Accessed Sept 2016
22. Sha K, Alatrash N, Wang Z (2017) A secure and efficient framework to read isolated smart grid devices. IEEE Trans Smart Grid 8(6):2519–2531
23. Li S, Da Xu L, Zhao S (2015) The internet of things: a survey. Inf Syst Front 17(2):243–259
24. Gokhale P, Bhat O, Bhat S (2018) Introduction to IOT. Int Adv Res J Sci Eng Technol 5(1):41–44
25. Gubbi J, Buyya R, Marusic S, Palaniswami M (2013) Internet of things (IoT): a vision, architectural elements, and future directions. Futur Gener Comput Syst 29(7):1645–1660
26. Stergiou C, Psannis KE, Kim BG, Gupta B (2018) Secure integration of IoT and cloud computing. Futur Gener Comput Syst 78:964–975
27. Atzori L, Iera A, Morabito G (2017) Understanding the internet of things: definition, potentials, and societal role of a fast-evolving paradigm. Ad Hoc Netw 56:122–140
28. Chahal RK, Kumar N, Batra S (2020) Trust management in social internet of things: a taxonomy, open issues, and challenges. Comput Commun 150:13–46

29. Srivastava G, Parizi RM, Dehghantanha A (2020) The future of blockchain technology in healthcare internet of things security. In: Blockchain cybersecurity, trust and privacy. Springer, Cham, pp 161–184
30. Shafique MN, Khurshid MM, Rahman H, Khanna A, Gupta D (2019) The role of big data predictive analytics and radio frequency identification in the pharmaceutical industry. IEEE Access 7:9013–9021
31. Manavalan E, Jayakrishna K (2019) A review of internet of things (IoT) embedded sustainable supply chain for industry 4.0 requirements. Comput Ind Eng 127:925–953
32. Georgakopoulos D, Jayaraman PP, Fazia M, Villari M, Ranjan R (2016) Internet of things and edge cloud computing roadmap for manufacturing. IEEE Cloud Comput 3(4):66–73
33. Bilal K, Khalid O, Erbad A, Khan SU (2018) Potentials, trends, and prospects in edge technologies: fog, cloudlet, mobile edge, and micro data centers. Comput Netw 130:94–120. https://doi.org/10.1016/j.comnet.2017.10.002
34. Yousefpour A, Fung C, Nguyen T, Kadiyala K, Jalali F, Niakanlahiji A, Kong J, Jue JP (2019) All one needs to know about fog computing and related edge computing paradigms: a complete survey. J Syst Archit 98:289–330. https://doi.org/10.1016/j.sysarc.2019.02.009
35. Wang Y, Meng W, Li W, Liu Z, Liu Y, Xue H (2019) Adaptive machine learning-based alarm reduction via edge computing for distributed intrusion detection systems. Concurr Comput Pract Exp 31(19). https://doi.org/10.1002/cpe.5101
36. Wang S, Zhao Y, Xu J, Yuan J, Hsu C-H (2019) Edge server placement in mobile edge computing. J Parallel Distrib Comput 127:160–168. https://doi.org/10.1016/j.jpdc.2018.06.008
37. Wang Y, Xie L, Li W, Meng W, Li J (2017) A privacy-preserving framework for collaborative intrusion detection networks through fog computing, pp 267–279. https://doi.org/10.1007/978-3-319-69471-9_20
38. Agarwal P, Alam M (2019) A lightweight deep learning model for human activity recognition on edge devices. Journal title Sensors and page 1–17
39. Makkar A (2020) Machine learning techniques. In: Machine learning in cognitive IoT. CRC Press, pp 67–85
40. Gope D, Dasika G, Mattina M (2019) Ternary hybrid neural-tree networks for highly constrained iot applications. arXiv preprint arXiv:1903.01531
41. Lin J, Yu W, Zhang N, Yang X, Zhang H, Zhao W (2017) A survey on internet of things: architecture, enabling technologies, security and privacy, and applications. IEEE Internet Things J 4(5):1125–1142
42. Guevara JC, Torres RDS, da Fonseca NL (2020) On the classification of fog computing applications: a machine learning perspective. J Netw Comp Appl 159:102596
43. Wang J, Pan J, Esposito F, Calyam P, Yang Z, Mohapatra P (2019) Edge cloud offloading algorithms: issues, methods, and perspectives. ACM Comput Surv (CSUR) 52(1):2
44. Aazam M, Zeadally S, Harras KA (2018) Offloading in fog computing for IoT: review, enabling technologies, and research opportunities. Futur Gener Comput Syst 87:278–289
45. Mach P, Becvar Z (2017) Mobile edge computing: a survey on architecture and computation offloading. IEEE Commun Surv Tutor 19(3):1628–1656
46. Boukerche A, Guan S, Grande RED (2019) Sustainable offloading in mobile cloud computing: algorithmic design and implementation. ACM Comput Surv (CSUR) 52(1):11
47. Peng K, Leung VC, Xu X, Zheng L, Wang J, Huang Q (2018) A survey on mobile edge computing: focusing on service adoption and provision. Wirel Commun Mob Comput 2018., Article ID: 8267838:1–16. https://doi.org/10.1155/2018/8267838
48. Shakarami A, Shahidinejad A, Ghobaei-Arani M (2020) A review on the computation offloading approaches in mobile edge computing: a game-theoretic perspective. Softw Pract Exp 50:1719–1759
49. Cao B, Zhang L, Li Y, Feng D, Cao W (2019) Intelligent offloading in multi-access edge computing: a state-of-the-art review and framework. IEEE Commun Mag 57(3):56–62

50. Taleb T, Samdanis K, Mada B, Flinck H, Dutta S, Sabella D (2017) On multi-access edge computing: a survey of the emerging 5G network edge cloud architecture and orchestration. IEEE Commun Surv Tutor 19(3) Third Quarter:1657–1681
51. Madakam S, Ramaswamy R, Tripathi S (2015) Internet of things (IoT): a literature review. J Comput Commun 3:164–173
52. Dolui K, Datta SK (2017) Comparison of edge computing implementations: fog computing, cloudlet and mobile edge computing. IEEE
53. Mao Y, You C, Zhang J, Huang K, Letaief KB (2017) A survey on mobile edge computing: the communication perspective. IEEE
54. Rodriguez-Zurrunero R, Ramiro U (2019) Dataset of process management in IoT operating systems: cross-influence between processing and communication tasks in end-devices. https://doi.org/10.17632/rxsdfg8ct9.1
55. Davis GA, Nihan NL (1991) Nonparametric regression and short-term freeway traffic forecasting. J Transp Eng 117(2):178–188
56. Mehrotra R, Sharma A (2006) Conditional resampling of hydrologic time series using multiple predictor variables: a K-nearest neighbour approach. Adv Water Resour 29(7):987–999
57. Bannayan M, Hoogenboom G (2008) Weather analogue: a tool for real-time prediction of daily weather data realizations based on a modified k-nearest neighbor approach. Environ Model Softw 23(6):703–713
58. Mangalova E, Agafonov E (2014) Wind power forecasting using the k-nearest neighbors algorithm. Int J Forecast 30(2):402–406
59. Cai P, Wang Y, Lu G, Chen P, Ding C, Sun J (2016) A spatiotemporal correlative k-nearest neighbor model for short-term traffic multistep forecasting. Transp Res Part C Emerg Technol 62:21–34
60. Cheng S, Lu F, Peng P, Wu S (2018) Short-term traffic forecasting: an adaptive ST-KNN model that considers spatial heterogeneity. Comput Environ Urban Syst 71:186–198
61. Martínez F, Frías MP, Pérez-Godoy MD, Rivera AJ (2018) Dealing with seasonality by narrowing the training set in time series forecasting with kNN. Expert Syst Appl 103:38–48
62. Fan GF, Guo YH, Zheng JM, Hong WC (2019) Application of the weighted k-nearest neighbor algorithm for short-term load forecasting. Energies 12(5):916
63. Xu D, Wang Y, Peng P, Beilun S, Deng Z, Guo H (2020) Real-time road traffic state prediction based on kernel-KNN. Transportmetrica A TranspSci 16(1):104–118
64. Kück M, Freitag M (2020) Forecasting of customer demands for production planning by local k-nearest neighbor models. Int J Prod Econ 231:107837
65. Hattori K, Takahashi M (1999) A new nearest-neighbor rule in the pattern classification problem. Pattern Recogn 32(3):425–432
66. Hattori K, Takahashi M (2000) A new edited k-nearest neighbor rule in the pattern classification problem. Pattern Recogn 33(3):521–528
67. Jiang S, Pang G, Wu M, Kuang L (2012a) An improved K-nearest-neighbor algorithm for text categorization. Expert Syst Appl 39(1):1503–1509
68. Jiang JY, Tsai SC, Lee SJ (2012b) FSKNN: multi-label text categorization based on fuzzy similarity and k nearest neighbors. Expert Syst Appl 39(3):2813–2821
69. Miao D, Duan Q, Zhang H, Jiao N (2009) Rough set based hybrid algorithm for text classification. Expert Syst Appl 36(5):9168–9174
70. Cui B, Shen HT, Shen J, Tan KL (2005, December) Exploring bit-di® erence for approximate KNN search in high-dimensional databases. In: Conferences in research and practice in information technology series, vol 39, pp 165–174
71. Tan S (2006) An effective refinement strategy for KNN text classifier. Expert Syst Appl 30(2):290–298
72. Wan CH, Lee LH, Rajkumar R, Isa D (2012) A hybrid text classification approach with low dependency on parameter by integrating K-nearest neighbor and support vector machine. Expert Syst Appl 39(15):11880–11888
73. Yoon JW, Friel N (2015) Efficient model selection for probabilistic K nearest neighbour classification. Neurocomputing 149:1098–1108

74. Zhang H, Berg AC, Maire M, Malik J (2006, June) SVM-KNN: discriminative nearest neighbor classification for visual category recognition. In: 2006 IEEE computer society conference on computer vision and pattern recognition (CVPR'06), vol 2. IEEE, pp 2126–2136
75. Zhang S, Cheng D, Deng Z, Zong M, Deng X (2018) A novel kNN algorithm with data-driven k parameter computation. Pattern Recogn Lett 109:44–54
76. Khateeb N, Usman M (2017). Efficient heart disease prediction system using K-nearest neighbor classification technique. In Proceedings of the international conference on big data and internet of thing, pp 21–26
77. Hashi EK, Zaman MSU, Hasan MR (2017) An expert clinical decision support system to predict disease using classification techniques. In: 2017 International conference on electrical, computer and communication engineering (ECCE). IEEE, pp 396–400

Big Data Analytics of IoT-Based Cloud System Framework: Smart Healthcare Monitoring Systems

Joseph Bamidele Awotunde, Rasheed Gbenga Jimoh, Roseline Oluwaseun Ogundokun, Sanjay Misra, and Oluwakemi Christiana Abikoye

1 Introduction

The rapid merging and technological expansions of micro-electro-mechanical digital electronics and wireless communication technologies resulted in the advent of the Internet of Things (IoT). The IoT has given birth to the recent growth in the number of sensors and devices connected, and an increase in the data collected from these devices is called big data. The consumption reflects how the growth of big data naturally overlays with IoT-based devices. The management of big data in an uninterrupted expanded network becomes nontrivial because of analytics, security, and data processing [1]. The effective utilization of IoT-based has been examined by experts in the areas of connection challenges, and this has been a concern to many researchers. Several opportunities have been created through the convergence of IoT-based devices, and various studies have been conducted in the area of big data analytics and IoT-based cloud for the utilization of big data analytics in an IoT system [1].

The number of objects connected within the Internet is many compared with human beings globally according to the report from Cisco [2, 3]. The PCs,

J. B. Awotunde (✉) · R. G. Jimoh · O. C. Abikoye
Department of Computer Science, University of Ilorin, Ilorin, Nigeria
e-mail: awotunde.jb@unilorin.edu.ng; jimoh_rasheed@unilorin.edu.ng; abikoye.o@unilorin.edu.ng

R. O. Ogundokun
Department of Computer Science, Landmark University, Omu Aran, Nigeria
e-mail: ogundokun.roseline@lmu.edu.ng

S. Misra
Department of Computer Science and Communication, Ostfold University College, Halden, Norway
e-mail: sanjay.misra@hiof.no

S. Misra et al. (eds.), *Artificial Intelligence for Cloud and Edge Computing*, Internet of Things, https://doi.org/10.1007/978-3-030-80821-1_9

Fig. 1 The sources of big data in the Internet of Things

smartphones, and tablets are examples of Internet-connected devices that form the IoT systems, coupled with wearable devices and Wi-Fi-enabled sensors and household appliances, among others as shown in Fig. 1. According to recent reports, the Internet-connected devices are expected to be double from 22.9 billion in 2016 to 50 billion by 2020 [4, 5].

The IoT systems are not only monitoring events in real-time but can also be used in mining information collected from IoT devices. The sensor-fitted devices are most data collection tools in the IoT setting that require data distribution service and message queue telemetry transport called custom protocols. The IoT devices are used in nearly all fields; therefore, IoT sensors are expected to produce a big amount of data. The data created from IoT objects can be used to investigate the impact of any events or decisions, treatment, diagnosis, and prediction of any illness and in finding potential research trends. The generated data are processed using different data analytic tools [6]. In the field of healthcare, IoT devices such as satellites could have diverse technologies, such as heart rate monitors, blood glucose monitoring, and endoscopic capsules [7–11].

Cloud computing enables businesses to easily handle enormous data and extract valuable data from the collected data, which are the fundamental obstacles facing big data technologies. To ensure the consistency of the software, the cloud offers enormous space for data storage and processing. However, cloud computing is impacted by the number of leaks and leakages. The structure should be built to protect data privacy. To avoid data theft by unauthorized users, encryption methods might be more useful [12]. To make the system more autonomous and effective, the distributed approach should include a separate partition for data and analysis. This improves the precision of analysis that ensures a stable data-cloud platform. In the cloud-based world, protection is the main challenge, and new technology can be developed to guarantee data security and privacy in the cloud environment [13].

It is possible to send the gathered patient data to their caregiver. Infusion pumps that connect to diagnostic steering wheels and waiting rooms with sensors that monitor vital patient signs are medical devices that can be incorporated or implemented as IoT technology. To promote proper coordination in an IoT-based system, and integrated communication networks need to interact effectively. It is not enough to collect a huge amount of data alone; it must be turned to be used as materials that can help organizations to revolutionize their business processes.

The use of IoT systems has reduced costs, increased revenue generations, and improved efficiencies in different fields; thus putting all the data into use becomes paramount. In generating cost-effective results from IoT systems, enterprises need to create a platform where the collection, managing, and analyzing of big data become scalable and reliable [14]. Therefore, it has become important to make use of big data platforms that make the process and analysis of big data reading from various sensors easier and facilitate meaningful results. Integration and review of data allow businesses to revolutionize their business processes. In particular, these companies can turn a large amount of sensor-collected data into useful insights using data analytics software. Therefore, this paper focuses on the BDA IoT-based cloud system for processing and analysis of real-time data generated from IoT sensors use during healthcare monitoring systems.

The rest of this paper is prearranged as follows: Section 2 discusses the IoT in the healthcare system. Section 3 presents an extensive discussion on BDA in IoT. Section 4 presents related work on big data analytics IoT-based cloud monitoring systems in healthcare. Section 5 presents the architecture of the BDA IoT-based cloud monitoring system. Section 6 presents results and a discussion of the proposed system. Lastly, Sect. 7 concludes the paper and discusses future works for the realization of efficient uses of big data analytics IoT-based cloud in healthcare monitoring systems.

2 Internet of Things in Healthcare System

Technology with the healthcare system is working hand-in-hand in recent years, and the expansion in the areas of IoT and big data has created new opportunities for experts in this area. Furthermore, with the adoption of wearable biosensors by people across the globe, there are newly emergėd applications for individualized telemedicine and mobile health. This new emergence of these technologies is a result of their high availability, being simple to personalize, and easy accessibility, thus creating cost-effectiveness on large scale, and experts were able to deliver individualized content. The next generation of mHealth and eHealth services has increasingly attracted popularity with the introduction of big data analytics and IoT in the healthcare system. Though the new fields are emerging speedily, they also have their challenges, particularly when the goal is providing healthcare systems with a complicated problem and experiencing difficulty in energy-efficiency safe, flexible, suitable, and consistent solutions. By 2020 in the healthcare system, IoT is projected to rise to a market size of $300B covering medical devices, systems, applications, and service sectors [15].

This desire for customized e-healthcare is also likely to be promoted by government initiatives. Many database clusters and resources are needed in general to store big data, and these have become a challenge. It is an important problem to get concrete trends from big data, such as patient diagnostic information. Nowadays, a variety of emerging applications for different environments are being developed. For insensitive systems, sensors are most commonly used for actual or near-future applications. Various body sensors have been developed for continuous monitoring of personal wellness and physical activity of a patient [16]. In recent years, experts have tried to develop different wearable devices that can be used for monitoring patients' health conditions in remote areas [17]. For instance, the wearable sensor is used for monitoring physiological patients' health challenges (e.g., exercises and food habits). Wearable sensors can continuously observe and store the health data of the patient in a data store during this period [18].

The use of laboratory tests can now be combined with big data generated from sensors to diagnose the health condition of patients by physicians to achieve better results. Sensor data is also most commonly used to take effective action for the recommendation of patients' well-being and care, lifestyle decisions, and early diagnosis for patients' health improvements. The old methods of storing data are no more helpful where the volume, speed, and variety of data are increasing in IoT-based sensor application domains. The creation of an appropriate storage system for the storage and processing of voluminous data needs this issue [19].

The accelerated growth of mobile phones, wireless networking, radar systems, and entrenched technologies has contributed to medical information schemes entering the age of omnipresent computers. This has contributed to a steady increase in technologies and apps (i.e., wearable gadgets) for the delivery of innovative, omnipresent health systems [20], which are being increasingly implemented worldwide [21]. On the other hand, the IoT [22] offers computing resources and web connectivity to ordinary objects that can access, interact, and archive data from the real world, and most IoT applications are very widespread [23].

More precisely, IoT implementations in the healthcare sector have contributed to the definition of the Internet of Medical Things (IoMT) [24] (also known as the IoT healthcare system). This applies to a digital network containing medical devices and Internet-interconnected applications. Within this framework, monitors for medical constraints (e.g., ECG, blood pressure, blood oxygen saturation, etc.), actions (e.g., pedometers, gyroscopes, GPS, etc.), and sick person awareness (e.g., relative air humidity, temperature, etc.) require constant measurement and control of parameters connected to the well-being. As a result, medical providers could deliver more reliable and cost-effective treatments outside interfering with the everyday lives of sick persons.

Big advancements in miniaturization, cellular communication, networking, and computer power accelerate progress in the healthcare industry and contribute to the production of connected medical devices. Such tools are capable of recording, processing, storing, and forwarding data. Along with the data itself, these tools build the IoT, a digital network comprising mobile applications, medical equipment, healthcare facilities, and computer networks, Medtech, and the IoT [25].

Smart systems improve patient care during health crises, for instance, asthma attacks, heart attacks, diabetes, collapse, and so on [26]. IoT-embedded apps will provide real-time tracking to save lives. Healthcare companies will streamline their clinical processes and process control from remote areas with smart tools and sensors to enhance patient care [25]. The IoT is a regional network composed of gathering medical instruments and technologies that are linked by ICT [24, 27]. IoT is as well acknowledged as IoT Health Treatment. The IoT employs an accelerometer tracker, a vision tracker, a temperature sensor, a carbon dioxide sensor, an ECG/EEG/EMG radar, a motion radar, a gyroscope radar, a saturation sensor for blood oxygen, a humidity sensor, a ventilation sensor, and a blood pressure radar to track and control the patient's well-being continuously. The IoT detects the health status of the patients and then sends the clinical data with the aid of remote cloud data centers to doctors and care holders [28].

Such results are utmost extensively employed in the analysis and medical management of sicknesses. Useful clinical software information is used to deter and protect patient safety during emergencies. Useful information mined from the medical record is employed in emergency circumstances to prevent and protect patient health. The biggest problem in IoT, though, is how to treat sensitive utilizations where a variety of associated gadgets produce a vast volume of therapeutic data [29]. This growing volume of data is also called big data, which conventional data analysis systems and software cannot handle. IoT can improve decision-making and early detection of illness by intelligently researching and gathering vast volumes of diagnostic data (big data). Therefore, accessible machine learning and intellectual procedures have been required that result in additional interoperable resolutions and that could result in successful verdicts in evolving IoTs.

Nevertheless, the sum of IoT gadgets is anticipated to rise significantly in the approaching years, and the complexity that exists in various IoT mechanisms (system crossing point, communication protocols, data structure, system semantics) would bring interoperability and confidentiality correlated difficulties [30]. A universal healthcare framework must be robust enough to address all of these principles in this way. The incorporation of IoT technologies in an interoperable setting and the creation of software for the collection, analysis, and extensive distribution of IoT data is now becoming important.

IoT-based systems will enable patients to feel self-confident about their health status [31]. In many cases, the intervention of healthcare professionals is even avoided or minimized. For instance, [32] suggested a cloud-centric IoT-centered system for tracking sickness prediction that robotically forecasts future sicknesses and their harshness rates outside the involvement of medical practitioners. Besides, these platforms authorize medical doctors with bits of knowledge that make simpler the medical procedure and provide them with instruments to backing medical verdicts. It has resulted in IoT commonly been employed to transform medicine and is today well thought out to be a cornerstone of modern universal healthcare systems [33].

3 The Edge Computing and Artificial Intelligence in Healthcare System

The IoT-based enabled with edge computing provided lower latency services, energy-effective, cost-effective, and maximum satisfaction in healthcare systems. Most IoT-based environments depend on a cloud platform for massive smart health systems [34]. The model can be used to forward capture data produced from IoT devices through the Internet to the cloud and thereby used for diagnosis to provide useful reports using learning algorithms like ML or deep learning. However, the IoT cloud system is inappropriate, especially where lower latency is necessary. Hence, IoT-based systems require a faster and low latency protection technique with delay-sensitive, smart, secure, stable smart healthcare management. Edge computing is the answer to this with a prolonged type of cloud computing where IoT-based data can be computed closer to the edge of the network where data are produced [35].

The edge computation reduced latency, data traffic, and data distance to the network since it is running at a local processing level close to the cloud database. Edge computing has become relevant and important since devices can recognize data instinctively and thus become useful in IoT-based systems to reduce the latency to a lower level. Figure 2 depicted the edge computing architecture; the first part of a network uses the IoT-based devices and sensors to collect data to be processed through a gateway using a radio access network that uses edge devices to compute data aggregated by the network locally. Once the data processing has been done, the full computing operations and memory storage have been processed to the cloud.

The edge layer is like a junction point where enough networking, computing, and storage resources are available to manage local data collection, which can be readily obtainable and deliver fast results. Low-power system-on-chip (SoC) systems are used in most situations because they are meant to preserve the trade-off between processing performance and power consumption. Cloud servers, on the other hand, have the power to conduct advanced analytics and machine learning jobs to combine time series generated by a variety of heterogeneous or mixed kinds of items [36].

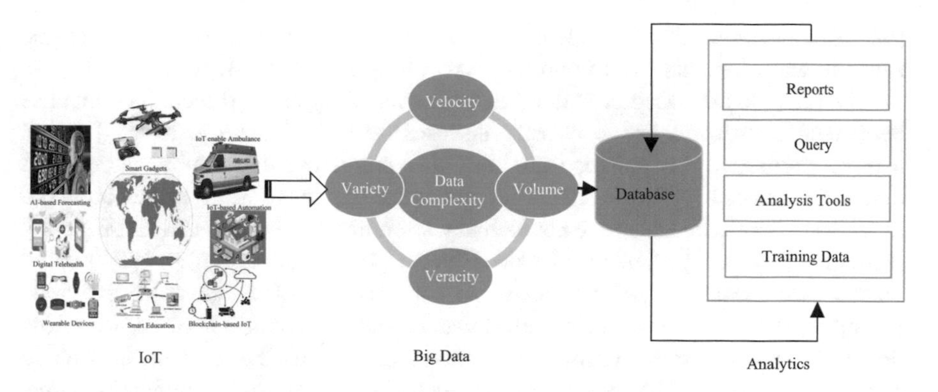

Fig. 2 The relationship between IoT and big data

With the rapid development in wireless technology, IoT can generate a huge amount of medical data and smart devices and customize enhanced services. Such big medical data can be of countless types, like text, multimedia, and image, which the cloud server needs to store, analyze, and process [37]. The high latency, security problems, and network traffic arise as a result of the handling of big cloud medical data. Fog computing was introduced to minimize the burden of the cloud being a new computing platform. The fog also helps in bringing the cloud service closer to the network edge; the layer acts as a flyover between the terminal device and the conventional cloud server. Hence, fog computing allows refined and secured healthcare services.

The AI methods focus on problems that, in theory, only people and animals can solve, problems that require intelligence. It is a branch of computer science that studies problems for which no operational computational algorithms exist. The term serves as patronage, allowing for the addition of new approaches over time [38]. As defined in many recent findings, machine learning approaches offer valuable detection accuracy in comparison with different data classification techniques [39–41]. Accomplishing conspicuous correctness in forecasting is significant because it could lead to an appropriate precaution programmer. Forecasting correctness may differ on diverse studying system methodologies. Hence, it is fundamental to recognize devices proficient in offering extreme accuracy of projection in a system like IoT-based.

Artificial intelligence approaches are the most effective policymaking tools for dealing with real-world and systemic problems. The goal is also to evaluate the efficacy of various AI approaches for categorizing various disease samples. The performance of AI approaches has been evaluated using a variety of categorization performance capabilities. Artificial neural network (ANN), support vector machine (SVM), linear regression, k-nearest neighbor (KNN), multilayer perceptron, and classification algorithm are six widely used artificial intelligence methods.

Machine learning has previously been used in the biomedical field [42–44], for the study of heart disease and diabetes [45] and for the study of diabetic proteins

[46], among other things. Academics have used ANN, SVM, fuzzy logic systems, k-means classifier, and other important AI techniques [42, 43, 47, 48].

The fuzzy logic classifier method is an alternative type of AI technique that has been used by researchers to classify diseases [49]. Al-Qaness et al. [49] unveil a novel predictive method for estimating and predicting the number of reported disease instances over the next ten days, due to the previous cases reported observed in China. An enriched adaptive neuro-fuzzy inference system (ANFIS) is proposed, which employs a significantly improved flower pollination algorithm (FPA) based on the Salp Swarm Algorithm (SSA). In comparison to the rate of misclassified examples, the accuracy of their studies was investigated, and it showed promising results in terms of mean absolute percentage error (MAPE), root mean squared relative error (RMSRE), determination coefficient (R2), and computation time. Also, we evaluated the proposed framework utilizing two separate datasets of reported cases of weekly influenza in two nations, namely, the USA and China. The results showed an even strong performance.

The adaptive neuro-fuzzy inference method (ANFIS) [50] is commonly used in analyzing and modeling issues in the time series and has shown strong success in several current applications. It provides versatility in evaluating time series data nonlinearity, as well as incorporating the features of both artificial neural networks (ANN) and fuzzy logic systems. It has been used in different applications for prediction. AI is the ability of systems that are typically in the computer hardware and software format to imitate or exceed human intelligence in everyday technological and strategic tasks involving identifying, cognitive, and execution. Because individual intelligence is multifaceted, AI encompasses a wide range of objectives, from knowledge representation and intellectual acquisition to ophthalmic observation and language comprehension.

4 Internet of Things and Big Data Analytics

Wearable sensor medical devices constantly produce massive data that is often referred to as big data either structured or unstructured data. The sophistication nature of data makes big data difficult to analyze and process in order to find relevant knowledge useful for decision-making. Originally, also, with a huge amount of data involved, big data processing with conventional techniques becomes practically impossible due to the slower speed and variety of information involved. In particular, big data plays a vital role in the healthcare system [51, 52]. Modern healthcare systems are now increasingly implementing clinical data, which is used to expand the online availability of clinical record sizes [53]. As a result, news technologies and tools were developed in this direction to help process big data generated from IoT-based sensors, and new market insights are gleaned toward those directions. As a result, a variety of options for lowering healthcare costs and diagnosing diseases are established using big data [54].

Integrating health smart systems (such as smart hospitals, smart clinic, etc.) and IoT devices (such as sensors, actuators, and smartphones) will play a vital role in improving healthcare services by creating a digital and smarter hospital and clinic. However, the interconnection of several IoT items to gather medical data over the Internet to launch smart digital healthcare, called big data, affects a massive amount of data generation. Research on big data in healthcare was performed by [55], and six big data use cases were identified to minimize healthcare costs. The use case in healthcare is furthermore divided into four parts by [56], namely, management and distribution, support for the clinical decision, support for services, and customer [57].

The era of big data impacted all parts of people's lifespan, including genetics and medicine in this twenty-first century [58]. The transition from paper patient reports to electronic health record (EHR) systems resulted in exponential data development [59]. Big data thus creates a tremendous incentive for physicians, epidemiologists, and public policy experts to take evidence-driven action that will inevitably strengthen patient safety [60]. For the biomedical scientist, big data is not only a technical fact but a necessity that needs to be thoroughly understood and used in the quest for new insights [61].

Big data has been categorized as conferring to five essential basics: quantity (data dimensions), diversity (related data categories from numerous bases), speed (data obtained in real-time), veracity (data vagueness), and importance (advantages for various manufacturing and academic fields). Besides, other research work like in [62] has really shown the important of big data in various field. Big data knowledge system in healthcare. In the Internet of things and big data technologies for next-generation healthcare. Springer, Cham. data processing), variation (data context), viscosity (latency data communication amid source and destination), virality (speed of data transmitted and obtained from various sources), and visualization; Notwithstanding big data's additional capabilities, the 5 V model lays the basic definition of the big data framework [63]. Big data science has recently undergone a significant transition from its analysis production to its high impact and deployment in numerous fields [64].

There are various sensors all over in the modern age that contribute greatly to the massive growth of big data. Current advances in the fields of computation, storage, and connectivity have created vast datasets; extracting valuable information from this immense volume of data would bring value to science, government, industry, and society. Through embracing biosensors, wearable devices, and mHealth, the volume of biological data collected has expanded [64]. According to a study by Cisco [65], the total amount of data produced by IoMT devices will hit 847 ZB (zettabytes) by 2021. The vast volume of data can be processed, collected, and controlled with conventional methods. Big data analytics is the method of analyzing broad datasets comprising a range of data types [66] to uncover invisible patterns, latent associations, industry dynamics, consumer desires, and other important business knowledge [67]. The ability to evaluate vast volumes of data can help a company cope with important knowledge that can impact the corporation [68].

The principal aim of big data analytics is, therefore, to assist industry organizations in strengthening their interpretation of data and thereby to make successful and knowledgeable choices. Big data analytics allow data mineworkers and experts to scrutinize a vast amount of data that cannot be connected employing conventional methods [67]. Big data analytics comprises technology and materials that can turn a vast quantity of organized, formless, and semi-organized data into an extra understandable framework for investigative procedures. The procedures employed in these computational methods have to recognize forms, tendencies, and associations in the data across several time prospects [69]. Such methods, after analyzing the results, illustrate the discoveries for effective verdict taking in tables, charts, and three-dimensional maps. Big data processing is also a major problem for many implementations owing to the sophistication of data and the scalability of fundamental procedures facilitating these systems [70].

Obtaining useful information from big data analysis is a serious issue requiring adaptable investigative procedures and methods to report timely outcomes, while existing methods and procedures are incompetent in handling big data analytics [71]. Therefore, there is a need for broad networks and external technologies to support parallel data. Besides, data foundations, for instance, very advanced data streams obtained from various data foundations, have diverse arrangements that make it essential to incorporate multiple analytics solutions [72].

The problem is thus based on the efficiency of existing procedures employed in big data processing, which does not grow uninterruptedly with the exponential growth in computing capital [73]. Big data analytics systems take a large amount of time to provide consumers with input and advice, although only a few applications [74] can handle massive datasets within a short processing period. By comparison, the greatest of the outstanding approaches employ a complex trial-and-error approach to tackle large volumes of datasets and data complexity [75]. There are big data analytics platforms.

For instance, the Investigative Data Examination System used big data visual analytics for exploratory earth system simulation analysis [76]. Computers & Geosciences, 61 [77–88], is a big data graphic analytics program that is employed to scrutinize complicated earth structure models of massive quantities of datasets. Big data volume is enormous, and thus conventional applications for database management cannot be used to accumulate and investigate huge data. The resolution comes from modern warehousing databases such as Apache Hadoop, which support distributed data processing.

Smart health tracking apps have evolved exponentially in recent years. These devices generate massive quantities of data. Therefore, adding data processing to data obtained from baby monitoring, electrocardiograms, temperature sensors, or blood glucose level monitoring will assist medical professionals in accurately determining patient's clinical conditions. Data analytics allow healthcare practitioners to detect dangerous diseases at their first steps to aid being rescued. Data analytics increases the medical excellence of treatment including ensuring patient's health. Furthermore, the background of the doctors can be checked by looking at past patient care, which can boost client loyalty, acquisition, and retention.

Enhanced competency: The criteria for handling and storing data from progressive analytics utilization have hindered their implementation in numerous regions. Such obstacles are thus starting to collapse due to IoT [89]. Big data technology, for instance, Hadoop and cloud-based mining tools, provide substantial cost-cutting benefits relative to conventional mining techniques. Besides, conventional analytical techniques involve data in a positive form, which is hard to do while employing IoT-based data. Using existing big data solutions that develop around less-cost group infrastructure, though, will help boost analytic capabilities and reduce computing costs.

Independence from data silos: The initiation of IoT including empowering technology, for instance, cloud computing, has enabled data storage towers to be replaced in various realms [89]. Typically, each data type is deemed only usable for its context, but cross-domain data has arisen as powerful resolutions to diverse glitches [90]. Different data types, for instance, runtime data, system metadata, business data, retail data, and corporate data, can today bc employed because of the numerous supporting technology that supports IoT, including big data, cloud, semantic web, and data storage.

Value-added applications: Deep learning [91], machine learning [92], and artificial intelligence are kcy innovations that offer value-added IoT and big data applications. Before IoT and cloud computing emerge, large volumes of data and processing resources remain inaccessible for many applications and hence prohibit them from employing such machinery. Various data analytics solutions [93], business intelligence systems [94], simulation frameworks [95], and analytics apps [96] have recently appeared and have helped companies and enterprises change their processes and improve their profitability and diagnostics and have incrust them. This amount of specificity had not been available until IoT appeared.

Decision-making: The explosion of IoT-based apps, mobile phones, and social networks presents decision-makers with an ability to collect useful knowledge about their customers, forecast potential patterns, and identify fraud. By rendering knowledge accessible and available to enterprises, big data will create tremendous value, thus allowing them to reveal uncertainty and improve their performance. Considerable data engendered through IoT and numerous analytical gadgets produce a wide range of healthcare system changes. These methods use statistical analysis, grouping, and clustering approaches to deliver diverse approaches to data mining [97, 98]. Mining IoT will also use big data to improve people's decision-making behaviors.

Big data analytics provides well-designed tools to analyze big data in IoT in real-time, generating accurate decision-making outcomes. Big data analytics focused on IoT demonstrate the complexity, growing scale, and capabilities of real-time data processing. Big data fusion with IoT is introducing new possibilities for creating a smart healthcare environment. Big data analytics focused on IoT has wide-ranging implementations in almost every industry. The key performance areas of healthcare analytics, however, are reduction in hospital backlog admissions, rating healthcare network, the accuracy of forecast and tests, enhanced decision-making, efficiency, decreased risk assessment, and improved patient segmentation.

5 Related Work

To obtain useful and realistic knowledge from the generated big data from different sources, various techniques have been used by most businesses to process their data globally. The approaches are called big data analytics using various algorithms, mathematical models, models of estimation and classification in processing decision-making, and many more. In the global world of Internet connectivity, in generating vast quantities of data, social media plays a significant role in our society. This is because corporate or small-scale organizations use social media to communicate with their staff and clients and to promote new goods or services, regardless of the size of their enterprise. In learning about new products and services from an organization, the clients depend on social media to make their choice [99]. Significant sensors, smartphones, IoT devices, and telemetry have also been used to generate quantities of data.

The creation of an increase in the huge size of data brought about the creation of big data analytics. Also, this has created a major challenge for the use of machine learning in the areas of restricting labeled data and highly distributed data, removing noisy data, and many more. Other techniques also face many difficulties due to data growth; such data analytical methods are indexing, storage, retrieval of information, and resources that are associated with the methods [77].

In addition to providing high-performance solutions, the techniques in conventional batch processing systems cannot be carried out. The implementation of distributed computing devices by experts is caused by large-scale performing techniques. The use of a highly complex model recruits the use of dispersed computer devices to scrutinize big data [78]. With the substantial increase in data generated by IoT-based devices, two prominent kinds of research have been created. The production of software and the associated research with other related to business that has big data analytic capabilities. The methods and resources associated with big data increase when the circumstances around it also increase.

The IoT is increasingly growing in healthcare. The new developments in healthcare using IoT include remote surveillance of patients and elderly people, home care. Other aspects of healthcare are the use of RFID, tracking, and monitoring. Health individuals, ordinary citizens, and hospitals will embrace these technologies. Interestingly, countless individuals have seen the need for these applications in many industries. The increased worries about the health-related problems by individuals have brought about the use of IoT devices to personally monitor their health. Also, elderly patients have immensely benefited from the use of IoT software to monitor their health challenges at home. In this regard, many powerful IoT applications are needed to be deployed to increase the services in these directions. The inventions in healthcare are still at the starting point when compared with the needs of medical experts in our hospitals globally with the ratio of the patient in each medical center. The overlapping of interdisciplinary workers and resources helps in the study of the consumption of health services extensively. The study of statistics and social medical government insurance with relevant healthcare together will help in understanding the need for IoT-based devices in a hospital around the globe. This will help balance the performance, cost, feasibility, consistency, and delivery analytically [79].

The city's smartness can be achieved when analyzing the IoT-based sensor results. To analyze huge data, for smart city applications, [80] proposed a multitier fog computing model. A multitier fog of dedicated and ad-hoc fog machine tools was used for dedicated and opportunistic computing system. The problems of the dedicated computing system and the slow cloud computing reaction were checked by the proposed model with functional modules. Therefore, when compared to both fog-based cloud computing data analytics in terms of service utility and block probability, the latter significantly boost the performance of data analytics of smart city services. Without fog computing architectures, the evaluation of data for QoS in cloud computing will be very difficult. The authors [81] claimed that, due to the application of a massive data stream, a huge number of data presented were generated everywhere. The inherently dynamic nature of big data makes it difficult to utilize data mining techniques and tools for the analysis of the data. So, analyzing massive data sources offers a methodical and systematic approach. It has a downside in terms of precision, anonymity, heterogeneity, scalability, and tolerance of faults.

Authors in [82] proposed the use of big data analytics in higher education. The uses of blogs, social media, learning management systems, research-based information, student information systems, and machine-generated data are some of the sources of big data in higher education. In segregating students at risk of failure or dropping from school, big data analytics have to play a dominant role through enrollment prediction when the data is analyzed and provide students with good stands. Hence, in higher education, BDA plays a crucial role in delivering competitive advantages. The RDBMS traditional methods for processing data and data storage are still not meeting the big data challenge. The use of educational mining techniques has helped students understand in a better way. The use of learning analytics enhanced learning and education in higher education globally. The learning management should not be designed separately for an effective way of incorporating the framework of learning analytics architecture.

A wireless wearable ECG monitoring device was designed by [83], focusing particularly on minimal power and cost [84] have developed a healthcare system that uses a smartphone to track diabetic patients [85] worked by finding the motion and pattern to monitor patients affected by Alzheimer's disease. Patient respiration was used to track the quality of sleep [86] in real-time by the author [86] using a monitoring system [87] surveyed the different types of technologies used to track Parkinson's disease-affected patients.

There are more data available in the generation of information in our hands to make decisions [88]. The variety and velocity are associated with big data as the datasets are very wide and high. The conventional approaches and software are very difficult to use in big data analytics. Nevertheless, solutions have to be given to derive information from the dataset, and trust and reliable studies have to be performed because of the exponential growth of data. The designed an advanced method for detecting fraud is very useful in data storage and management within organizations.

Access to the latest remote sensing images is becoming very costly, notwithstanding the growth of this application in recent years, due to over high imagery data processing and storage; thus the data produced has grown rapidly. The remote sensing images and image processing tools like the learning curve are becoming extremely costly [100]. Cloud computing was recommended for the analyses of a huge amount of image data. ArcGIS is one of the common cloud computing techniques used in processing, storage, and management, leading to next generation in this area, and the model is used in other areas. It is a downside that developing a remote sensing application model is not that easy. It is still very difficult to analyze broad imaging data for remote sensing. The problem of data-intensive and computational systems is very common in the remote sensing application model.

Efficient smart devices such as cell phones and sensors have resulted in a greater degree in the exponential growth of apps for data streaming; they are immersive gaming, event tracking, and augmented reality. The IoT has been identified as a major source of big data due to the immense data streams generated by the attached devices and sensors. There more interest by the industries and academia due to the creation of an IoT environment [101] due to the provision of latency and cost-effectiveness. The use of variables of four significant dimensions are used to analyze fog streaming's design space. The fog streaming design space analyzes data, device, and person for proper optimization. The embracement of fog architecture has significantly been useful in information stream processes and resulted in high-quality computing resources communications. These techniques and fog data streaming will lead to industry and academic growth in the network edge in the future.

A large amount of data is produced by the IoT that has grown exponentially dependent on eternal operational states [102] indicated that IoT devices produce massive data that has been a problem for cloud-controlled planned data analytics and processing since the IoT explosion. The creation of micro clouds' closest edge of data sources can greatly help the fog structure with perfect cloud system functionality. Research is more needed to make a smart decision since the fog computing approach carried out by IoT-based big data analytics is still evolving. Fog together with cloud computing plays a vital role in many applications like healthcare monitoring, smart grids, and smart cities. The downside of cloud computing is that it only operates at the edge of the IoT [103, 104] suggested a hybrid model using certain optimization strategies to evaluate vast quantities of data.

The analysis and processing of real-time data are very difficult, and an efficient solution should be given. Where the data is being generated and processed is very vital in data analytics to address this problem [105] suggested a solution known as the fog engine that integrates close-to-the-ground IoT and enables data analytics to transfer large quantities of data easily and quickly. With fog cloud, where information is generated, data analytics can be conducted very close, thereby minimizing data communication and data processing time together. Several parameters, such as network bandwidth, processing data speed, and data transmission size, evaluate the output. Lower latency, throughput, and lower network bandwidth consumption are some of the benefits of fog cloud computing with the disadvantages of expensive data and less data protection.

6 The Architecture of Big Data Analytics of IoT-Based Cloud Healthcare Monitoring System

In general, in the years ahead, the volume of healthcare data is projected to continue to rise significantly. In practice, the use of recent developments in ICT to easily evaluate and use such large data will provide substantial benefits in multiple use cases and implementation scenarios for healthcare organizations ranging from single-physician offices and multi-provider groups to large hospital networks. In particular, in many applications, healthcare analytics can be leveraged to turn vast volumes of data into actionable information that can be used to recognize needs, provide resources, anticipate issues, and avoid emergencies for the patient population.

6.1 The Flowchart of the Proposed System

The four layers made up of the architecture of the BDA IoT-based cloud were discussed as follows:

The objective of the framework is divided into two: (i) the data generation for health monitoring sensors represented by the conceptual framework and (ii) the decision management after the data processing for healthcare. Different patient physiological features were harvesting using diverse human body sensors, and the Cloudera platform was used for the processing of the data capture based on the MapReduce instrument. Figure 3 displayed the flowchart of the framework.

These sensors were attached to the human body, and the physiological result is gathered using wearable sensors planted in the human body. The acquired data was possible using the data acquisition system from different sensors. The data is carefully analyzed and inspected to formulate a resourceful framework and filter the data for preprocessing by transform, aggregate, and cleaning. Hadoop two-node cluster was used for the processing of the data after the first phase which is the preprocessing, and the data is preserved and stored using distributed storage mechanisms called Hadoop Distributed File System (HDFS). Also, the dataset is evaluated by setting different threshold limit values (TLVs) based on specifying different rules, and the rules engine is maintained.

The HDFS was used to store and hold the data generated and extract data for data analytics, ingestion, visualization, and smart healthcare monitoring control in real-time. Lastly, for decision-making and management, the processed data and

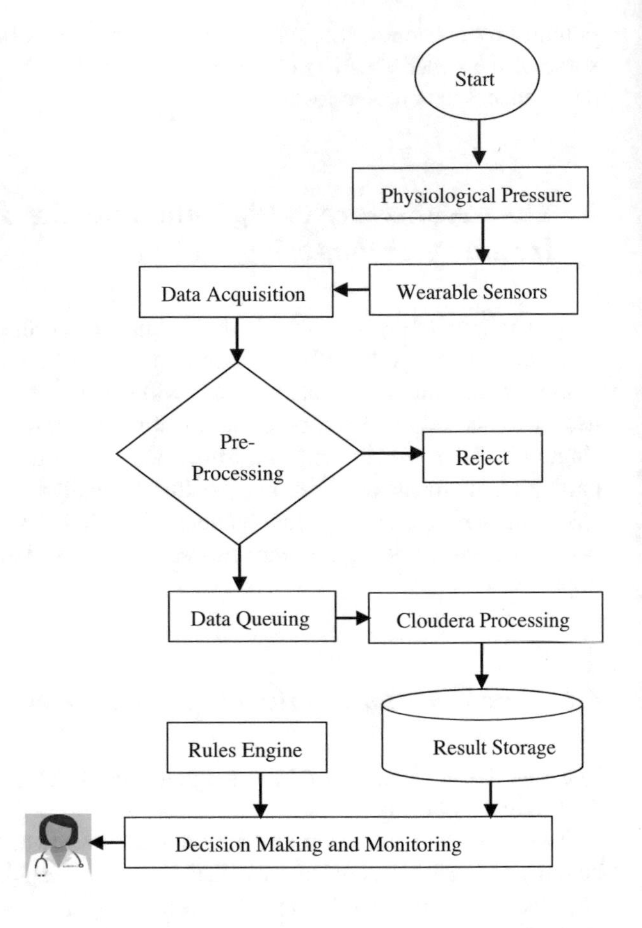

Fig. 3 Flowchart for the big data analytics of IoT-based cloud healthcare monitoring system

rules are used to notify the physicians, medical experts, and users. Big data is incorporating into the framework to be able to realize the analytics into smart health monitoring. This was done to provide real-time decision-making in smart healthcare to improve the data processing efficacy, and not using conventional big data-embedded healthcare monitoring system acts upon different filters and algorithms.

6.2 The Architecture of the Healthcare Monitoring System

The architecture is made up of four layers: (i) the data capture layer, (ii) the pre-processing layer, (iii) the data processing, and (iv) the application and management layer. The four parts are displayed in Fig. 4. These layers were extensively explained below.

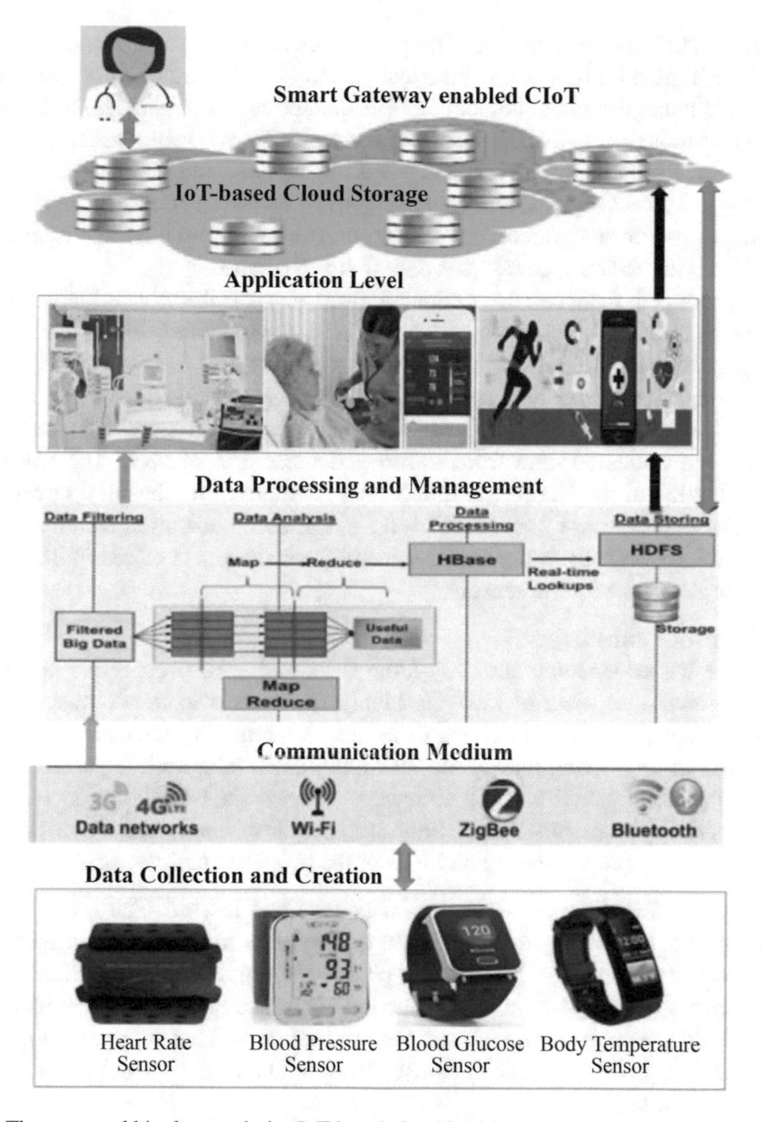

Fig. 4 The proposed big data analytics IoT-based cloud healthcare monitoring system

6.3 The Data Generation Layer

The first layer is used for data capture; it contains different sensors and devices that can be used for data generation; and the analysis and utilization of the wearable devices take place at the layer. The importance of this layer is that it is a virtual layer. In a healthcare monitoring system that needs to function 24/7, the layer displays the significance and importance in the big data analytics IoT-based cloud with zero maintenance tolerance. The physiological devices can be attached to any part of the human body that this information can be collected from certain pressure areas. This helps the physiological devices collect useful details about any patient who wears those sensors. The data capture from the wearable health monitoring sensors was used by the physicians, medical experts, and the patient. Different physiological effects can be collected using diverse sensors effect. To generate signs and symptoms for any patient, any of the sensors can be applied and implanted in the human body to produce relevant data for each disease.

The produced data can be collected from any of the physiological devices attached to humans; such sensors can be body temperature, blood glucose, and blood pressure, among others. The layer was responsible for the capturing of the physiological data from any patient and then converts the data using countless microcontrollers. The IoT-based cloud database is used to store and maintain the captured and collected data from different devices and sensors. The stored data are then loaded to the healthcare monitoring database at the hospital or clinic and any database as the case may be. With the aid of communication technologies and installed microcontrollers, the data generated from sensors is stored in the personal storage embedded with the sensors.

Data Preprocessing Layer

Due to the heterogeneous, naturally large scale and their prospective origin from various sources, big data of today is highly susceptible to noise, missing values diverse formats, and contradictions in natures. Low-quality data can contribute to low quality in processing outcomes. When implemented before the actual processing, data processing methods will noticeably increase the overall quality of the time needed for data processing. Data filtration, aggregation, and data transformation had encompassed the preprocessing methods of the proposed architecture.

Data Aggregation For data reduction, data accretion is used. The impractical and infeasible analysis of sophisticated data on the huge volume of data may take a long time to interpret and process. To produce a simplified demonstration of the dataset with a smaller length, data reduction techniques may also be implemented, but carefully preserve the quality of the original data. That is, processing should be more effective on the reduced dataset but should still deliver the exact or nearly analytical results.

Data Transformation Data is converted into a suitable format for mining at this stage. By scaling information to a specific and limited range, the min-max normalization technique is used to process the data. Min-max preprocessing is used in the specific range $[x - y]$ changing M value to N value.

$$N = \frac{M - X_{min}}{M_{max} - X_{min}} * (y - x) + x$$

Where the normalized value is M, the minimum value is X_{min}, the maximum value is represented by M_{max}, and the upper and lower limit is represented by y and x, respectively. Algorithm 1 is used for the said perseverance.

Data Filtration For inconsistencies and incompleteness and to remove noise, data filtration was applied. The potentially incorrect data is observed in the real-world data because these data are really dirty. The data that is not valuable always causes noise and in most cases does not affect the real-time processing. Incomplete data means lacking attribute values, inconsistencies and discrepancies in data, or certain characteristics of interest. To eliminate data noise, KF is the most advantageous filter technique.

6.4 Data Processing

The overall processing and data transformation are the responsibility of the data processing stages. The data processing contains the following units: the training unit, storage, Cloudera waitron, and rules engine unit. The queue that does not require a prompt response will be monitored by the queuing process by providing a way to position the data on top of the queue to process results. The processing stops and interruptions are the aims of the queue unit. The D data item is received by the queue functions one item at a time and is then forwarded according to the explanation stated and is being handled by a handler H [106]. The M/M/1 queuing model is used for effective functioning with [106] in the framework. For central processing, the Cloudera server is used. This carries data in the server form and serves as the foremost unit of the processing unit. The Cloudera server is used to process and store large datasets in distributed form. With thousands of nodes, the Cloudera platform can make it possible to execute requests by keeping thousands of terabytes of data and facilitating high data transfer speeds between nodes. The Cloudera platform is empowered by the MapReduce method to analyze the data that is processed in two steps: first, by mapping processes where the data is converted into another data collection and second by reducing results generated during the process by combining the mapped data.

The processed results were stored in the IoT cloud storage, and this result is used for decision-making in the last stage of the proposed system. Data storage plays a vital role in the healthcare big data analytics system. Hence, for easy storage, HDFS was used for the proposed framework. In Cloudera and Hadoop platform, HDFS is very useful because it is a distributed storage device. To simplify the scalability requirement of big data processing, the MapReduce implementation on smaller subsets of a larger data cluster is very necessary. The threshold is a precise value denoted by the threshold limit value (TLV) defined based on data. The TLV was used to evaluate the process data, and the rules engine is used to maintain the various rules. For instance, TLV can be used for alarming high glucose in the human body. For different measurements in the proposed system, the rules were defined based on predefined thresholds.

6.5 Application and Monitoring Management Layer

The overall decision-making is being performed in this phase. The stage gives the detailed result by which the physicians, caregivers, medical experts, and users can make a useful decision. The application and monitoring management phase receive the processing data that have been transformed into useful results for further use by physicians and other experts. TLV-based rules with if/then statements are used for decision-making. To make an intelligent decision based on the processing data, the application and monitoring phase utilized the results, and the physicians, patient caregivers, experts, and citizens are advised accordingly based on the generated results. It categorizes the generated data to produce the decisions and serves as the intermediary for the users using the system. This defines the ontological decision used to unicast the monitoring, and high-level and low-level incidents are differentiated by identical units or users.

7 Results and Discussions

The Smart Health Management device results are greatly in use by patients and physicians. From the comfort of their home, the patient can check their health condition in real-time and attend hospitals only when they need to. This can be achieved by the proposed method; the product of which is brought about the use of big data analytics IoT-based cloud healthcare monitoring system and can be used by any patient from anywhere in the world. The system displays the almost real-time values of different health parameters as it is a prototype model and emulates how the same can be applied in the real world. Doctors may also use the patient's body condition record to analyze and assess the impact of medication or other such products.

Figure 3 displayed the flowchart of the proposed model; the body temperature, blood glucose, and blood pressure sensors are controlled and initially shown on the Cloudera processing unit. The captured data from the temperature and blood pressure sensors are stored in the IoT-based cloud storage system. The range used for the body temperature is in Table 1. The temperature range membership feature is possible to clarify as:

$$Low = \begin{cases} 1, x < 36^{\circ}\mathrm{C} \\ 0, x > 36^{\circ}\mathrm{C} \end{cases}$$

$$Normal = \begin{cases} 1, 36^{\circ}\mathrm{C} \le x \le 37.5^{\circ}\mathrm{C} \\ 0, x > 37.5^{\circ}\mathrm{C}\ and < 36^{\circ}\mathrm{C} \end{cases}$$

$$High = \begin{cases} 1, x > 37.5^{\circ}\mathrm{C} \\ 0, x < 37.5^{\circ}\mathrm{C} \end{cases}$$

Table 1 The body temperature measurement range

Range	State
36.0–37.5 °C	Normal
>37.5 °C	High
<36.0 °C	Low

Table 2 The blood pressure measurement range

Range	State
60 BPM–100 BPM	Normal
>100 BPM	High
<60 BPM	Low

Likewise, various ranges of blood pressure readings are often considered to assess the patient's health status, as in Table 2. The blood pressure membership feature is listed as follows:

$$Low = \begin{cases} 1, x < 36 \text{ BPM} \\ 0, x > 60 \text{ BPM} \end{cases}$$

$$Normal = \begin{cases} 1, 36 \text{ BPM} \leq x \leq 100 \text{ BPM} \\ 0, x > 100 \text{ BPM } and\ x < 60 \text{ BPM} \end{cases}$$

$$Normal = \begin{cases} 1, x > 100 \text{ BPM} \\ 0, x < 100 \text{ BPM} \end{cases}$$

The rules for diagnosing the patient's health status are carried out based on these various ranges of values. The following membership feature is diagnosed with the output health state: strong, sick, hypothermia, fever, and diabetes mellitus as required (Table 3). The overall functions of the state are as follows:

$$Checkup = \begin{cases} 1, x < 20 \\ 0, x > 20 \end{cases}$$

$$Unwell = \begin{cases} 1, 20 \leq x \leq 40 \\ 0, x > 40\ and\ x < 20 \end{cases}$$

$$Hypothermia = \begin{cases} 1, 40 \leq x \leq 60 \\ 0, x > 60\ and\ x < 40 \end{cases}$$

$$Healthy = \begin{cases} 1, x > 80 \\ 0, x < 80 \end{cases}$$

Table 3 Classification of blood pressure, diabetes, and obesity

Risk factors		Normal	Pre-stage	Stage I	Stage II	Stage III
Blood pressure	SBP	<110	110–138	139–160	≥160	
	DBP	<70	70–89	90–99	≥100	
Glucose	PGC	≤71–100	100–121	≥122		
	INS	≤97–138	138–190	≥191		
BMI		≤30	30–39.9	40–44.9	45.0–49.9	≥ 50
Waist circumference		≤40, ≤30	>40, >30			

SBP systolic blood pressure, *DBP* diastolic blood pressure, *PGC* plasma glucose concentration, *INS* 2-hour serum insulin, and *BMI* body mass index

Table 4 Rules for diagnosing disease

	Body temperature		
Blood pressure	Low	Normal	High
Low	Health review	Sick	Health review
Normal	Hypothermia	Strong	Fever
High	Health review	Sick	Health review

As shown in Table 4, the rules for the diagnosis of performance health status are:

The period of diabetes is a strong predictor of diabetes complications. Patients with diabetes over 15 years of age believe that they need to obey the prescribed treatment of the doctor and the nutritional restrictions on diabetes prevention, manage their blood glucose, and monitor blood sugar level and that the effects on their vision problems have been affected to a far higher degree than their quality of living over the last month. However, patients with a relatively short to medium duration of diabetes (less than 5 years or 5–10 years after diagnosis) tend to be slightly more confident with the time required to control insulin.

These diagnosis rules can be summarized by taking into account all the combinations.

The body temperature, blood glucose, blood pressure, room temperature, and humidity sensors' values, respectively, are regulated using the microcontroller. The full framework of the sensor health monitoring system is illustrated in Fig. 4. The output values of the sensors measured and shown on the physicians and experts are shown in Fig. 4. These sensor values are then sent to the IoT-based cloud server in the database. The data can be accessed from the cloud by registered users using the platform of the IoT application. The patient's illness is diagnosed based on these values obtained by succeeding the rules set out. The health disorder diagnosis is performed by the medical practitioner and expert as shown in the system framework. The medicines can be administered, and necessary action can be recommended even from a distance by the physicians and medical experts. The entire proposed work is autonomous, so the medical staff ratio requirement can be minimized, and if patients can use this monitoring device at home, it would also eliminate the need for physical accompaniment to monitor them. Hospital staffing costs are also exponentially decreased.

8 Conclusion and Future Research Direction

The basic concepts of big data analytics IoT-based cloud healthcare monitoring healthcare system have been elaborately discussed in this paper. With the advent of infectious and other associated diseases, healthcare has recently been given extreme importance globally. So, in this respect, the best option for such an emergency is an IoT-based health monitoring device. The IoT is the latest Internet revolution that is a growing field of study, particularly in healthcare. Such remote healthcare tracking has developed at such a rate with the rise in the usage of wearable sensors and mobile phones. IoT health surveillance helps to avoid the spread of the disease and to allow a correct evaluation of the state of health, even though the doctor is far away. In healthcare, IoT is the main player in supplying patients with better medical services and also facilitates doctors and hospitals. A portable physiological checking system is proposed in this paper, which can continuously screen the body temperature, blood glucose, blood pressure, and other specific room parameters of the patient. The big data analytics IoT-based cloud system for the storage and analysis of real-time data generated from IoT sensors deployed for the healthcare monitoring system using Wi-Fi module-based remote communication; the proposed system provided a nonstop checking and control tool to track the patient condition and store the patient information on the IoT-based cloud server to generate big data. Various problems relating to information security and confidentiality, mobility control, and applications have been identified during design processes in previous studies. As a result, it is necessary to focus on such challenges, particularly as they relate to IoT-based cloud with big data analytics about healthcare prediction, diagnosis, examinations, and tracking of patient disease. In this case, an intelligent security model should be thoroughly investigated in order to reduce the identified risks. While a variety of IoT systems are widely available on the Internet, they suffer from a number of flaws, including a lack of trust and privacy, as well as effectiveness and acceptability. The issue of load imbalance and information dissemination over cloud servers is a major concern for future research. To name a few, more efficient security algorithms such as DNA encryption, fully homomorphic encryption, and cloud decryption must be adopted. It is because cybersecurity is a primary concern for IoT-based medical system surveillance. This is mostly because the cloud's recorded healthcare information may be vulnerable to a variety of cyberattacks. Moreover, since sensors produce multifactorial data on clinical tests, monitoring, and affordable healthcare, it is often impossible to match or connect data from multiple sensor devices. To achieve a better diagnostic result, it will be necessary to determine which of the IoT sensor data performs best for a collection of biomarkers. The proposed system will be improved further by integrating components of the artificial intelligence system to assist both doctors and patients in making decisions.

References

1. ur Rehman MH, Yaqoob I, Salah K, Imran M, Jayaraman PP, Perera C (2019) The role of big data analytics in the industrial internet of things. Futur Gener Comput Syst 99:247–259
2. Ahmed E, Yaqoob I, Gani A, Imran M, Guizani M (2016) Internet-of-things-based smart environments: state of the art, taxonomy, and open research challenges. IEEE Wirel Commun 23(5):10–16
3. Kashyap R (2020) Applications of wireless sensor networks in healthcare. In: IoT and WSN applications for modern agricultural advancements: emerging research and opportunities. IGI Global, Hershey, pp 8–40
4. Vijay P (2015) Evolution of internet of things go-to-market strategies for semiconductor companies. Doctoral dissertation, Massachusetts Institute of Technology
5. Priyanka EB, Thangavel S (2020) Influence of internet of things (IoT) in association of data mining towards the development smart cities-a review analysis. J Eng Sci Technol Rev 13(4):1–21
6. Yaqoob I, Hashem IAT, Gani A, Mokhtar S, Ahmed E, Anuar NB, Vasilakos AV (2016) Big data: from beginning to future. Int J Inf Manag 36(6):1231–1247
7. Pramanik PKD, Upadhyaya BK, Pal S, Pal T (2019) Internet of things, smart sensors, and pervasive systems: enabling connected and pervasive healthcare. In: Healthcare data analytics and management. Academic Press, London, pp 1–58
8. Darwish A, Ismail Sayed G, Ella Hassanien A (2019) The impact of implantable sensors in biomedical technology on the future of healthcare systems. In: Intelligent pervasive computing systems for smarter healthcare. Wiley, Hoboken, pp 67–89
9. Awotunde JB, Bhoi AK, Barsocchi P (2021) Hybrid Cloud/Fog Environment for Healthcare: An Exploratory Study, Opportunities, Challenges, and Future Prospects. Intelligent Systems Reference Library, 2021, 209, pp. 1–20.
10. Manogaran G, Chilamkurti N, Hsu CH (2018) Emerging trends, issues, and challenges on the internet of medical things and wireless networks. Pers Ubiquit Comput 22(5–6):879–882
11. Qadri YA, Nauman A, Zikria YB, Vasilakos AV, Kim SW (2020) The future of healthcare internet of things: a survey of emerging technologies. IEEE Commun Surv Tutor 22(2):1121–1167
12. Akhtar P, Khan Z, Rao-Nicholson R, Zhang M (2019) Building relationship innovation in global collaborative partnerships: big data analytics and traditional organizational powers. R&D Manag 49(1):7–20
13. Chan MM, Plata RB, Medina JA, Alario-Hoyos C, Rizzardini RH, de la Roca M (2018) Analysis of behavioral intention to use cloud-based tools in a MOOC: a technology acceptance model approach. J UCS 24(8):1072–1089
14. Riggins FJ, Wamba SF (2015) Research directions on the adoption, usage, and impact of the internet of things through the use of big data analytics. In: 2015 48th Hawaii international conference on system sciences. IEEE, pp 1531–1540
15. Firouzi F, Rahmani AM, Mankodiya K, Badaroglu M, Merrett GV, Wong P, Farahani B (2018) Internet-of-Things and big data for smarter healthcare: from device to architecture, applications, and analytics. Futur Gener Comput Syst 78:583–586
16. Ayeni F, Omogbadegun Z, Omoregbe NA, Misra S, Garg L (2018) Overcoming barriers to healthcare access and delivery. EAI Endorsed Trans Pervasive Health Technol 4(15):e2
17. Paradiso R, Loriga G, Taccini N (2005) A wearable health care system based on knitted integrated sensors. IEEE Trans Inf Technol Biomed 9(3):337–344
18. Lorincz K, Malan DJ, Fulford-Jones TR, Nawoj A, Clavel A, Shnayder V et al (2004) Sensor networks for emergency response: challenges and opportunities. IEEE Pervasive Comput 3(4):16–23
19. Ng JW, Lo BP, Wells O, Sloman M, Peters N, Darzi A et al (2004) Ubiquitous monitoring environment for wearable and implantable sensors (UbiMon). In: International conference on ubiquitous computing (Ubicomp)

20. Connelly K, Mayora O, Favela J, Jacobs M, Matic A, Nugent C, Wagner S (2017) The future of pervasive health. IEEE Pervasive Comput 16(1):16–20
21. Kutia S, Chauhdary SH, Iwendi C, Liu L, Yong W, Bashir AK (2019) Socio-technological factors affecting user's adoption of eHealth functionalities: a case study of China and Ukraine eHealth systems. IEEE Access 7:90777–90788
22. Minerva R, Biru A, Rotondi D (2015) Towards a definition of the internet of things (IoT). IEEE Internet Initiat 1(1):1–86
23. Zahoor S, Mir RN (2018) Resource management in pervasive internet of things: a survey. J King Saud Univ Comput Inf Sci 33(8):921–935
24. Manogaran G, Chilamkurti N, Hsu CH (2018) Emerging trends, issues, and challenges on the medical internet of things and wireless networks. Pers Ubiquit Comput 22(5–6):879–882
25. Haughey J, Taylor K, Dohrmann M, Snyder G (2018) Medtech and the medical internet of things: how connected medical devices are transforming health care. Deloitte
26. Patel N (2017) Internet of things in healthcare: applications, benefits, and challenges. Internet: https://www.permits.com/blog/internet-of-things-healthcare-applications-benefits-andchallenges. HTML. Accessed 21 Oct 2020
27. Manogaran G, Varatharajan R, Priyan MK (2018) Hybrid recommendation system for heart disease diagnosis based on multiple kernel learning with an adaptive neuro-fuzzy inference system. Multimed Tools Appl 77(4):4379–4399
28. Aceto G, Persico V, Pescapé A (2020) Industry 4.0 and health: the internet of things, big data, and cloud computing for healthcare 4.0. J Ind Inf Integr 18:100129
29. Al-Turjman F, Nawaz MH, Ulusar UD (2020) Intelligence on the medical internet of things era: a systematic review of current and future trends. Comput Commun 150:644–660
30. Awotunde JB, Jimoh RG, Folorunso SO, Adeniyi EA, Abiodun KM, Banjo OO (2021) Privacy and security concerns in IoT-based healthcare systems. Internet of Things, 2021, pp. 105–134
31. Orsini M, Pacchioni M, Malagoli A, Guaraldi G (2017) My smart age with HIV: an innovative mobile and MIoT framework for patient empowerment. In: 2017 IEEE 3rd international forum on research and technologies for society and industry (RTSI). IEEE, pp 1–6
32. Verma P, Sood SK (2018) Cloud-centric IoT based disease diagnosis healthcare framework. J Parallel Distrib Comput 116:27–38
33. Guarda T, Augusto MF, Barrionuevo O, Pinto FM (2018) Internet of things in pervasive healthcare systems. In: Next-generation mobile and pervasive healthcare solutions. IGI Global, Hershey, pp 22–31
34. Janet B, Raj P (2019) Smart city applications: the smart leverage of the internet of things (IoT) paradigm. In: Novel practices and trends in grid and cloud computing. IGI Global, Hershey, pp 274–305
35. Raj P, Pushpa J (2018) Expounding the edge/fog computing infrastructures for data science. In: Handbook of research on cloud and fog computing infrastructures for data science. IGI Global, Hershey, pp 1–32
36. Rehman HU, Khan A, Habib U (2020) Fog computing for bioinformatics applications. In: Fog computing: theory and practice. Wiley, Hoboken, pp 529–546
37. Devarajan M, Subramaniyaswamy V, Vijayakumar V, Ravi L (2019) Fog-assisted personalized healthcare-support system for remote patients with diabetes. J Ambient Intell Humaniz Comput 10(10):3747–3760
38. Kourou K, Exarchos TP, Exarchos KP, Karamouzis MV, Fotiadis DI (2015) Machine learning applications in cancer prognosis and prediction. Comput Struct Biotechnol J 13:8–17
39. Ayo FE, Awotunde JB, Ogundokun RO, Folorunso SO, Adekunle AO (2020) A decision support system for multi-target disease diagnosis: a bioinformatics approach. Heliyon 6(3):e03657
40. Sundararajan K, Georgievska S, Te Lindert BH, Gehrman PR, Ramautar J, Mazzotti DR et al (2021) Sleep classification from wrist-worn accelerometer data using random forests. Sci Rep 11(1):1–10

41. Xu M, Ouyang L, Han L, Sun K, Yu T, Li Q et al (2021) Accurately differentiating between patients with COVID-19, patients with other viral infections, and healthy individuals: multimodal late fusion learning approach. J Med Internet Res 23(1):e25535
42. Awotunde JB, Ogundokun RO, Misra S (2021) Cloud and IoMT-based Big Data Analytics system during COVID-19 pandemic. Internet of Things, 2021, pp. 181–201
43. Liberti L, Lavor C, Maculan N, Mucherino A (2014) Euclidean distance geometry and applications. SIAM Rev 56(1):3–69
44. Caballero-Ruiz E, García-Sáez G, Rigla M, Balsells M, Pons B, Morillo M et al (2014) Automatic blood glucose classification for gestational diabetes with feature selection: decision trees vs. neural networks. In: Paper presented at the XIII Mediterranean conference on medical and biological engineering and computing 2013
45. Feizollah A, Anuar NB, Salleh R, Wahab AWA (2015) A review on feature selection in mobile malware detection. Digit Investig 13:22–37
46. Berglund E, Sitte J (2006) The parameterless self-organizing map algorithm. IEEE Trans Neural Netw 17(2):305–316
47. Awotunde JB, Folorunso SO, Jimoh RG, Adeniyi EA, Abiodun KM, Ajamu GJ (2021) Application of Artificial Intelligence for COVID-19 Epidemic: An Exploratory Study, Opportunities, Challenges, and Future Prospects. Studies in Systems, Decision and Control, 2021, 358, pp. 47–61
48. Amato F, López A, Peña-Méndez EM, Vaňhara P, Hampl A, Havel J (2013) Artificial neural networks in medical diagnosis. Elsevier
49. Al-Qaness MA, Ewees AA, Fan H, Abd El Aziz M (2020) Optimization method for forecasting confirmed cases of COVID-19 in China. J Clin Med 9(3):674
50. Jang JS (1993) ANFIS: adaptive-network-based fuzzy inference system. IEEE Trans Syst Man Cybern 23(3):665–685
51. Adeniyi EA, Ogundokun RO, Awotunde JB (2021) IoMT-based wearable body sensors network healthcare monitoring system. In: IoT in healthcare and ambient assisted living. Springer, Singapore, pp 103–121
52. Yang P, Stankevicius D, Marozas V, Deng Z, Liu E, Lukosevicius A et al (2016) The lifelogging data validation model for the internet of things enabled personalized healthcare. IEEE Trans Syst Man Cybern Syst 48(1):50–64
53. Samuel V, Adewumi A, Dada B, Omoregbe N, Misra S, Odusami M (2019) Design and development of a cloud-based electronic medical records (EMR) system. In: Data, engineering, and applications. Springer, Singapore, pp 25–31
54. Lopez D, Sekaran G (2016) Climate change and disease dynamics-a big data perspective. Int J Infect Dis 45:23–24
55. Bates DW, Saria S, Ohno-Machado L, Shah A, Escobar G (2014) Big data in health care: using analytics to identify and manage high-risk and high-cost patients. Health Aff 33(7):1123–1131
56. Ajayi P, Omoregbe N, Misra S, Adeloye D (2017) Evaluation of a cloud-based health information system. In: Innovation and interdisciplinary solutions for underserved areas. Springer, Cham, pp 165–176
57. Yan H, Xu LD, Bi Z, Pang Z, Zhang J, Chen Y (2015) An emerging technology–wearable wireless sensor networks with applications in human health condition monitoring. J Manage Anal 2(2):121–137
58. Riba M, Sala C, Toniolo D, Tonon G (2019) Big data in medicine, the present, and hopefully the future. Front Med 6
59. Dinh-Le C, Chuang R, Chokshi S, Mann D (2019) Wearable health technology and electronic health record integration: scoping review and future directions. JMIR Mhealth Uhealth 7(9):e12861
60. McCue ME, McCoy AM (2017) The scope of big data in one medicine: unprecedented opportunities and challenges. Front Vet Sci 4:194
61. Lacroix P (2019) Big data privacy and ethical challenges. In: Big data, big challenges: a healthcare perspective. Springer, Cham, pp 101–111

62. Manogaran G, Thota C, Lopez D, Vijayakumar V, Abbas KM, Sundarsekar R (2017) Big data knowledge system in healthcare. In: Internet of things and big data technologies for next-generation healthcare. Springer, Cham, pp 133–157
63. Kitchin R (2017) Big data-hype or revolution. In: The SAGE handbook of social media research methods. Sage Publications, Los Angeles/Thousand Oaks, pp 27–39
64. Ge M, Bangui H, Buhnova B (2018) Big data for the internet of things: a survey. Futur Gener Comput Syst 87:601–614
65. Hofdijk J, Séroussi B, Lovis C, Sieverink F, Ehrler F, Ugon A (2016) Transforming healthcare with the internet of things. In: Proceedings of the EFMI special topic conference 2016
66. Mital R, Coughlin J, Canaday M (2015) Using big data technologies and analytics to predict sensor anomalies. Amos 84
67. Berman E, Felter JH, Shapiro JN (2020) Small wars, big data: the information revolution in modern conflict. Princeton University Press, Princeton
68. Raghupathi W, Raghupathi V (2014) Big data analytics in healthcare: promise and potential. Health Inf Sci Syst 2(1):3
69. Teijeiro D, Pardo XC, González P, Banga JR, Doallo R (2018) Towards cloud-based parallel metaheuristics: a case study in computational biology with differential evolution and spark. Int J High-Perform Comput Appl 32(5):693–705
70. Candela L, Castelli D, Pagano P (2012) Managing big data through hybrid data infrastructures. ERCIM News 89:37–38
71. Marjani M, Nasaruddin F, Gani A, Karim A, Hashem IAT, Siddiqa A, Yaqoob I (2017) Big IoT data analytics: architecture, opportunities, and open research challenges. IEEE Access 5:5247–5261
72. Assuncao MD, Calheiros RN, Bianchi S, Netto MA, Buyya R (2013) Big data computing and clouds: challenges, solutions, and future directions. arXiv preprint arXiv:1312.4722, 10
73. Hashem IAT, Yaqoob I, Anuar NB, Mokhtar S, Gani A, Khan SU (2015) The rise of "big data" on cloud computing: review and open research issues. Inf Syst 47:98–115
74. Ajayi P, Omoregbe NA, Adeloye D, Misra S (2016) Development of a secure cloud-based health information system for antenatal and postnatal Clinic in an African Country. In: ICADIWT, pp 197–210
75. Siddiqa A, Hashem IAT, Yaqoob I, Marjani M, Shamshirband S, Gani A, Nasaruddin F (2016) A survey of big data management: taxonomy and state-of-the-art. J Netw Comput Appl 71:151–166
76. Steed CA, Ricciuto DM, Shipman G, Smith B, Thornton PE, Wang D et al (2013) Big data visual analytics for exploratory earth system simulation analysis. Comput Geosci 61:71–82
77. Hammou BA, Lahcen AA, Mouline S (2020) Towards a real-time processing framework based on improved distributed recurrent neural network variants with fast text for social big data analytics. Inf Process Manag 57(1):102122
78. Lozada N, Arias-Pérez J, Perdomo-Charry G (2019) Big data analytics capability and co-innovation: an empirical study. Heliyon 5(10):e02541
79. Xiao X, Hou X, Chen X, Liu C, Li Y (2019) Quantitative analysis for capabilities of vehicular fog computing. Inf Sci 501:742–760
80. He J, Wei J, Chen K, Tang Z, Zhou Y, Zhang Y (2017) Multitier fog computing with large-scale iot data analytics for smart cities. IEEE Internet Things J 5(2):677–686
81. Kolajo T, Daramola O, Adebiyi A (2019) Big data stream analysis: a systematic literature review. J Big Data 6(1):47
82. Matsebula F, Mnkandla E (2017) A big data architecture for learning analytics in higher education. In: 2017 IEEE AFRICON. IEEE, pp 951–956
83. Spanò E, Di Pascoli S, Iannaccone G (2016) Low-power wearable ECG monitoring system for multiple-patient remote monitoring. IEEE Sensors J 16(13):5452–5462
84. Chang SH, Chiang RD, Wu SJ, Chang WT (2016) A context-aware, interactive M-health system for diabetics. IT Prof 18(3):14–22
85. Cheng HT, Zhuang W (2010) Bluetooth-enabled in-home patient monitoring system: early detection of Alzheimer's disease. IEEE Wirel Commun 17(1):74–79

86. Milici S, Lázaro A, Villarino R, Girbau D, Magnarosa M (2018) Wireless wearable magnetometer-based sensor for sleep quality monitoring. IEEE Sensors J 18(5):2145–2152
87. Pasluosta CF, Gassner H, Winkler J, Klucken J, Eskofier BM (2015) An emerging era in the management of Parkinson's disease: wearable technologies and the internet of things. IEEE J Biomed Health Inform 19(6):1873–1881
88. Elgendy N, Elragal A (2014) Big data analytics: a literature review paper. In: Industrial conference on data mining. Springer, Cham, pp 214–227
89. Ahmed E, Yaqoob I, Hashem IAT, Khan I, Ahmed AIA, Imran M, Vasilakos AV (2017) The role of big data analytics in the internet of things. Comput Netw 129:459–471
90. Bröring A, Schmid S, Schindhelm CK, Khelil A, Käbisch S, Kramer D et al (2017) Enabling IoT ecosystems through platform interoperability. IEEE Softw 34(1):54–61
91. Chen XW, Lin X (2014) Big data deep learning: challenges and perspectives. IEEE Access 2:514–525
92. Qiu J, Wu Q, Ding G, Xu Y, Feng S (2016) A survey of machine learning for big data processing. EURASIP J Adv Signal Process 2016(1):67
93. Subiksha KP, Ramakrishnan M (2021) Smart healthcare analytics solutions using deep learning AI. In: Proceedings of international conference on recent trends in machine learning, IoT, smart cities and applications. Springer, Singapore, pp 707–714
94. Vidal-García J, Vidal M, Barros RH (2019) Computational business intelligence, big data, and their role in business decisions in the age of the internet of things. In: Web services: concepts, methodologies, tools, and applications. IGI Global, Hershey, pp 1048–1067
95. Jeong Y, Joo H, Hong G, Shin D, Lee S (2015) AVIoT: web-based interactive authoring and visualization of indoor internet of things. IEEE Trans Consum Electron 61(3):295–301
96. Strohbach M, Ziekow H, Gazis V, Akiva N (2015) Towards a big data analytics framework for IoT and smart city applications. In: Modeling and processing for next-generation big-data technologies. Springer, Cham, pp 257–282
97. Schorn MA, Verhoeven S, Ridder L, Huber F, Acharya DD, Aksenov AA et al (2021) A community resource for paired genomic and metabolomic data mining. Nat Chem Biol 17:1–6
98. Okuda M, Yasuda A, Tsumoto S (2021) An approach to exploring associations between hospital structural measures and patient satisfaction by distance-based analysis. BMC Health Serv Res 21(1):1–13
99. Ramírez-Gallego S, Fernández A, García S, Chen M, Herrera F (2018) Big data: tutorial and guidelines on information and process fusion for analytics algorithms with MapReduce. Inf Fusion 42:51–61
100. Huang Y, Gao P, Zhang Y, Zhang J (2018) A cloud computing solution for big imagery data analytics. In: 2018 international workshop on big geospatial data and data science (BGDDS). IEEE, pp 1–4
101. Yang S (2017) IoT stream processing and analytics in the fog. IEEE Commun Mag 55(8):21–27
102. Anawar MR, Wang S, Azam Zia M, Jadoon AK, Akram U, Raza S (2018) Fog computing: an overview of big IoT data analytics. Wirel Commun Mob Comput 2018
103. Nagarajan SM, Gandhi UD (2019) Classifying streaming of twitter data based on sentiment analysis using hybridization. Neural Comput & Applic 31(5):1425–1433
104. Murugan NS, Devi GU (2019) Feature extraction using LR-PCA hybridization on twitter data and classification accuracy using machine learning algorithms. Clust Comput 22(6):13965–13974
105. Mehdipour F, Javadi B, Mahanti A (2016) FOG-engine: towards big data analytics in the fog. In: 2016 IEEE 14th Intl Conf on dependable, autonomic and secure computing, 14th Intl Conf on pervasive intelligence and computing, 2nd Intl Conf on big data intelligence and computing and cyber science and technology congress (DASC/PiCom/DataCom/CyberSciTech). IEEE, pp 640–646
106. Din S, Paul A (2019) Smart health monitoring and management system: toward autonomous wearable sensing for internet of things using big data analytics. Futur Gener Comput Syst 91:611–619

Genetic Algorithm-Based Pseudo Random Number Generation for Cloud Security

Sudeepa Keregadde Balakrishna, Sannidhan Manjaya Shetty, Jason Elroy Martis, and Balasubramani Ramasamy

1 Introduction

The cloud's rationale is well delineated in the United States of America by the National Institute of Standards and Technology (NIST) as follows: "Cloud Computing is a model for enabling convenient, on-demand network access to a shared pool of configurable computing resources (e.g., networks, servers, storage, applications, and services) that can be rapidly provisioned and released with a minimal management effort or service provider interaction". It has a very great scope of solidification, being consistent, encompassing versatility, being easily reachable, characterizing agility, achieving higher performance, having multi-inhabitancy and being secure and easy to sustain [1].

Cloud computing is the next huge thing when it comes to the revolution in the progression of things to originate as the Internet of Services that offers various frameworks, software, and platforms as the legitimate services that can be made available via enrollment-based organizations through a pay per usage method to its end users. Cloud provides storage of information and computations on high from a low-end computer sitting anywhere in the world. We can broadly categorize cloud services into two types—the primary limits a single organization termed as an enterprise, and the latter is available to all known as public clouds. Cloud providers provide services to the user in a "use and pay" model, which is cost-friendly and reliable. Cloud storage allows users to store data on the go anytime, anywhere. Once the data is on the cloud, the user and any other authorized user can access the data across many devices. This leads to specific worries over providing safekeeping, as it can lead to a breach of the information.

S. K. Balakrishna · S. M. Shetty (✉) · J. E. Martis · B. Ramasamy
NMAM Institute of Technology, Nitte, Karnataka, India
e-mail: sudeepa@nitte.edu.in; sannidhan@nitte.edu.in; jason1987martis@nitte.edu.in; balasubramani.r@nitte.edu.in

S. Misra et al. (eds.), *Artificial Intelligence for Cloud and Edge Computing*, Internet of Things, https://doi.org/10.1007/978-3-030-80821-1_10

Over the years, cloud computing has hypothetically evolved in providing cutting-edge services to the end user, and the appeal on security-entitled information has extensively escalated because of the information's sensitivity transferred over the public transmission networks. Henceforth, preserving the information's secrecy is a primary objective to evade illegal access [2].

Previously utilized cloud infrastructure services rely on one-time authorization techniques that identify the user initial to services usage. These can cause certain data breaches as authorization can be imitative or replicated to masquerade the intended user. Current modern low-end machines can even perform brute force breakage in a matter of minutes by using high time computing power. Hence, there arises a need to authenticate the user, which can be time-consuming repeatedly and can cause performance degradations due to repetitive authentication scenarios. It is also necessary to rethink our authentication algorithms that can adapt to fraudsters' conditions. Genetic algorithm (GA) fit perfectly in this category as they are adaptable to various needs. In other terms, the algorithm evolves by selecting the best apposite self-motivated authorization keys, which are hastily elastic and tough to forge. One of the extensively applied fundamental techniques in the area of information security is linear finite state machines (LFSMs). Among many of them, various applications related to cryptography rely on linear feedback shift registers (LFSRs) involving the composition of pseudo-random number generators (PRNGs) or stream cipher systems [3–4]. The length of the LFSR key is least and cannot afford higher confidentiality levels from the unlicensed users. Hence, to overcome the fallacies of existing systems, we propose a novel scheme for generating a non-binary sequence of pseudo-random numbers through an adaptable LFSR.

The novelty of the prospective technique lies in the sequence length of the key that is overextended by scheming a hybrid model employing LFSR and a GA. To further enhance the entire cryptographic model, the anticipated system also aims to achieve a higher-level confusion on the cryptographic procedure by integrating one-time padding. We have also conducted prescribed statistical examinations to appraise the randomness properties of the engendered sequence of keys obtained from the designed hybrid model. Later, to test the correctness of cryptographic procedures, the engendered series of keys obtained from the model were practically imposed on some cryptographic applications. The security for cryptography applications is simulated on a cloud.

The principal aim of our work is focused on achieving the following objectives:

1. We design an LFSR to produce a non-binary PRNG for engendering a series of the pseudo-random key.
2. We increase the key series' span through a hybrid model designed via the amalgamation of LFSR and the GA.
3. Generate a non-binary key sequence model for the cryptographic application.
4. To achieve image encryption and decryption operations over the cloud environments using a broader length of random key sequence engendered from the designed hybrid model.

5. We accomplish statistical testing to test and validate the randomness of the generated number sequence.

The remainder of this chapter is further ordered into subsequent sections: Sect. 2 discusses the background study and related work carried out in connection to the implementation of the proposed system, Sect. 3 incorporates the detailed information about the design of the proposed system, Sect. 4 presents the implementation details in connection to the achievement of practical outcomes through the proposed system, and Sect. 5 debates on the analysis and validation of the practical outcomes that attest the strength of the endangered key along with the assurance of cryptographic procedure. The final section, i.e., Sect. 6, presents the conclusion and future work corresponding to the proposed system.

2 Background and Related Work

An exhaustive literature survey has been conducted to realize various breaches and analyze several security aspects concerning cloud applications. Numerous relevant websites have been visited to assimilate valuable information. Besides, textbooks have been referred to as understanding the advanced mathematical concepts required for understanding the GA and LFSR issues. This section has highlighted a few articles that have shown remarkable results in the scope of the study.

A widespread, diverse application uses random numbers, including encryption systems, real-time testing of digital circuits, simulation systems, etc. A good random number is essential to provide a factual foundation of randomness in applications where one must model a physical process. Since it is ideally possible to generate truly random sequences, we opt for an alternate randomness approach called pseudo-randomness. Pseudo-random number systems are imperative for specific algorithms like Monte Carlo (MC) systems and to perform various simulations. An ideal pseudo-random number generator engenders a number sequence that is not discrete compared to the series of accurate random numbers in a brief span of computation period if the kernel is unknown.

The method that engenders the sequence of random numbers and the engendered sequence not representing any definite order is called the random number generator (RNG). The resulting number sequence from the RNG may produce a binary or a non-binary series. We broadly classify an RNG into two distinct types in connection to its level of originality:

1. A true random number generator (TRNG)
2. A pseudo-random number generator (PRNG)

Any entropy source using TRNG makes use of the available sequence of random numbers that are previously existing instead of formulating a new one. The expanse of random numbers' volatility depends on the chosen entropy aspect, which is also arbitrary. One of the significant identified disadvantages of the TRNG system is the lowest rate of production. Apart from this, TRNGs have another considerable downside, as they are highly dependent on the genus of hardware systems used. Research investigations conducted under TRNG proved that there are primarily four standard TRNGs meant for generating truly random numbers. They are (1) Random.org, a website that generates true random numbers, (2) Hot Bits, (3) lasers that use inert gases like argon and neon, and (4) oscillators present in hardware devices [5–9].

PRNG is utterly opposite to that of TRNG that does not hinge on any physically existing unit. PRNG engenders a sequence of random numbers based on the initial value known as the "seed" value. Research conducted has upheld PRNG to be a lot more secure than the TRNG systems as it is hard to break from the attacker's prospect. A PRNG ensures the security aspect, as each number in the engendered sequence can be obtained and recalculated only if the PRNG has the information regarding the "seed" value. Any PRNG system to produce a series of random numbers utilizes a precomputed mathematical model [10–12].

In the early 1980s, the outstanding achievement in the area of security is done by Shamir [13], Blum and Micali [14], Yao [15], and Goldreich and Levin [16] by establishing the theory of pseudo-random number generator, which is suitable for cryptographic applications. A pseudo-random number generator would be considered as an essential part of the stream cipher. It enhances the initial seed value into a sequence of numbers so that if the initial input value is confidential and random, then the output is not much different from the perfect random number sequence. Håstad et al. [17] showed that security issues of stream cipher systems do not only depend upon the initial values of PRNS; also, it depends upon the one-way function (OWF). Blum et al. [18] proposed a design derived from the benefit of the quadratic residual modular integer to produce pseudo-random sequences. The principles of RSA were introduced by Alexi, Chor, Goldreich, and Schnorr to enhance security. Impagliazzo and Naor [19], Fischer and Stern [20], and Boneh et al. [21] introduced a design depending on the subset sum problem's complexity, which proves more secure.

Abu-Almash has proposed applying the GA approach for designing a PRNG in their research artifact. The proposed work was considered to be a novel research work with the utilization of genetic procedures to develop a PRNG. Genetic procedures were applied to generate keys and obtain the best possible sequence through many generations. In this approach, many parameters have been used for each chromosome as a complete solution for a specific problem [22].

Poorghanad et al. steered a research work in the area of stream cipher system for scrutinizing the efficacy of pseudo-random series of the lengthier period. Based on the investigations, researchers have proposed a novel generator system built on a genetic algorithm. The proposed approach furnished a complex architecture in generating lengthier series of the key [23].

Alireza et al., in their artifact, have proposed the utilization of evolutionary approaches for generating a PRN of a higher quality. In their article, the authors have stressed the importance of a random key sequence's strength to achieve a cryptosystem of higher quality. They provide a very sophisticated safety for the systems by encompassing a very complicated procedure for engendering random sequence values that are hard to predict in usual cases. Apart from this, the research work has also presented LFSR's design that performs the generation of binary polynomial value. However, the sequence generated does not guarantee the test of impossible predictability. Hence, the research work proves that LFSR alone is insufficient for generating a random sequence of higher strength [24].

Goyat has proposed a public key cryptosystem using the approach of a genetic key generation system. The authors have discussed cryptography and Vernam ciphers. In the proposed work, Vernam cipher is used to create a robust key sequence and has also ensured to avoid the re-utilization of the randomly chosen key by using a cipher system termed as a one-time pad. The key sequence of random numbers is engendered using a GA for cryptographic applications in this work. They have also conducted many standard operations to ensure the strength of the key sequence [25].

Ashalatha et al., in their research artifact, have presented a work relating to the usage and analysis of different tools utilized for simulation in the cloud computing framework. In their entire article, authors have covered the architectural aspects of standard simulation tools pertaining to the cloud, along with their advantages and disadvantages. As an outcome of the artifact, on comparing different simulation tools, researchers have concluded that a cloud simulator is the best simulator compared to grid computing in large-scale environments [26].

Nichatand Sikchi, in one of their research articles, has discussed about a technique for the encryption of an image via the hybrid genetic procedure. A hybrid model is designed out of chaotic function for encrypting the data with the combination of the genetic approach. In its proposed model, GA is utilized to acquire ideal outcomes. Also, during the final stage of an entire process, a superior cipher image is chosen dependent on the estimation of the coefficient of relationship and entropy. The excellent cipher image selection mainly depends on the lowest coefficient value of correlation and the highest entropy value connected to an image. Outcomes of the artifact achieved a better encryption system to generate cipher images using the designed hybrid model and uphold that the genetic procedure boosts up the cryptographic process [27].

Verma and Singh has designed an LFSR-based unique random sequence generator system. The proposed system concentrated on implementing a unique generator for engendering a series of random values by further improving an ordinary random sequence generator by imbibing an additional property that involves ensuring that a number generated via arbitrary number generator cannot be copied. The authors have used LFSR to implement the unique random number generator. They have successfully proved that LFSR's utilization boosts random number generation speed as it operates on bitwise X-OR operation and shift registers. On accomplishing their proposed system, authors have concluded that LFSR is the best option for any cryptographic system to generate a series of random keys [28].

Singh et al. have proposed an innovative approach to model a genetic procedure to enhance security in cryptographic applications. In the proposed algorithm, the GA technique's application has been merged along with the cryptographic operation to achieve an enhanced result within a short span of probable time. The combined cryptographic procedure was designed to achieve an optimized level of excellence, productivity, and efficacy. In analyzing the practical outcomes achieved from the proposed work, it is well understood that the designed cryptographic procedure has accomplished the target objectives and has outperformed when compared with previous systems [29].

In their article, Calheiros et al. presented work on utilizing the CloudSim toolkit for the purpose of simulation and design of the cloud service environment. Apart from these, authors have also relied on the toolkit even to assess different algorithms related to the provisioning of resources. In this research artifact, researchers have focused on simulating the cloud environment connected to the IT systems usage. It was identified in the artifact that required infrastructure and various applications are being offered as "services" to the end users with the policy of pay on use model. In connection to the facts mentioned earlier, real-time assessment of the performance factor of provisioning methodologies, workload models, etc., is tough to accomplish under fluctuating conditions of the applications under usage. Basically, to rule out the challenge, authors in their artifact have proposed using a CloudSim toolkit capable of performing the modeling and simulation of the real-time cloud environment based on the requirement [30].

In their research artifact, Agarwal has developed a novel encryption procedure using a secret key system by applying genetic procedures. Researchers have exploited crossover and mutation maneuvers' prevailing genetic procedure features to implement an innovative encryption system. The authors have demonstrated the importance of preserving information security via primary factors concerned with network systems. To validate the cryptosystem's working in connection to the secret key system, they have adopted the image's usage for encryption and decryption process. Experimental outcomes achieved in the research paper confirm the validation of both encryption and decryption process in connection to the utilization of genetic procedures [31].

Malhotra and Jain have presented work related to various cloud simulators' study and their comparison to evaluate the best simulator's different aspects. From their past research work, authors have identified that it is very much essential to evaluate the performance and security aspect of the cloud environment with an increased amount of deployment and adoption. With this concern, researchers have found out that simulation tools are ideally suited for this aspect. Hence, as a contribution to the research community, the authors of the article designed the simulator for performance evaluation and security aspects using distinct standard simulation tools. On carrying out the comparative analysis of all the devices concerning their working nature, it was identified that CloudSim outperformed all other simulators by extending continuous services without any break [32].

Atiewi and Yussof, in their research artifact, conducted the specific comparative tests between two simulators: CloudSim and GreenCloud, for the measurement of the amount of energy consumed in the cloud systems. The research work was carried out with the motivation of decreasing the consumption of power by replacing the server systems of an organization with cloud infrastructures. To test the real-time results, the authors have employed two simulation tools: CloudSim and GreenCloud.

On comparing the achieved outcomes, it was observed that both the simulators perform equally well, with the only difference where GreenCloud was explicitly meant for network simulation, which can even grasp the details of each packet on transportation [33].

3 Proposed System

In this section of the artifact, we elucidate about the working concept of the proposed hybrid model that is utilized for engendering a protracted length of the non-binary sequence of random value. The foremost aim of the proposed methodology is to enhance the cryptographic process' strength by spreading the key span. It is also very significant to design a complicated encryption procedure with a substantial breakthrough in deciding upon cryptosystems' quality. In this regard, the following subsections illustrate the hybrid model's design with a combination of genetic procedure and LFSR.

3.1 Hybrid Model with a Combination of a Genetic Algorithm and LFSR

The working methodology mainly functions by spreading the extent of the non-binary sequence values. Considering the circumstances for a non-binary arrangement, we measure the supremacy cryptosystems by assessing the complexity involved in encryption procedure as well as randomness properties engaged in generating the secret key. Based on the utilization of the maximum span of key sequence, the approach chosen for encryption will offer a strong cryptosystem. Considering the discussed advantage, in this section, we have emphasized selecting a procedure that could engender the most significant span of the sequence of non-binary values and an encryption procedure that could boost the cryptosystem quality. Based on the conducted literature review, we have evidently identified that LFSR is one of the better PRNG that has the capability of engendering the key values of periodic sequence, which are also limited in nature. Recent studies have also demonstrated that the inclusion of genetic procedures yielded a strong crypto structure. The following subsections elucidate the complex working concept of different modules embodied in designing a hybrid model.

3.1.1 Algorithm for Generating Pseudo-Random Numbers

To generate a random number that acts as a truly random number, we have to develop a pseudo-random number taken from a seed value initially from the LFSR. We

Algorithm 1 Algorithm showing the functioning of the hybrid model

Step1:	Generate a pseudo-random number using LFSR.
Step2:	Consider the output key as input to GA (selection)
Step3:	Perform mutation operation to swap 2 bits
Step4:	Perform crossover operation
Step5:	Perform mutation operation
Step6:	Compare the last key and first key
	If (newkey == firstkey) terminate,
	Clock for the next input
Step7:	Repeat until end generation of the LFSR

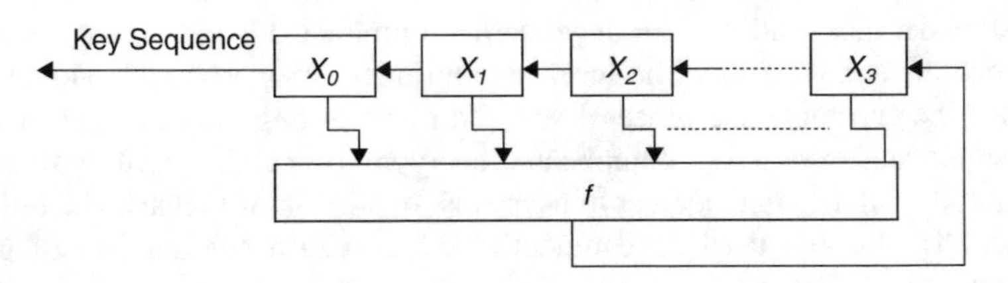

Fig. 1 Figure showing the inner functioning of the linear feedback shift register

then perform slight mutations and crossovers using genetic randomness mapping. The process repeats until LFSR completes its operation. Algorithm 1 presents the sequential set of procedures carried out in generating PRNG.

3.1.2 Design of LFSR

This section of the chapter discusses the design of LFSR in detail. In general, an LFSR encompasses "N" users of register entities, which are also entitled to the LFSR stages. Each stage displays the real-time station of the register at an instant in time. The status will be updated based on the current station of the register applied to the defined feedback function '*f*'. The engendered series of values will postulate the modular value corresponding to the defined feedback function. Let us assume that the achieved modular value of LFSR to be '*m*' and the total amount of dimensions to be '*n*,' then the highest range of the sequence of the key is represented as $m^n - 1$. Here for the designed LFSR, to outspread the shift register's span, we utilize the methodology of the GA. Figure 1 illustrates the general structure of the LFSR.

Figure 2 depicts the design of a very simple LFSR corresponding to the proposed hybrid system design. For our problem, we have embodied an LFSR comprising of three distinct stages entitled X_0, X_1, X_2, and a corresponding feedback function represented as $(X_0 + X_1)\ \%\ 256$ is utilized.

As it is clearly discussed, the maximum length is fixed in any linear feedback shift register, with an intention to extend the maximum sequence length; the utilization of the genetic procedure accomplishes the length expansion. The detailed working concept of the GA in correspondence to the proposed methodology is illustrated in the chapter's trailing subsections.

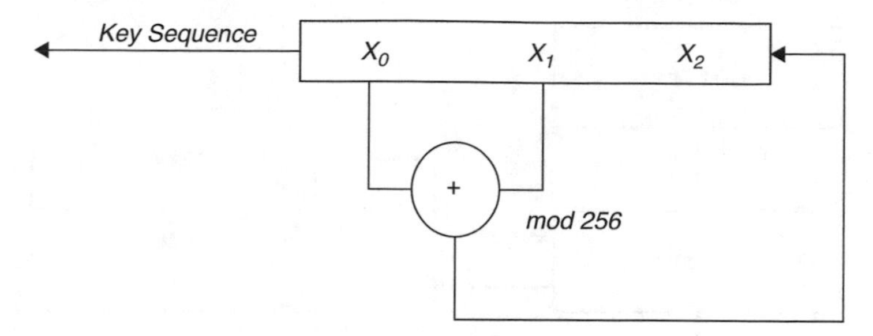

Fig. 2 Figure showing the LFSR used in the proposed hybrid model

Algorithm 2 General working procedure of the GA

Step1:	Planning a distinct amount of solutions for a problem under study
Step2:	Analysis of different solutions planned and performing the appropriate decision and labeling of good and bad solutions.
Step3:	Defining a standard procedure for mingling the fragments of improved solutions to produce an innovative solution.
Step4:	Application of mutation operator to evade everlasting loss of variety in the solution

3.1.3 Genetic Algorithm (GA)

GAs are typical procedures related to an optimization problem that is adopted for the purpose of achieving an optimal solution for any situation that is given to it. The general working concept of a GA is presented as a sequence of distinct steps in Algorithm 2.

Any genetic procedure makes use of three standard operators to accomplish the functionality: (1) selection, (2) crossover, and (3) mutation. Each operator has its significance and ably contributes to the process of achieving the solution using the designed genetic procedure. Choosing the best result for the utilization in the subsequent generation is accomplished by the selection operator. Executing the merger of information generated by two distinct solutions is performed by the crossover operator. Finally, the spinning of the values in the information is taken care of by the mutation operator. On accomplishing the operations by all the operators, an innovative solution for a problem under study is generated. As mentioned earlier by the involved operators, procedures are repetitive until the achievement of an optimal solution. An abstract working concept of the genetic approach is portrayed in Fig. 3.

For our proposed model, we have utilized LFSR for engendering the pseudo-random sequence of key values. From individual new key values, a new series of the key is engendered via a GA. In this case, the production of a new key-value via LFSR is thought of as an operation of selection. Once after this, to engender a series of random values in the next level, we have adopted GA operations corresponding to mutation and crossover. Here for the system proposed, to spawn an efficient sequence of random values, we have imbibed mutation operation twice: mutation-1 and mutation-2. The first level of mutation, mutation-1, is applied immediately after

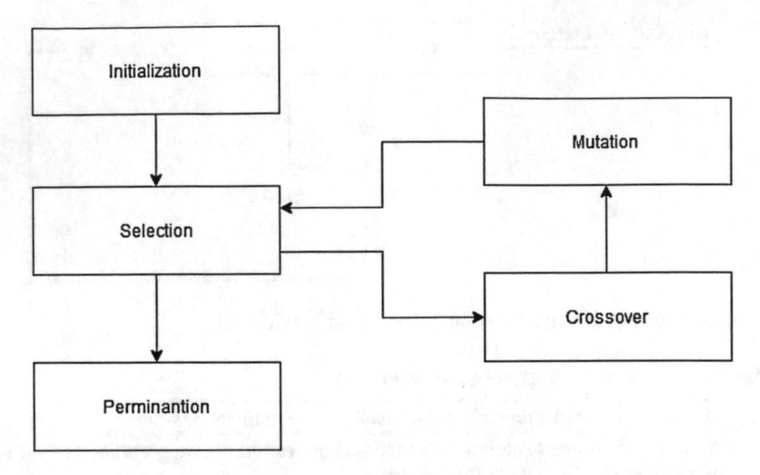

Fig. 3 Abstract working procedure of genetic algorithm

Fig. 4 Process of key sequence generation using a GA process and LFSR

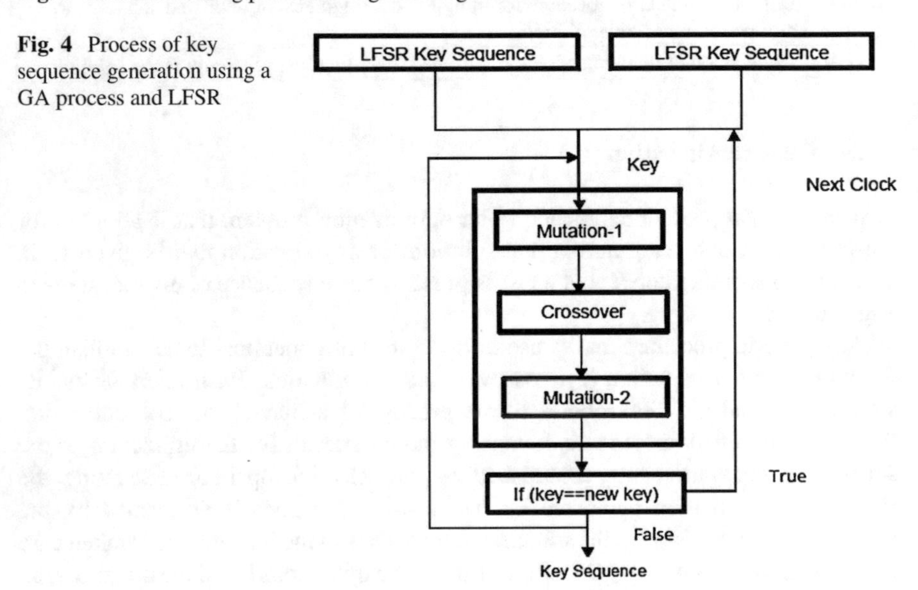

receiving the sequence of keys via LFSR. The next level of mutation, mutation-2, is applied after the crossover operation on merging the results. The entire process is depicted in Fig. 4.

The mutation-1 operation is achieved by considering two-bit values of the output that is engendered from LFSR. LFSR engenders the production as a series of 8-bit values, which are considered as $B_0B_1B_2B_3B_4B_5B_6B_7$. Later the engendered series of bits achieved through LFSR is fed as input into the designed genetic procedure. In our problem, we have considered the output attained via mutation-1 as $B_0B_1B_2B_3B_4B_5B_6B_7$. The process of the same is presented in Fig. 5

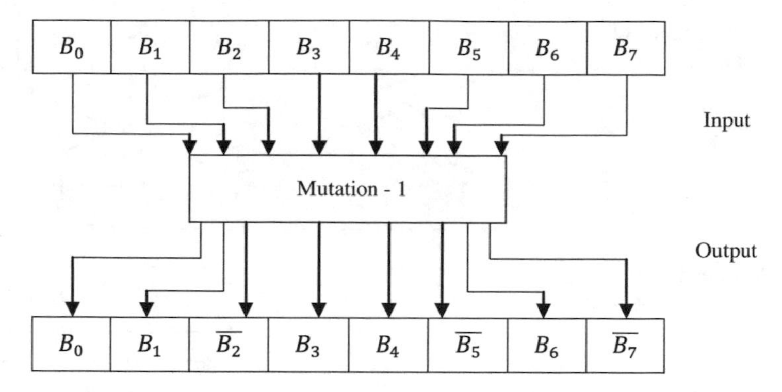

Fig. 5 Schematic representation of mutation-1operation of genetic process

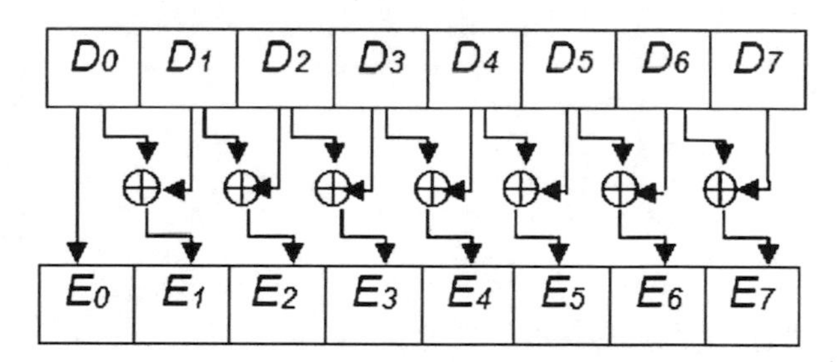

Fig. 6 Schematic representation of mutation-2 operation of genetic process

Once after achieving the result through mutation-1, it is then supplied into the next standard crossover operator of the genetic procedure. On the able utilization of crossover operation, it is possible to mingle two distinct strings to achieve a new string. In that case, the strings utilized for mingling are considered parent strings, and the engendered new string is labeled as a child string. In this proposed artifact, crossover's operation is accomplished through the application of shift operation to the least significant bit in the series. The output achieved through the crossover operator in the next step is supplied to the process of mutation-2. Under the operation of mutation-2, each bit of the input provided is sequentially XORed with the preceding bit of the sequence, excluding the very initial bit of the arrangement. The overall process of the crossover operation is presented in Fig. 6, where the series of bits D_0 *to* D_7 represents the supplied input values and E_0 *to* E_7 represents the achieved output series.

The crossover operator recombines the two different strings representing parents to achieve a new string of better quality defined as a child. Basically, three different variations correspond to crossover operators: (1) one-point crossover, (2) two-point crossover, and (3) uniform crossover. In the proposed work, we apply the operation by shifting a nibble designated as least significant to most significant and most significant to least significant. The overall process of the procedure performed is outlined in Fig. 7. In the model, we have considered the series $K_0K_1K_2K_3K_4K_5K_6K_7$ as inputs and $K_4K_5K_6K_7K_0K_1K_2K_3$ series as output.

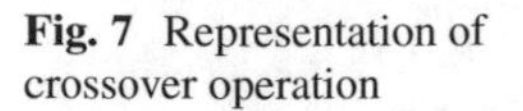

Fig. 7 Representation of crossover operation

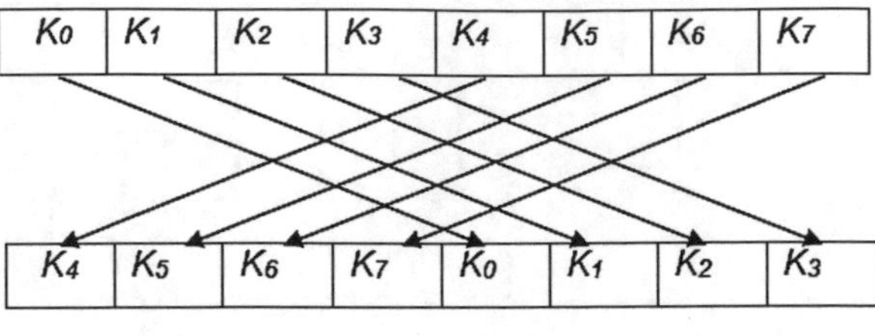

Fig. 8 Organization of different layers of CloudSim components

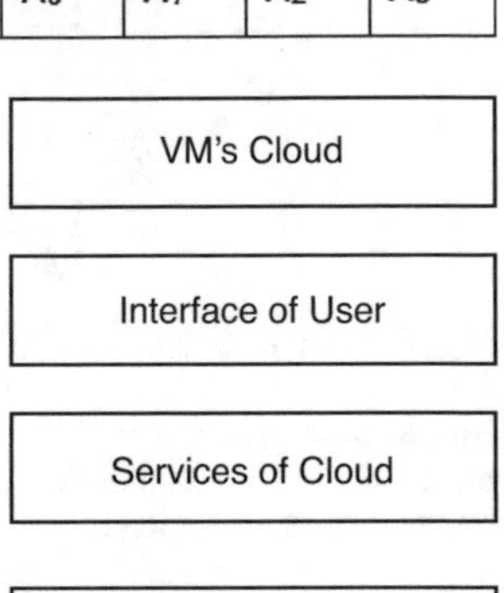

3.2 CloudSim

CloudSim basically refers to a toolkit for cloud simulation that is established by the Department of Computer Science and Engineering under Clouds Research Center at the University of Melbourne. CloudSim is a toolbox that offers various essential classes of services meant for delineating data center, different applications, number of virtual machines, and collection of assets corresponding to the computations, group of clients, and several distinct methods for the administrators in connection to the arrangement of infrastructure. Various extensions considered for the CloudSim toolkit are CloudsimEx, CloudAnalyst, Cloud Auction, WorkflowSim, RealCloudSim, SimpleWorkFlow, CloudReports, CloudMIX Xpress, and EMUSim. The prime benefits of employing the CloudSim toolkit for required execution and testing squeeze time adequacy and malleability. Figure 8 presents an abstract organization of CloudSim toolkit encompassing different components.

Following are the most commonly adopted service entities of the CloudSim that can accomplish specific actions related to the cloud infrastructure

(i) *DataCenter (DC)*: This service segment pertaining to the cloud framework operates as a center to equip consumers with necessary application tools and programming resources. A datacenter can encompass multiple hosts.
(ii) *Cloud Information Service (CIS)*: This entity of the CloudSim acts as an archive that comprises the resources necessary prevailing in the cloud.
(iii) *Broker*: As according to its name, this entity acts like a mediator that is used to succumb the information into the required virtual machine (VM) belonging to

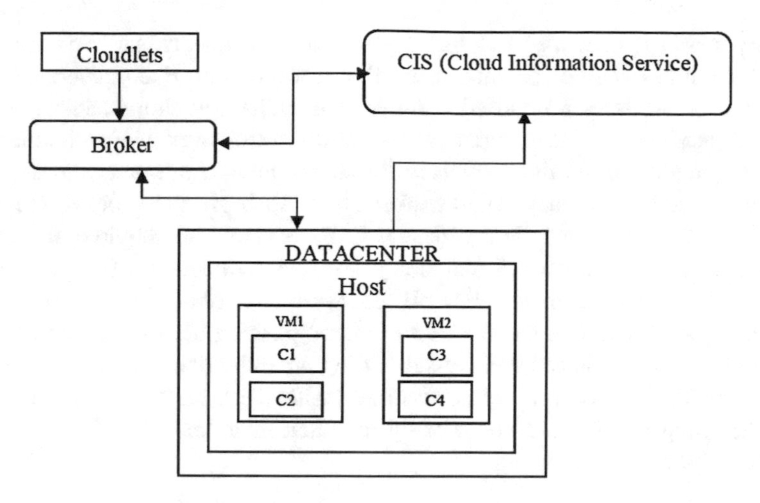

Fig. 9 Abstract representation CloudSim model

a specific host and later associates with applications and cloudlets to achieve the task. Once executing the complete set of cloudlets, the entity also takes the responsibility of extinguishing the unoccupied virtual machines.

(iv) *DataCenter Broker*: This entity is also a mediator for succumbing the required set of applications into the data center and accomplishes the obligation of acting as a custodian to convey the necessary data among the virtual machine and datacenter.

(v) *Cloudlet*: These act as a set of applications running on the VMs.

(vi) *Cloudlet generator*: An entity of the CloudSim toolkit, which is meant for spawning the cloudlets that are succumbed into the broker entity.

(vii) *Virtual machine (VM)*: The name virtual corresponds to being logical, and these entities are the rational machines meant for the purpose of running the spawned applications.

(viii) *VM generator*: An entity responsible for spawning the required VMs in the simulator and provides the new VM to the broker.

(ix) *Host*: This corresponds to the machines where physically existing VMs are being embodied. The host encompasses a type of hardware configuration (HC).

(x) *VM allocation policy*: This is referred to as an abstract class instigated by the components corresponding to the host.

Figure 9 projects a high-level representation of the CloudSim model encompassing a few entities corresponding to CIS, datacenter, VM, cloudlets, host, and broker.

In Fig. 9 presented, CIS serves as a record maintaining register that comprises the entire set of resources prevailing in the cloud. Each asset existing in the cloud is compulsorily made to register under CIS. The datacenter entity encompasses host modules, which basically represent the components of hardware pertaining to the primary memory, network bandwidth, processor components, etc. Any number of virtual machines can be spawned using the host modules. Each generated virtual machine encompasses a required number of cloudlets running exclusively inside them. Broker entities have direct access and control over every cloudlet. This simulator module takes the responsibility of assigning the jobs to the datacenter through its dedicated class called broker class. To begin the process, the broker builds up the communication with the CIS to stem the required information pertaining to the resources. Upon doing so, CIS reestablishes the potentials of the datacenter. Once after receiving all the information necessary connected to the datacenter, it then carries out the task of accomplishing the submission of the set of cloudlets into the datacenter. On submitting an entire list of cloudlets, they start running inside the virtual machine residing inside the host. The virtual machines' job scheduling will be taken care of by the datacenter, and virtual machines will maintain cloudlets' scheduling.

3.3 Proposed Hybrid Model Design

The hybrid model's design is constructed using an LFSR of non-binary sequence and a GA to engender a series of pseudo-random values. The proposed hybrid design further strengthens the process of key generation via LFSR by producing an expanded key series span. Output produced from each iteration of the LFSR is supplied into the GA. The abstract working concept of the proposed model with a combination of LFSR and GA is presented in Fig. 10. The sequence of procedures depicted in the figure works by comparing each generated sequence of the key with the GA's initial input and the designed LFSR. If it is equal, then terminate the arrangement LFSR is clocked to engender the next input sequence to the GA. The process keeps repeating until LFSR reaches the end of the series generation.

The mathematical formulation presented in Eq. (1) describes calculation pertaining to the individual series' size engendered via hybrid design.

$$LHM = (m * n) \tag{1}$$

Parameter m corresponds to the length of the series engendered via LFSR, and n represents the total number of non-repeating random values engendered. LHM represents the individual length of the input series corresponding to the hybrid model.

3.4 Key Sequence Generated Using GA

The PRNG portrayed in Fig. 10 is hybrid and is simulated for producing the random key sequences. The fusion model is an amalgamation of GA and LFSR; the

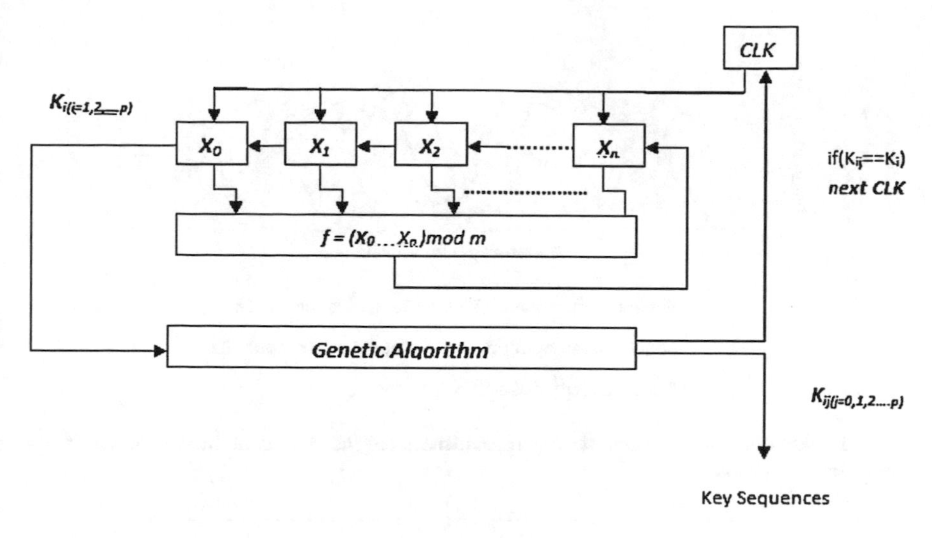

Fig. 10 Hybrid model to generate the random sequence

Table 1 Key sequence generated from hybrid model

122	73	181	157	94	116	135	47	224	106	150	190	125	87	164	12	**122**
5	83	196	9	179	205	217	56	33	112	231	42	144	238	250	27	**5**
68	247	171	140	207	233	186	29	82	212	136	175	236	202	153	62	**68**
132	25	50	209	248	43	128	111	230	58	17	242	219	8	163	76	**132**
120	199	41	176	237	218	24	34	80	228	10	147	206	249	59	1	**120**

LFSR generates key sequence with preliminary kernel standards $X_o = 122$, $X_l = 5$, $X_2 = 68$. Each integer output from LFSR is subjected to GA as per Fig. 2. The first column of Table 1 is a sequence generated from LFSR. The process is then passed on to the GA, and a novel key arrangement is produced until the key sequence recurrences. The latter column characterizes the recurrence of the key created by GA.

The graph plotted displays the sequence of sovereign keys as engendered by the genetic procedure via the individual key generated from LFSR as illustrated in Fig. 11. The series of random values engendered from the proposed hybrid model is depicted in Fig. 12.

It is evident from the above two methods that the sequence's length can be increased to a certain limit. Work is basically to think about a generator that will have no limits for the generation without changing much in the architecture of the system. This kind of generation is elaborated in the next section.

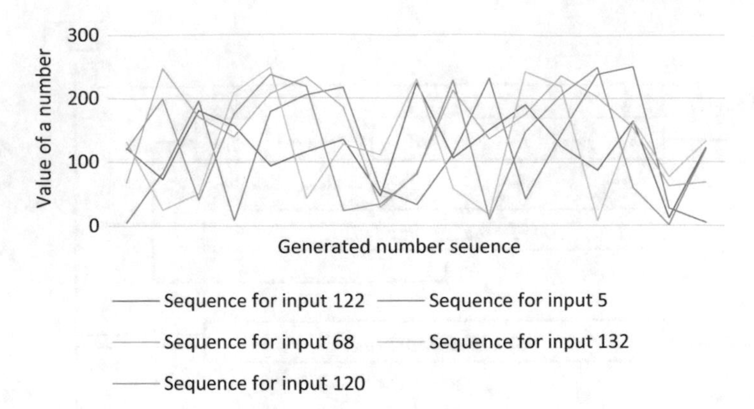

Fig. 11 Graph of GA extension of random sequences engendered via an individual number of LFSR shown in Table 1

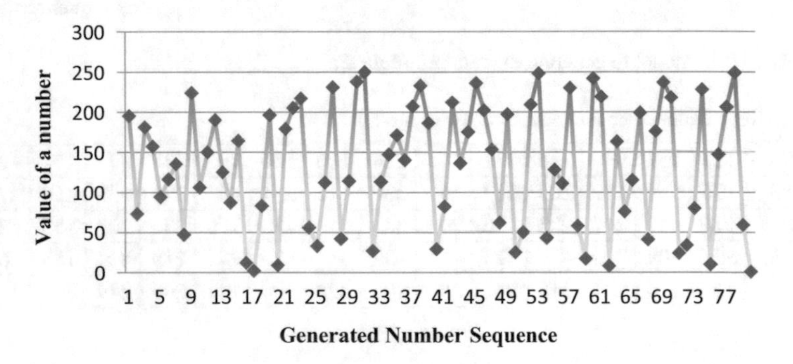

Fig. 12 Graph standing for the series of random values engendered via the hybrid model

3.5 Statistical Test Results of Hybrid Model PRNG

The generated pseudo-random sequence has to be tested for the probabilistic property, which describes the randomness. A function of data is computed and used to decide whether to reject the null hypothesis or not. The various statistical tests are grouped into an empirical test and theoretical tests [34, 35]. Empirical Empirical tests include chi-square test, autocorrelation test, gap test, serial test, run test which are hypothetical in nature. These tests are conducted on a sequence generated by PRNG and require no knowledge about how the sequence is produced. However, theoretical test requires knowledge about how the sequence is generated. The number generated by LFSR, together with the subsequence, is concatenated and used in cryptographic applications as a key sequence. This sequence is as shown in Fig. 12.

Table 2 Result of maximum length obtained by simulating three- stage LFSR and its hybrid model

Function	Type of PRNG	Period of key sequence	Initial seed value
$(x_0 + x_1)$ *mod* 256	LFSR	895	$X_o = 122, X_l = 5, X_2 = 68$
Same LFSR with same initial condition and GA	Hybrid model	13,425	$X_o = 122, X_l = 5, X_2 = 68$

3.5.1 Maximum Length

In this example, a three-stage LFSR is considered with the seed values $X_o = 122$, $X_l = 5$, $X_2 = 68$ and modular 256; for this, the length of the sequence generated is 895. These values are chosen arbitrarily to prove that the addition of GA to this LFSR will lead to a generation of higher length sequences. The generated sequence from the LFSR is used further to generate a higher length sequence by processing it through GA. The same LFSR has been adopted in the hybrid model and simulated, which shows that each LFSR sequence produces a subsequence of 15 more random numbers.

If p is the entire LFSR length and q is the sequence length of the GA, the key sequence for each input, then the span of a proposed hybrid system is as depicted in Eq. (2).

$$Length = p * q \tag{2}$$

The maximum length of GA depends upon the number of bits of the key. If b is the number of bits of key the $q = 2^b - 1$.

Based on Eq. (2), the length of the hybrid model's key sequence with the same LFSR condition is 13425. The simulation result shows that the length of the hybrid model is 15 times more than the LFSR length in this case. The comparison of the key sequence length of both generators is shown in Table 2. It indicates definitely that the extent of the sequence generated by the proposed model is more than the length of the sequence generated by the LFSR.

It is clear from the observation that length of hybrid model augments beyond the extreme measurement of LFSR.

3.5.2 Test for Uniformity

Uniformity testing is carried out under initial conditions for the sequence of pseudo-random values engendered via the proposed hybrid model. This is accomplished through the utilization of the chi-square test. The complete set of samples is classified into a class of numbers "c" and class interval range: [0,64), [64,128), [128,192), [192,255]. Uniform number distribution is predicted from each class, i.e., the expected value of the class is E_i. The value predicted is calculated by Eq. (4).

Table 3 Chi-square value corresponding to the proposed hybrid model and LFSR for the considered intervals of the class: [0,64), [64,128), [128,192), [192,255)

Level of significance	Critical value	Type of PRNG	X_0^2	Degree of freedom	Accept/reject
0.05	7.81	LFSR	4.458	3	Accepted
0.05	7.81	Hybrid Model	2.862	3	Accepted

Let the observed value O_i be the count of observations belonging to the category "*i*" with $O_1 + O_2 + \ldots + O_c = x$ and p_i be the probability that an observation falls into a category such that $p_1 + p_2 + \ldots p_c = 1$.Then it would be expected such that:

$$O_i \approx xp_i \tag{3}$$

for large values of x. The expected value category i is given by:

$$E_i = xp_i = {}^{x}\big/_{c} \tag{4}$$

The statistic value for $§^2$ is defined as:

$$§^2 = (O_1 - E_1)^2\big/_{E_1} + (O_2 - E_2)^2\big/_{E_2} + \cdots + (O_c - E_c)^2\big/_{E_c} \tag{5}$$

For the purpose of statistical test, the likelihood of incorrectly discarding the null hypothesis, i.e., the percentage regarding the possibility of concluding that the null hypothesis is false even though it is true, has to be considered. The likelihood of such incorrect decisions occurring in the process of testing is considered as the level of significance. Usually, the testers consider the level of significance as 0.05, 0.01, or 0.001, and rarely, lesser values such as 0.0001 are considered. Degrees of freedom h shows the amount of standards measured for scheming the arithmetic value, which is lesser than one number of categories, i.e., denoted by $h = numberofclasses - 1$.$§_0^2$ is designed based on Eq. (6) and is associated with critical value. Compared with a hypothetical study against a point on a test distribution, the value to be tested is termed to be a critical value. Here, a null hypothesis is accepted only if the essential value is higher than the statistical value; otherwise, the hypothesis is considered to determinate.

The actual value corresponding to the class i is O_i; the total number of classes prevailing in the set of random values is considered as c. Considering 200 samples of keys generated from the hybrid model, the significance level a is *0.05*, h is considered to be the degree of freedom and is represented as *number of classes – 1,* and based on these values, $§_0^2$ is calculated and compared against the critical value given in Table 3. Since the $§_0^2$ is lesser than the mentioned critical value, null hypothesis corresponding to the uniform distribution is passed.

3.5.3 Test for Independency/Autocorrelation

One of the important properties of randomness is being independent in nature, and hence independent nature of a generated pseudo-random series can be obtained via autocorrelation test. The test actually defines if the engendered sequence is independent/not independent. Under this test, the dependency between the numbers occurring in the engendered series may be well-defined. It decides whether or not the generated number is autonomous. On the consideration of the numbers chosen in the series, the test estimates association between the numbers. Depending on the consideration of the very first number under test and the difference among the numbers, the numbers are chosen. As the starting number, we recognize "d" as distance and number "i" as a preliminary number.

Autocorrelation term ρid has to be determined for the series of numbers R_i, $R_{i+d}, R_{i+2d}, \ldots\ldots R_{i+(L+1)d}$. Here integer L is considered as the largest value of integer according to the form $i + (1 + L)d \leq N$, the number integers occurring in a sequence is given by N, d space between the numbers in a sequence; i is considered as starting digit. The statistical form of the test for the relationship between the series of number $R_i, R_{i+d}, R_{i+2d}, \ldots\ldots\ldots R_{i+(L+1)d}$ is computed according to the following mathematical formulations:

$$Z_0 = \frac{\hat{\sigma}_{id}}{\rho_{\hat{\sigma}_{id}}} \tag{6}$$

The mathematical formulation related to distribution estimator $\hat{\rho}_{id}$ and standard deviation concerning the estimator $\sigma_{\hat{\rho}_{id}}$ are presented in the subsequent equations:

$$\hat{\rho}_{id} = \frac{1}{L+1}\left[\sum_{j=0}^{L} R_{i+jd} R_{(j+1)d})\right] - 0.25 \tag{7}$$

and

$$\sigma_{\hat{\rho}_{id}} = \frac{\sqrt{13L+7}}{12\,(L+1)} \tag{8}$$

The Z_0 statistical value is calculated on the basis of Eq. (6) and compared with the critical value for the approval judgment. If $-Z_{\alpha/2} \leq Z_0 \leq Z_{\alpha/2}$, the numbers generated are independent and agree with the series.

Here, we have considered the autocorrelation test series of 500 samples produced via the proposed model. The 5-number difference is taken into account in the series, and the degree of significance α is 0.05. The test demands the computation of autocorrelation ρ_{id} among the series of numerical values $R_i, R_{i+d}, R_{i+2d}, \ldots\ldots$. $R_{i+(L+1)d}$. The identification of the integer value L representing the largest value is necessary as according to the formulation $i + (1 + L)d \leq N$, where N denotes the total count of numbers in the series, i represents the starting digit, and d denotes the

Table 4 Autocorrelation test result of hybrid model and LFSR

Generator	l	m	M	N	Z_0
LFSR	1	5	98	500	1.2305688
Hybrid model	1	5	98	500	−1.1255204

difference among the numbers under evaluation. Equations (7) and (8) were used to calculate the distribution and the standard deviation connected to the estimator. The test figures for the similarity of the number series under consideration are estimated from Eq. (6), and the results are shown in Table 4. Since the Z_0 value is between the critical value of 1.96 and −1.96, the main series fulfills the autocorrelation test.

3.5.4 Run Test

The pattern of sequence generating by the hybrid model has to be tested based on the run test because the randomness properties of the sequence also depend upon the pattern of how it occurs. Two types of run tests are the following:

1. Runs up and down
2. Run above and below the mean

- *Runs up and runs down*

Under this, we count the types of run; the type would run up and run down. A set of figures each was accomplished by a greater amount. Likewise, a down run refers to the set of numerical values, each of which is achieved via smaller numbers. If N denotes the count of numbers in a series, *a* represents the total count of runs, and then the corresponding mean value and variance value are depicted as according to the subsequent equations.

$$\mu_a = \frac{2N - 1}{3} \tag{9}$$

and

$$\sigma_a^2 = \frac{16N - 29}{90} \tag{10}$$

For >20, distribution of a is considered roughly to be normal; the statistic test is depicted as follows

$$Z_0 = \frac{a - \mu_a}{\sigma_a} \tag{11}$$

The hypothesis is rejected when $-z_{\alpha/2} \leq z_0 \leq z_{\alpha/2}$, where the significance level is selected as α.

Table 5 The statistical outcomes of run-up and run-down test for the proposed hybrid model

Generator type	Total runs (a)	Total numbers N	σ_a^2	Z_0	μ_a	Accept/reject
LFSR	341	500	88	0.86	333	Accepted
Hybrid model	317	500	88	−1.72	333	Accepted

Taking into account, the primary series of 500 key values is produced using the hybrid model for the up-down test by considering a = 0.5 level of significance. If N represents the total count of numerical values in a series, and a is the cumulative count of runs, in that case, Eqs. (9) and (10), respectively, give the mean and variance. The test statistics are determined using Eq. (11) for the pseudo-random main series dependent on mean and variance. In Table 5, the outcome of the run-up and the run-down test is given.

The result depicted in Table 5 approves that $-1.96 < Z_0 < 1.96$ on level of significance 0.05. Hencc, it is evidently proved that engendered series is a sequence of pseudo-random.

- *Runs above and below the mean*

With the term considered as the consolidated count of runs, variable n_1 and n_2, respectively, represent the total count of observations above and below the mean values [(0.99 + 0.00)/2 = 0.495], and then mean μ_b and variance computation of b for a truly independent series are depicted as according to the following equations:

$$\mu_b = \left(\frac{2n_1n_2}{N}\right) + \left(\frac{1}{2}\right) \tag{12}$$

and

$$\sigma_b^2 = 2n_1n_2\left(2n_1n_2 - N\right)/N^2\left(N - 1\right) \tag{13}$$

The statistical test for the same is depicted by formulation in Eq. (14):

$$Z_0 = \frac{b - \mu_b}{\sigma_b} \tag{14}$$

The hypothesis is rejected when $-z_{\alpha/2} \leq z_0 \leq z_{\alpha/2}$, where the significance level is selected as α.

For conducting the statistical test run above and below, the mean for the validation of randomness with significance level 0.05 and the main series of 500 key values engendered via the proposed hybrid model are considered for calculation. On utilizing Eq. (14), required test values are computed, and the outcomes obtained are depicted in Table 6.

Table 6 Statistical test values corresponding to run above and below for the proposed model

Generator	n_2	n_1	μ_b	Run	Z_0	σ_b^2	Accept/reject
Hybrid model	290	210	244.1	241	−0.2848	118.43	Accepted
LFSR	274	226	248.19	259	0.982	122.4	Accepted

The generated sequence satisfies randomness because the outcomes depicted in the table validate that $-1.96 < Z_0 < 1.96$ is at the level of significance 0.05.

4 Implementation Details

Here, we have considered a mixed model to procreate a pseudo-random number sequence utilizing an LFSR and GA. The number generated is random and uses a hybrid model for cryptographic applications like image encryption and image decryption [36]. The implementation of the cryptographic operations will be done in a cloud environment using CloudSim.

4.1 Employment of Hybrid Model for the Encryption of Image Data over the Cloud System

The pseudo-random number engendered via the proposed hybrid model is utilized as a key for encrypting/decrypting an image on stream ciphers. The operation of encryption/decryption algorithms is described by considering r_i as a plain text obtained via the stream of plain text with i ranging from 1 to n with n being the total number of plain text elements, and k_i denotes the corresponding hybrid model's generated key. The operation of encryption is additive and is demarcated according to Eq. (15).

$$Ciphertext\ \{c_i\} = \{r_i\} \oplus \{k_i\} \bmod m, \text{ where } i = 0, 1, 2, 3 \ldots .n \text{ and } m \text{ modulus.} \quad (15)$$

The output obtained from an encryption process represents part concerning the stream of ciphertext. The cipher data are received on the recipient side, and the initial plain text data must be decrypted. The method of decryption is set accordingly.

$$Plain\ text\ \{r_i\} = \{c_i \Theta\ k_i\}\ modm, \text{ where } i \text{ ranges between 0 to } n. \quad (16)$$

The techniques are replicated via Lenna's contribution image as depicted in Fig. 13, and the final picture from the cryptographic techniques is shown in Fig. 13b, c.

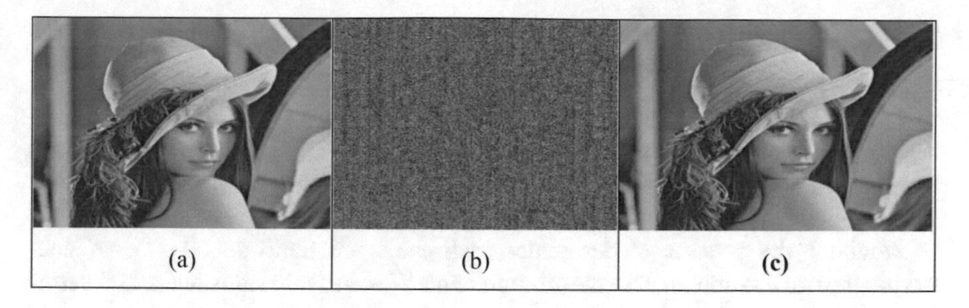

Fig. 13 Process of encryption and decryption of an image via additive cipher technique: (**a**). actual image, (**b**) encrypted image, (**c**) decrypted image

5 Result Analysis of the Cipher System Based on Security Parameters

The analysis of the outcomes achieved via the proposed system is performed by the utilization of the following tests over the values of encrypted data and also on the original data:

1. Standard deviation (SD)
2. Entropy

5.1 Security Constraints

5.1.1 Computation of Standard Deviation on the Occurrence of Ciphertext Elements

The standard deviation usually represents the distribution of the average data for a sampling dataset. More data is distributed closely on average by the slight standard deviation. A higher value reflects the more data distributed over the mean. Contemplate *N* to be the amount corresponding to the ciphertext component and the term n_i related to the existing total of cryptographic component "*i*," which ranges between $0 \leq i < m$. According to Eq. (17), the σ representing the normal deviation of cryptographic text existence is denoted (17):

$$\sigma = \frac{1}{m}\sqrt{\sum_{i=0}^{m-1} (n_i - \overline{n})^2} \tag{17}$$

where $\overline{n}$ is denoting the mean value corresponding to N cryptographic text components and is intended as per Eq. (18).

$$\overline{n} = \frac{1}{m}\sum_{i=0}^{m-1} n_i \tag{18}$$

Minimal the rate of σ designates, estimates are equivalent being of each cryptogram text component in the arrangement. We can avoid statistical attacks from them.

5.1.2 Entropy Computation of Elements of Ciphertext

Entropy represents the intensity to be determined by the instability of the cipher variable. The maximum entropy limit is the total amount of bits per pixel of the ciphertext. An entropy that is considered close to the upper limit's value signifies that the ciphertext is volatile and is also proved to be secure. The entropy computation of the ciphertext variable is assessed as according to Eq. (19):

$$Entropy = \sum_{i=0}^{M-1} p_i \log_2 \frac{1}{p_i} \tag{19}$$

Here, pi corresponds to the probability of ciphertext with i ranging between 0 to m-1 and is computed as per Eq. (20)

$$p_i = \frac{n_i}{N} \tag{20}$$

The higher the entropy valuc, thc higher the degree of protection since pattern uncertainty increases. The maximal entropy value of a sequence of 8 bits of cipher data is 8 because the proposed work is done.

5.2 *Result Analysis of the Cipher System Based on Security Constraints*

The strength of any security entity is measured based on the constraints corresponding to the security. The standard deviation and entropy values corresponding to the proposed system are determined using Eqs. (17) and (19); the result achieved from the entropy is compared against other standard systems, as depicted in Fig. 14.

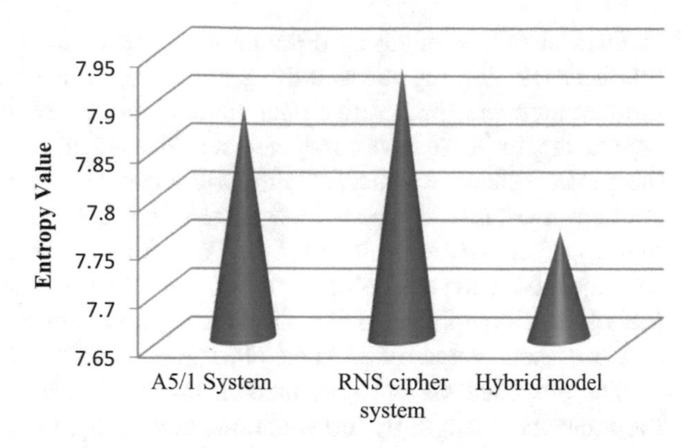

Fig. 14 Entropy comparison of the hybrid model against the standard systems

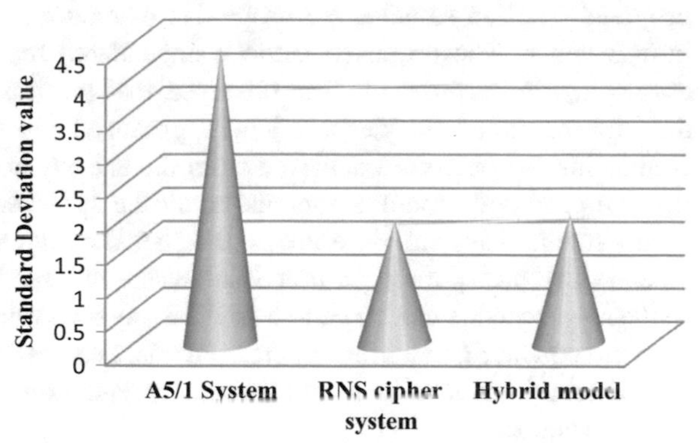

Fig. 15 Comparison of standard deviation for the proposed system's ciphertext with other standard systems

The findings undoubtedly confirm that the proposed scheme's entropy is equal to roughly normal systems and as strong as RNS with a non-maximum LFSR series of main lengths. The default variance outcome reveals the standard deviation of the hybrid model decreases significantly, and Fig. 15 reveals its relation. The standard deviation for the hybrid model is greatly reduced.

6 Conclusions and Future Scope

This research artifact has efficaciously implemented a novel model for engendering the series of pseudo-random numbers. The proposed work incessantly aims to optimize the pseudo-random number series generation's length based on the designed feedback shift register. With the progression of the research exertion, it was also well discovered that there is a limitation in connection to the series' length, or it is required to have a fixed value of maximum length for a series. Based on the conducted research study, it was also observed that it becomes impossible to outspread the length of a series until and unless there is a change in the generator's component level. This downside of the system enforces the extension of the sequence's length by utilizing the genetic procedure. The amalgamation of the feedback shift register and the genetic algorithm engenders a series of pseudo-random numbers that gratify randomness and the extension of the series' length beyond the limit. We have conducted out various statistical tests to assess the forte of the generated key sequences. Our hybrid model generated a higher range of 13,425 subsequences; the chi-square test generated a value of 2.862 which is way enhanced than a critical average value of 7.81. We have also confirmed our system by using runs up down and runs below and above the mean test. As per the tests, our hybrid model has accomplished a hypothesis acceptance region for a run length of 317 and 241 with mean values of 333 and 244.1, respectively.

The proposed system presented in this artifact the efforts to further enhance the sequence's length by the able utilization of the genetic algorithm. By this, the proposed exertion satisfies the properties of randomness, such as uniformity and independence. The proposed model's engendered sequence was also applied for cryptographic purposes. It was observed that the outcomes are relatively better than the maximum sequence's length produced via the feedback shift register. Comparing the outcomes achieved from the entropy and standard deviation attests that the proposed model accomplishes all the specified objectives and is as good as any standard system. As a future prospect, the implemented work can be further extended by incorporating a procedure where the maximum length corresponding to the key sequence can be generalized for any non-word or word-oriented pseudo-random sequence generation. Apart from this, as a future enhancement, the designed hybrid model can also be generalized to engender any non-word-oriented pseudo-random sequence.

References

1. Aishwarya C, Sannidhan MS, Rajendran B (2014) DNS security: need and role in the context of cloud computing. In: 2014 3rd international conference on eco-friendly computing and communication systems. IEEE, pp 229–232
2. Marston S, Li Z, Bandyopadhyay S, Zhang J, Ghalsasi A (2011) Cloud computing—the business perspective. Decis Support Syst 51(1):176–189
3. Arnault F, Berger T, Minier M, Pousse B (2011) Revisiting LFSRs for cryptographic applications. IEEE Trans Inf Theory 57(12):8095–8113
4. Panneton F, L'ecuyer P (2005) On the xorshift random number generators. ACM Trans Model Comput Simul (TOMACS) 15(4):346–361
5. Haahr M. Random.org. Retrieved 10/16/2011 from http://www.random.org
6. Walker J (2006) HotBits: Genuine random numbers generated by radioactive decay. http://www.fourmilab.ch/hotbits/
7. Stefanov A, Gisin N, Zbinden H (2008) Optical quantum random number generator. J Mod Opt 47:595–598
8. Li P, Wang Y, Zhang J (2010) All-optical fast random number generator. Opt Express 18(19):20360–20369
9. Sunar B, Martin W, Stinson D (2006) A provably secure true random number generator with built-in tolerance to active attacks. IEEE Trans Comput 56(1):109–119

10. Lehmer DH (1951) Mathematical method in large-scale computing unit. In: Proceeding of the second symposium on large-scale digital computing machinery. Harvard University Press, Cambridge, pp 141–146
11. Knuth DD (2006) Art of computer programming semi numerical algorithm, vol 2, 3rd edn. Pearson Education Inc. Publication
12. Blum L, Blum M, Shub M (1986) A simple unpredictable pseudo random number generator. SIAM J Comput 15(2):850–864
13. Shamir A (1981) On the generation of cryptographically strong pseudo-random sequences. In: ICALP, pp 544–550
14. Blum M, Micali S (1984) How to generate cryptographically strong sequences of pseudo-random bits. SIAM J Comput 13(4):850–864
15. Yao AC (1982, November) Theory and application of trapdoor functions. In: 23rd Annual symposium on foundations of computer science (SFCS 1982). IEEE, pp 80–91
16. Goldreich O, Levin LA (1989) A hard-core predicate for all one-way functions. In: Johnson DS (ed) 21th ACM symposium on theory of computing-STOC '89. ACM Press, pp 25–32
17. Håstad J, Impagliazzo R, Levin LA, Luby M (1999) A pseudorandom generator from any one-way function. SIAM J Comput 28(4):1364–1396
18. Blum L, Blum M, Shub M (1986) A simple unpredictable pseudo-random number generator. SIAM J Comput 15(2):364–383
19. Impagliazzo R, Naor M (1996) Efficient cryptographic schemes provably as secure as subset sum. J Crypt 9(4):199–216
20. Fischer JB, Stern J (1996) An efficient pseudo-random generator provably as secure as syndrome decoding. In: EUROCRYPT, pp 245–255
21. Boneh D, Halevi S, Howgrave-Graham N (2001) The modular inversion hidden number problem. In: ASIACRYPT, pp 36–51
22. Abu-Almash FS (2016) Apply genetic algorithm for pseudo random number generator. Int J Adv Res Computr Sci Softw Eng 6(8):8–19
23. Poorghanad A, Sadr A, Kashanipour A (2008) Generating high quality pseudo random number using evolutionary methods. In: 2008 international conference on computational intelligence and security, vol 1. IEEE, pp 331–335
24. Poorghanad A, Sadr A, Kashanipour A (2008) Generating high quality pseudo random number using evolutionary methods. In: International conference on computational intelligence and security, pp 331–335. https://doi.org/10.1109/cis.2008.220
25. Goyat S (2012) Genetic key generation for public key cryptography. Int J Soft Comput Eng (IJSCE) 2(3):231–233
26. Ashalatha R, Agarkhed J, Patil S (2016) Analysis of simulation tools in cloud computing. In: IEEE WiSPNET, pp 748–751
27. Nichat SP, Sikchi SS (2013) Image encryption using hybrid genetic algorithm. Int J Adv Res Comput Sci Softw Eng 3(1):427–431
28. Verma HK, Singh RK (2012) Linear feedback shift register based unique random number generator. Int J Comput Sci Inform 2(4):77–82
29. Singh D, Rani P, Kumar R (2013) To design a genetic algorithm for cryptography to enhance the security. Int J Innov Eng Technol (IJIET) 2(2):380–385
30. Calheiros RN, Ranjan R, Beloglazov A, De Rose CAF, Buyya R (2010) CloudSim: a toolkit for modeling and simulation of cloud computing environments and evaluation of resource provisioning algorithms. Wiley Online Library, pp 23–50. https://doi.org/10.1002/spe.995
31. Agarwal A (2012) Secret key encryption algorithm using genetic algorithm. Int J Adv Res Comput Sci Softw Eng 2(4):216–218
32. Malhotra R, Jain P (2013) Study and comparison of CloudSim simulators in the cloud computing. SIJ Trans Comput Sci Eng Appl (CSEA) 1(4):111–115
33. Atiewi S, Yussof S (2014) Comparison between CloudSim and GreenCloud in measuring energy consumption in a cloud environment. In: 3rd international conference on advanced computer science applications and technologies, 2014, pp 9–14. https://doi.org/10.1109/ACSAT.2014.9

34. Biebighauser D (2000), Testing random number generators. University of Minnesota – Twin Cities REU
35. Rukhin A, Soto J. et al (2010) A statistical test suite for random and pseudo random number generators for cryptographic applications. NIST Special Publication, 800–22 Revision 1a
36. Jambhekar ND, Misra S, Dhawale CA (2016) Cloud computing security with collaborating encryption. Indian J Sci Technol 9(21):95293

Anomaly Detection in IoT Using Machine Learning

Saadat Hasan Khan, Aritro Roy Arko, and Amitabha Chakrabarty

1 Introduction

Internet of Things (IoT), one of the most recent developing technology that holds the potential to change our lifestyle completely [5, 6, 27]. It is already being used for household purposes, industrially and commercially. The "Things" in IoT refers to the sensors [15]. These sensors can collect data of itself or the environment it is placed in. Depending on the kind of the object, it can either perceive data or act on the surrounding with other objects that are able to communicate with each other through the "Internet." However, IoT devices are not only used to communicate between themselves. The extent to which their use goes is very large. They help people in making decisions too. For example, a heart monitoring system uses sensors to collect data regarding an individual's heart and analyze the data to give a recommendation regarding the person's visit to a doctor [18, 21].

With IoT systems extensive popularity, the number of malicious attacks and newer, sophisticated tools to attack have also increased [1, 19]. According to Gartner Research [25], IoT devices will hit a total of 26 billion within 2020. Logical controls are used as software shields to ensure that data is accessed by the authorized individuals and systems. Intrusion detection and prevention system, passwords, access control, etc. includes logical controls [2]. An Intrusion Detection System (IDS) is one of the logical controls to guard the data [10, 22]. If there is a security breach or an attack on the IoT system, the IDS should be able to identify that an attack has taken place [31]. Moreover, an IDS is able to detect a fault using two techniques. They are: (1) Signature based detection (2) Anomaly-based detection [4, 20].

S. H. Khan (✉) · A. R. Arko · A. Chakrabarty
BRAC University, Dhaka, Bangladesh

S. Misra et al. (eds.), *Artificial Intelligence for Cloud and Edge Computing*,
Internet of Things, https://doi.org/10.1007/978-3-030-80821-1_11

Signature based Intrusion Detection System (IDS) is mostly used to examine packets in a network which come from various locations of the network to find those specific packets which are anomalous in nature. However, with regard to heavy false alarms, the computational process has become a heavy drawback in which takes a toll on their performance [34]. Signature based detection methods are useful when detecting known attacks. They can be either coded with concrete if-else blocks or supervised machine learning methods can be used to achieve their goal. Another disadvantage to the system is that, supervised learning cannot always be used to detect anomalies as many anomalous data can take new forms which might not exist in the data-set. Either way, they use a lot of computational power and can be troubling at detecting anomalies Signature based detection is powerful in cases where the pattern of the breach is known beforehand [29].

On the contrary anomaly-based detection techniques get handy against attacks which are new or do not follow a specific pattern [12, 16]. With new types and forms of data being in the offing, it becomes difficult to trace the pattern of anomalous data. Anomaly-based detection systems generally run-on AI techniques. The target is to make the system learn and be as accurate as possible at classifying the real data from the fake ones. Despite, many notable algorithms present, the field of anomaly detection in many places is vastly growing. With newer security tools appearing every now and then to solve newer challenges of anomaly detection. Anomaly-based detection IDS is growing to cover many sectors where anomalies need to be identified, as each anomaly can take a disparate form. Thus, compared to signature based detection, anomaly-based detection has a wider prospect to cover.

Therefore, with careful attention to previous knowledge, our system is designed to detect anomalous data from the set of data exchanged by a given system, where anomalous data can be in various forms and do not follow a given signature. Hence, we are to follow the direction of Anomaly-based detection. In order to achieve such an intelligent system, ML algorithms are used. Furthermore, it is tough to get a data-set of the best fit for our purpose, and thus, a data-set of environmental characteristics had been generated to be used for anomaly detection. Different sensors, each capable of sensing different environmental characteristics were set up in different places of Dhaka City and they were sent to a central cloud Server. This created the network that a simple IoT would possess. The data were then retrieved from the cloud server, analyzed, pre-processed, labeled, scaled, transformed and a few synthetic data were added to give a hint to the ML algorithms about how some anomalous data might look like, during the training phase. The data was then fit to different ML models to determine anomalous data inside the network. The best performing model can then be used in several IoT systems which depend on ML algorithms to detect anomalies.

The rest of the paper is planned to give a literature review first, then explaining the methodology of our entire research, followed by implementation of different ML algorithms with there optimised parameters and analyses on these algorithms' results. Lastly, the paper is finished with a conclusion where the best performing algorithm is selected keeping time of execution and accuracy as the main parameters of success.

2 Literature Review

There have been many works on anomaly detection of IoT. However, most of the works are on network intrusion detection. These papers mainly focus on machine learning algorithms like SVM, bagging, boosting algorithm sets and their performance comparison [33]. Another paper is [9], which talks about network intrusion detection based on learning techniques. It mainly analyses the existing network intrusion detection tools and finds the best solution. In the paper [8], a detailed review on the Intrusion Detection and Prevention System (IDPS) is presented till the year 2019.

An intrusion detection system is basically of two types, which are—anomaly detection IDS and Signature based IDS. The work [7] shows how intrusion detection system can be implemented using a Hashing Based Apriori Algorithm with Hadoop MapReduce. In present times, most Anomaly detection techniques rely on ML algorithms as these malicious data can take new shapes at any time and thus lack a signature. In the past Anomaly Detection techniques have provided a significant number of steps to detect anomalous attacks beforehand in internet. However, in "Undermining an Anomaly-Based Intrusion Detection System Using Common Exploits," the weaknesses of an anomaly detection based IDS are highlighted [31].

Signature based IDS have also used supervised Machine Learning algorithms to detect aberrant data. In "A survey and taxonomy of the fuzzy signature-based Intrusion Detection Systems," the authors discuss how IDS along with firewall in networks perform as a more robust security check [24]. This paper uses fuzzy misuse detection schemes along with different ML and data mining techniques to deal with intrusions. In [20], Decision Trees and clustering techniques were used for a Signature based IDS.

Obtaining labeled data in IoT for intrusion detection is quite difficult. KDD (Knowledge Discovery and Data Mining) Cup 1999 data-set published by MoD are one of the freely available non-IoT data-set for intrusion detection [11]. But after a long period of researching and analyzing, it has been marked as full of errors and outdated[32].

There are also different areas anomaly detection techniques can be used. For example, In the paper [28], differences between host and network-based intrusion detection techniques to demonstrate how they can work with each other for providing an extra effective intrusion detection and protection. For that purpose, the idea points to examine the role of a set of definite ML algorithms in a way to build robust defensive methods for IoT environments that depend on data-set/flows for following two layers:

1. Application Layer (Host based)
2. Network Layer (Network Based)

Mostly, machine learning, data mining, data processing are used to understand information from the respective system and provide an output in the application layer. To work on the intrusion detection in application layer, the paper titled "A

Model for Anomalies Detection in Internet of Things (IoT) Using Inverse Weight Clustering and Decision Tree" used a freely available unlabeled IoT data-set from Intel Berkeley Research Lab [3]. The data-set contains more than 2.3 million readings collected from 54 sensors in Intel Berkeley Research lab in 2004 [23]. To use the data-set further in anomaly detection research, it requires processing, as it is not primarily aimed for intrusion detection. The paper [14] discusses about attack on IoT devices using machine learning algorithms. They also used algorithms like Decision Tree for their data-set.

Even though cyber-security research is almost in its peak, there is a scarcity of proper data-set. Recently IoT eligible UNSW NB15 data-set has come to rescue for modern intrusion based researches of the network layer. This data-set has a huge number collection of benign and malignant network traffic instances. It integrates normal behavior and latest attack instances in traffic. The malignant data traffic is generated keeping many types of attacks under consideration. The lower number of malicious attack instances than the large number of normal traffic instances makes the UNSW-NB15 data-set unbalanced. This data-set is providing a good support for a wide range of networking IDS analysis [26]. In the paper [32], researchers have applied and evaluated the performance of many supervised learning algorithms, which include SVM (Support Vector Machine) and a range of other ensemble classifiers. These ensemble algorithms accumulate fundamental classifiers, that have been trained on different subsets of data-set. IDS (Intrusion Detection System) is capable of differentiating between abnormal and normal behaviors using the classification model inside the IDS. An IDS is based on SVM (Support Vector Machine) and C5.0, which is a widely used implementation of Decision tree with each node, aim to provide an answer with regards to the specific attributes in the input record. Generally existing IDS fails to detect the attacks in critical cases because it is unfamiliar with such new attacks. Using the combination of C5.0 and SVM, data can be divided into classes and after being trained; SVM will be able to map data into high dimensional space [13].

3 Methodology

3.1 Proposed Model

Figure 1 represents the overall steps in our research. Our research starts from collecting data, adding synthetic and wrong data to applying machine learning algorithms and then finally comparing which algorithm performs the best.

We used a total of 4 sensors to collect environment data. This includes light sensor, temperature sensor, humidity sensor and 3 V batteries. We set up 10 boards each consisting of a ESP32 (Node-MCU micro-controller) and the 4 sensors. These 10 boards are placed in different areas to collect data with greatest variation possible from which data is sent to Ada fruit server. The data-set is then clustered to add

Fig. 1 Flowchart of the system

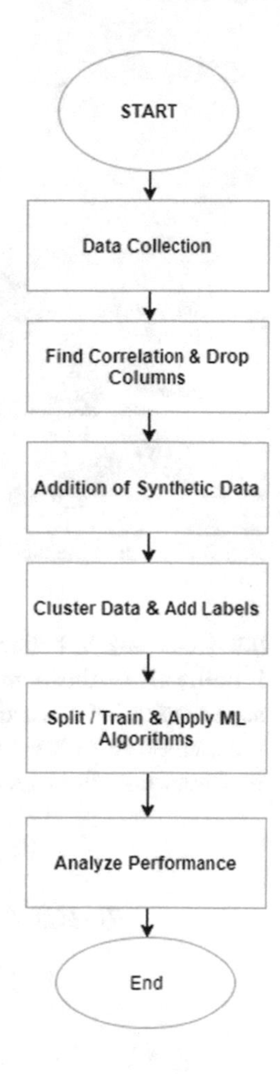

labels and different machine learning algorithms like Decision tree, Random forest, Logistic Regression are applied afterwards.

3.2 Hardware Setup

Figure 2 represents the overall hardware setup. The system data as input from the sensors and then passes the data to the Node-MCU (ESP32), respectively. The micro-controller gets electrical signals from all the sensors and then computes and converts the received electrical signals to scaled values like LUX, humidity (percentage), temperature (degree Celsius) and voltage (Volts) which are understandable by humans. For measuring the temperature and humidity, we have used the DHT11 module [17]. We have used a LDR based sensor for finding the light intensity and

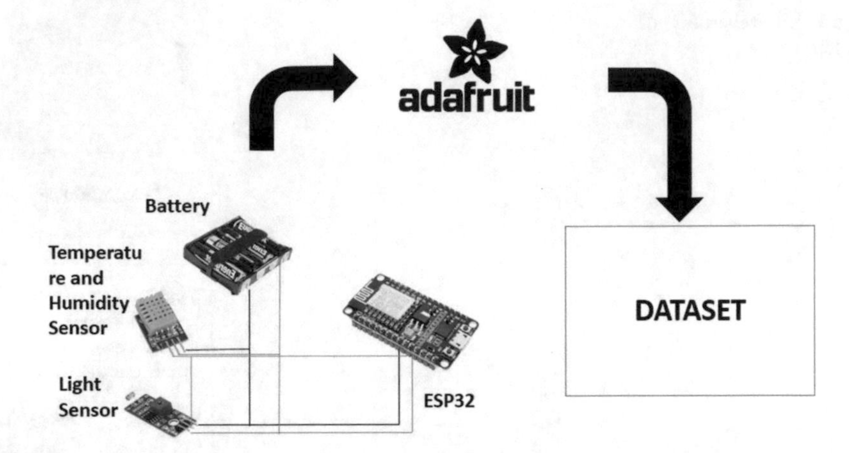

Fig. 2 Sketch of the hardware connections

connected a 3 V battery directly to the micro-controller to read voltage readings. Finally, all of these are sent to the Ada Fruit server. The data gets stored in four separate feeds from a single board for the four features at the same time. All the data are merged in a CSV file after 2 months of data collection. The following equations represent how the LDR voltage gets converted to appropriate LUX units.

$$VoltLDR = 0.0008057 * analogReading \tag{1}$$

$$ResistLDR = (10000 * (3.3 - VoltLDR))/(VoltLDR) \tag{2}$$

$$LUX = (500 * ResistLDR)/64881.00 \tag{3}$$

3.3 Data Preparation

Our data-set has five features. They are: (1) Temperature (2) Humidity (3) Light Intensity (4) Voltage and (5) Controller Ids. Table 1 shows an extract of our data-set. Before any other work is done, the features are tested for their importance. The feature importance was analyzed using Pearson Correlation. The correlation between features in our data-set is shown in Fig. 3. From the figure it has been clear that "Controller ID" has almost negligible correlation with other features of data. Therefore, it has been discarded as a feature for further work.

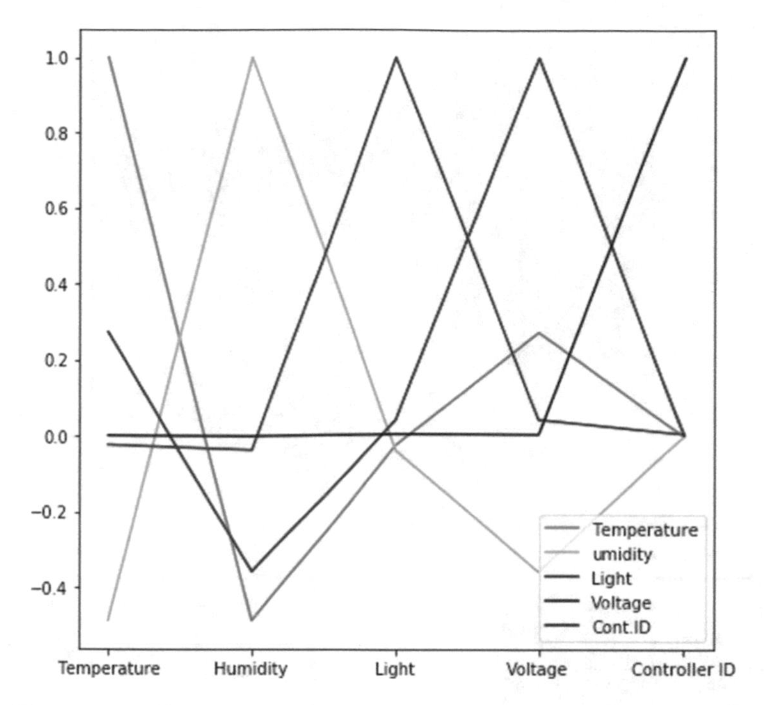

Fig. 3 Feature correlation

3.4 Labelling Data

Before applying any sort of Machine Learning Algorithms, the data-set needs to be labeled so that supervised ML algorithms can be applied. The underlying idea of labelling the data is to create different clusters for data that are exchanged in the IoT network and data that are maliciously injected into the network. Since the data we collected from the server was actually the data that belonged to the network, the data that have been synthetically added will be treated as malicious data. The following figures (Figs. 4 and 5) show an extract of data and spread of data.

Once synthetic data have been added to the data-set, they are analyzed and pictured. K-means clustering algorithm has then been applied to the data-set to label them. Manually labelling this many data would be tough and time consuming. Hence an unsupervised algorithm has been applied to form clusters and label data. In simple words, K means clustering calculates the distance between each data point and its K nearest data points, and then labels each data point to its according, closest cluster. The Purple set in the scatter diagram represents synthetic data. The yellow set is the original set of data (Table 1).

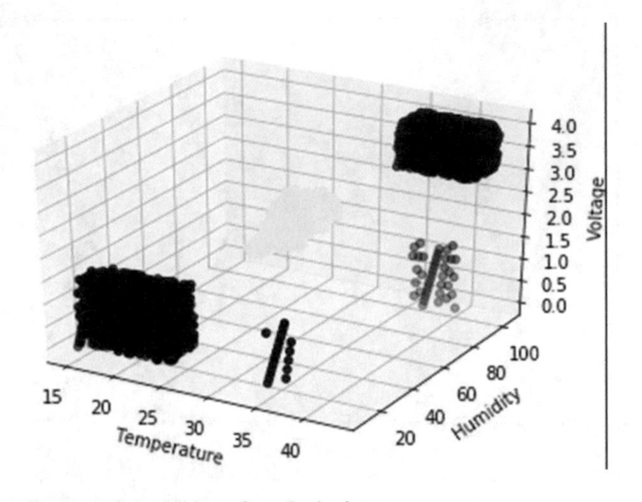

Fig. 4 Scatter diagram after addition of synthetic data

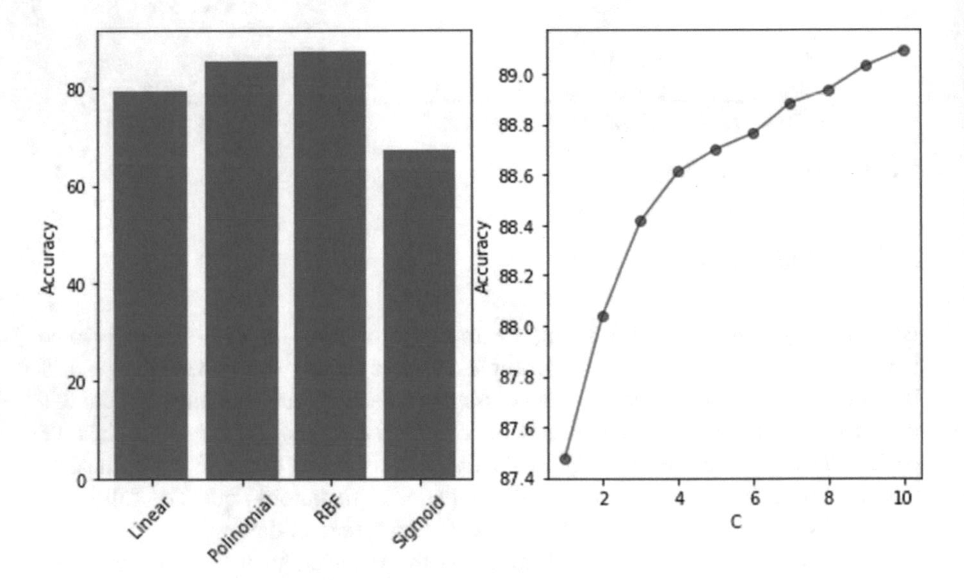

Fig. 5 Accuracy of different SVM kernels

Table 1 Data-set after addition of synthetic data

Temperature	Humidity	Light	Voltage	Controller Id	Output
31.7	47	227.92	2.52	3	1
31.7	47	227.92	2.52	1	1
31.7	47	235.78	2.53	1	1
40.1	108.8	424.4	3.4	Nan	0
40.4	108.9	424.8	3.8	Nan	0

4 Algorithm Implementation and Parameter Tuning

This section describes how different Machine Learning algorithms have been applied to our data-set and how they perform in classifying malicious/benign data instances. For all algorithms that have been applied, a random state of 0 is initialized while fitting and splitting the data. Moreover, the data is split in an 80/20 ratio for evaluation of algorithms. Each algorithm is applied and the parameters are tuned to get the best results. Any analysis made is also presented.

4.1 SVM Implementation

SVM (Support Vector Machines) is a classification model which is usually done to place 2 or more classes separately and then identify them. If we consider two classes only, the algorithm takes a line (which is also called the hyper-plane) and places it in between the classes, separating them from each other in such a way that the distance between each of the two nearest points of different classes have the maximum distance from the hyper-plane. The equation of such a hyper-plane is

$$w * x - b = 0 \tag{4}$$

where w is the normal vector to the hyper-plane and x can vary for each set of data, b is the distance between the nearest point to the hyper-plane and the hyper-plane itself. If the value of this equation is positive or negative, then it represents data to be in either of the two respective classes. The SVM model is fit into our data-set and different shapes of hyper-planes are tested to get different results. Not all kernels fit the model very well. Figure 5 shows how each kernel performs when fit to the data-set. When seen, RBF kernel performs the best, and the graph on the right shows the variation of accuracy with variation of Regularization Parameter, "C" for the model.

4.2 Decision Tree Implementation

First, the root of the tree is built by placing the best attribute. For our data-set, it can be seen that humidity is the best attribute. The training set is broken into subsets. We repeated the process over and over again until we reach the leaf node which contains the exact classification. The attribute with highest priority is chosen as the root and the priority of attribute decreases as we go down the tree. This priority is based on

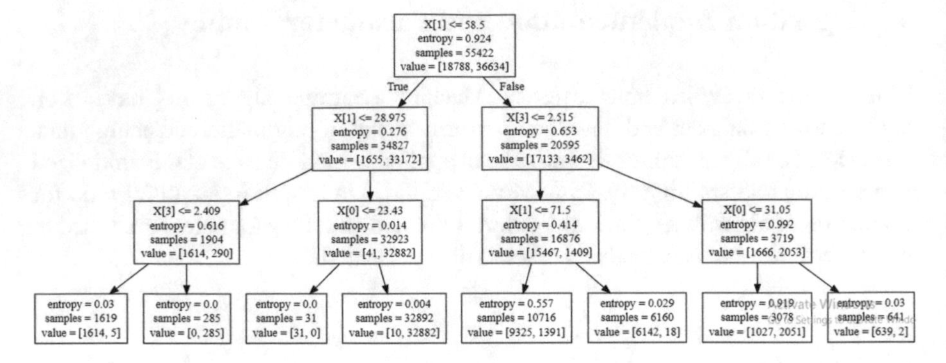

Fig. 6 Decision tree

the information gain which is dependent on the entropy. The entropy is calculated using the following formula:

$$Entropy(S) = \sum_{i=1}^{c} -p(i)logp(i) \tag{5}$$

Here, "c" is the number of classes of an attribute. It is two in our case "pi" is the fraction of examples of the class "i" [30].

$$Gain(S, A) = Entropy(S) - Entropy(A) \tag{6}$$

Here, S is the parent node and A is the attribute that we want to split. Information gain is basically the decrease in entropy due to partitioning of data-set based on an attribute. Figure 6 shows our decision tree based on our features. Initially, we got the highest information gain for the attribute—humidity. As a result, the algorithm starts building the tree using humidity as the root node. The "max height" for the diagram was set to 4. However, the decision tree would be too big to fit if we want to put the tree with a complete split.

4.3 K Nearest Neighbor Implementation

k Nearest Neighbor algorithm assumes that similar things exist in close proximity and works based on the shortest distance the algorithm finds from the instance to the training samples to determine its nearest neighbors. Then counts the highest number of neighbors belonging in a particular class and places the query instance as the member of that class. The value of K plays a significant role in determining the accuracy of the model. For our data-set, we first found the value of k that best suit our case. We initially classified with k-NN using k as 5. However, this had no

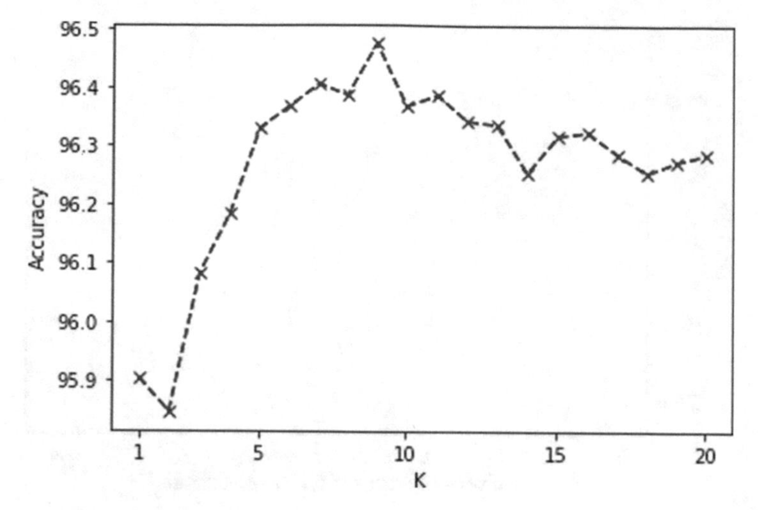

Fig. 7 k-NN's accuracy with variation in k

real logic behind it. As a result, a graph of accuracy against K, ranging from 1–20 values was plotted, which is shown in Fig. 7. As per Fig. 7, K was set to 9 while being fit to the data-set.

4.4 Random Forest Implementation

For our paper, the algorithm first grows multiple trees as opposed to a single tree. This gives better generalization and eliminates the problem faced by a decision tree. Since our training set contains 6480 entries, 6480 bootstrap samples are created with random "pick and replace" of data. Furthermore, the "number of features estimator" parameter is set to 2 to get the best result as shown by Fig. 8. To classify whether the data is normal or abnormal based on the attributes, each tree gives a classification. The model chooses the classification which is most supported (highest in number) over all the Decision Trees.

4.5 Logistic Regression Implementation

Logistic Regression is a classification algorithm used to class data according to different data types. It is used to accurately represent a data-set which has more than one feature. The logistic Regression uses a logit function which helps to build

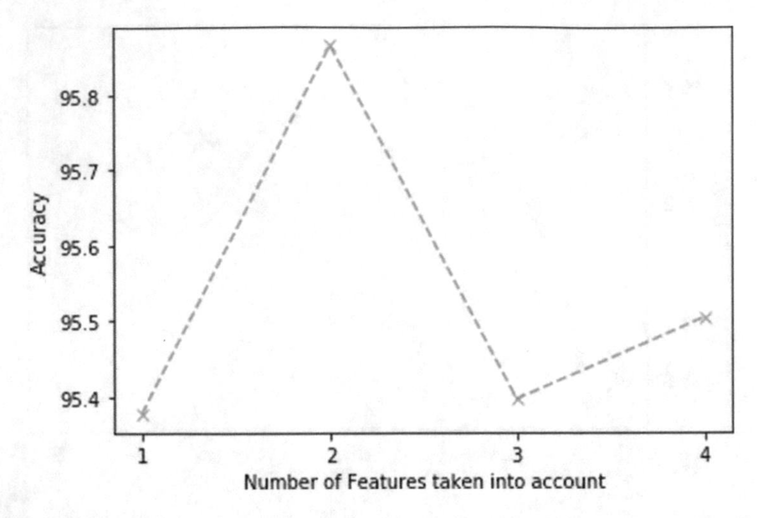

Fig. 8 Random forest's accuracy with variation in feature numbers

a sigmoid function to classify the two groups. The equation for logit function is as follows:

$$logit(Y) = \pi/1 - \pi \tag{7}$$

The function calculates the probability of an event (pi). If the probability is below a threshold it is fit to 0, or else 1.

The model is fit to our data-set and the regularization parameter (C), which helps the model to generalize better and make it less prone to over-fitting, is varied. Lower the value of C, higher is the regularization, lower is the chance of over-fitting. Figure 9 shows the relation of regularization parameter and accuracy. The value of C has been varied from 1 to 10 and the corresponding accuracy is plotted. It is seen that there is no change in accuracy over the different values of C, hence producing the straight horizontal line. Therefore, it can be deduced the model cannot be further tuned to give us a better accuracy than 81%.

5 Result and Analyses

The research is carried using Spyder (Python 3.6) in Windows environment with 16GB RAM and a 3.6 GHz Ryzen 3600X processor. After carrying out several supervised algorithms, we present all our findings here. We fit five disparate Machine Learning algorithms, which are: Logistic Regression, SVM, k-NN, Random Forest, and Decision Tree algorithms to our data-set. The results for the optimal parameters for each algorithm after thorough analysis are shown below in Table 2.

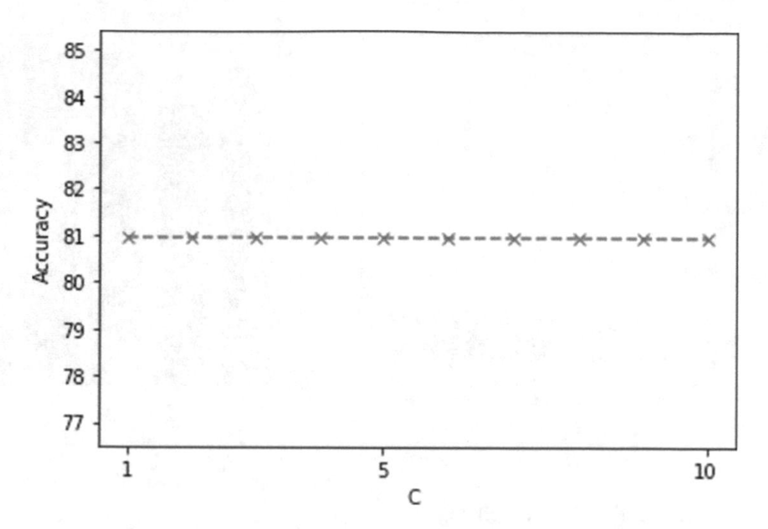

Fig. 9 Relation between regularization parameter and accuracy for logistic regression

Table 2 Summary of optimal parameters for each algorithm

Algorithm	Optimal parameters
SVM	RBF kernel with $C = 10$
Logistic regression	$C = 1$
Decision tree	No tuning needed
k-NN	$K = 9$
Random forest	No. of features = 2

5.1 Accuracy and Relative Time to Run for Each Model

Figure 10 shows the accuracy and the time it takes to run the best version of each model. All algorithms performed very well in terms of accuracy. The highest accuracy out of all these models comes from k-NN with a stellar accuracy of 96.5%. Both Decision Tree and Random Forests have accuracies that are close to k-NN's. Random forest is an improvised version of Decision Trees which eliminates the shortcomings of Decision Trees. Therefore, Random Forest's accuracy is slightly a bit higher than Decision Trees'. SVM and Logistic Regression have performed less accurately in detecting anomalies with accuracies of 89.1% and 81%, respectively. Moreover, SVM algorithm takes the highest time to run compared to others, which all run very fast and less than 1 s for our given data-set. The fastest algorithm is Logistic Regression which took 0.176 s to fit in our data-set and in determining the accuracy. However, despite its swiftness in carrying out the entire process, Logistic Regression provides us with the lowest accuracy, as already said. Therefore considering accuracies and time take to run, the best three algorithms are: Decision Tree, Random Forest, and k-NN.

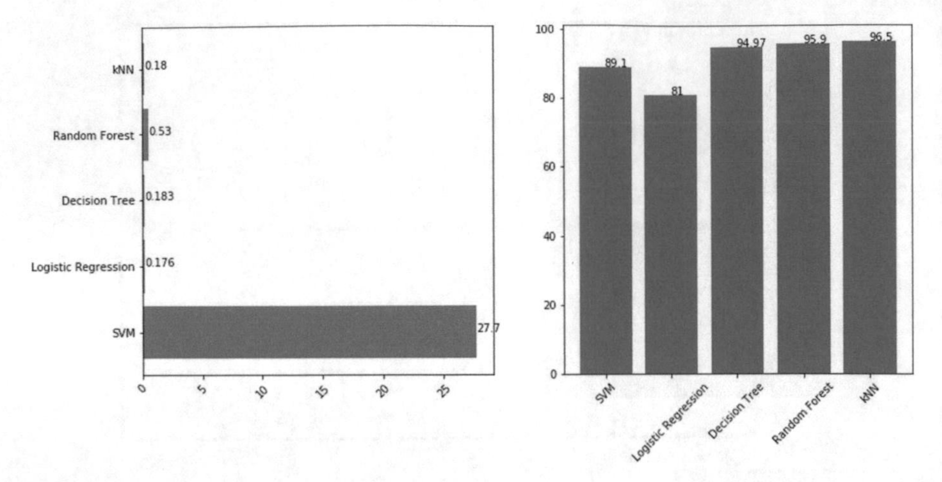

Fig. 10 Accuracy and timing of each model

5.2 *Confusion Matrices of Each Model*

The confusion matrix of all the models is found out. In Fig. 11, it can be seen that the highest true positive vale comes from Logistic Regression but it has a very low true negative value. The false positive and false negative values are also high for Logistic Regression which makes the algorithm not suitable. SVM has a very high true negative value, but its true positive value is not very impressive. Moreover, the false positive and false negative values are high for SVM, which again makes it unsatisfactory as the other algorithms performed way better.

Random Forest, k-NN and Decision Trees have similar confusion matrices. They are all very efficient with high true positive and true negative values, which means the predicted labels and the actual labels matched. Their high true positive and negative values justification is also bolstered by high accuracies obtained from testing these models.

Figure 11 shows the confusion matrix of all the models in their most fit version for the applied data-set.

5.3 *Area of ROC Curve for Each Model*

The area of the ROC curve shows the ratio between the true positive rate and the false positive rate. The better the model, the higher is the true positive rate's steepness because it can detect true positive values more correctly. Therefore, the better the model, steeper would be the gradient of the curve, and higher will be the area under the curve. In Fig. 12, we can see, Random Forest has the highest area under its ROC curve. Therefore, it classifies the true positive rates very efficiently,

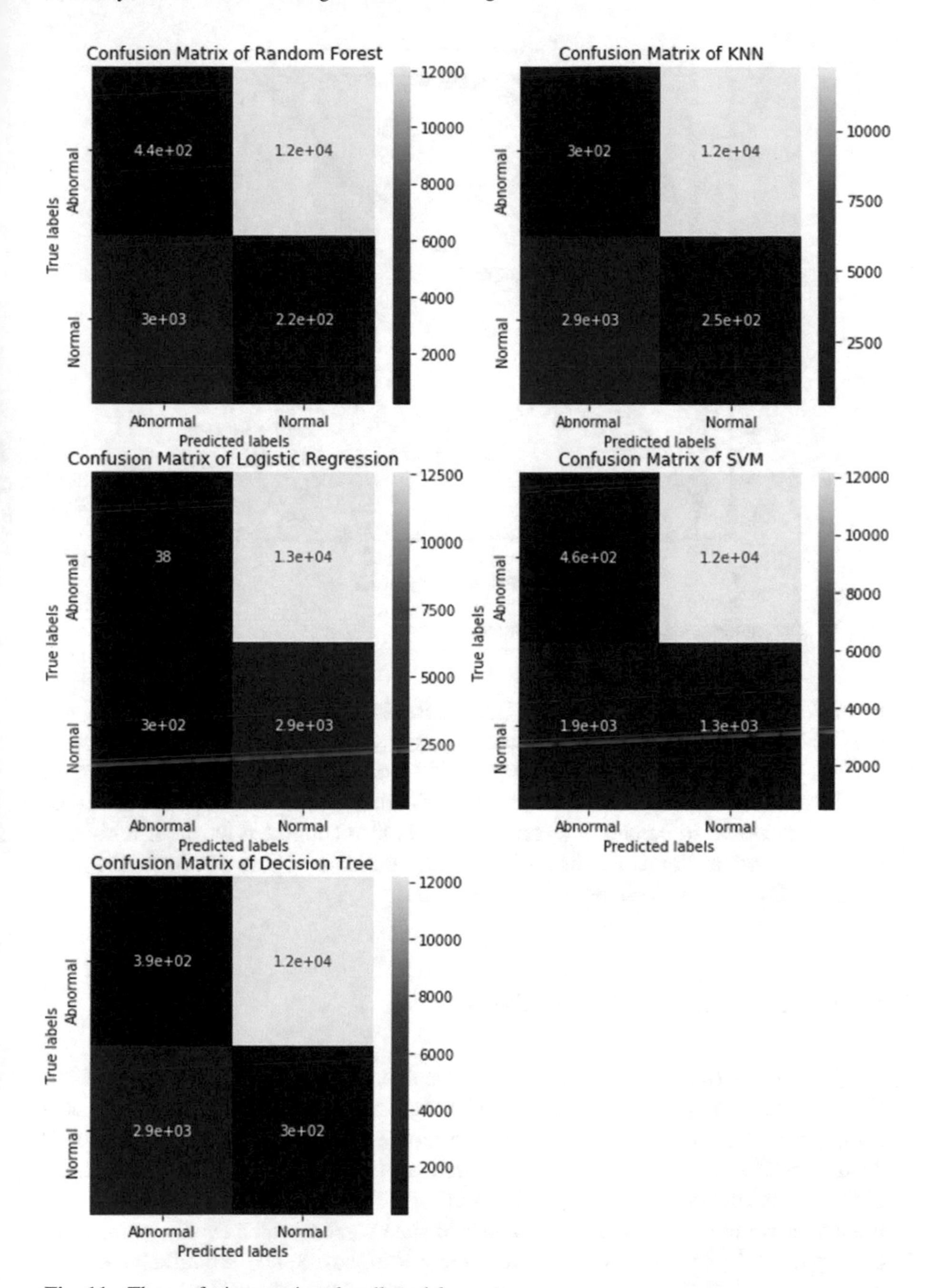

Fig. 11 The confusion matrices for all models

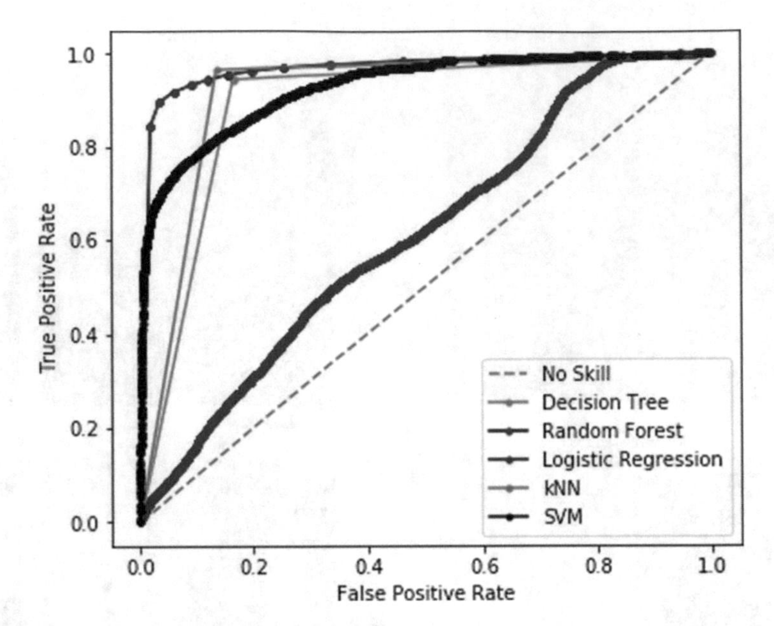

Fig. 12 Area under ROC curve for each model

which leads to this high area. Similarly, Logistic Regression has the lowest area and thus it is the least accurate. The fact that Logistic Regression has such low accuracy, derived from the area under ROC curve, is also supported by the accuracy and confusion matrix of the models. So, from ROC curve's perspective, Random Forest is the best model to be chosen. The decision is also supported by the astonishing accuracy Random Forest reaches while testing the data. Figure 12 shows the area under ROC curve for each model's best version

6 Conclusion and Future Work

Finally, after performing all the analyses, we can see that, SVM takes a very long time despite its considerable accuracy. Making a hyper-plane for such a large data-set is very time consuming. Therefore the model is discarded due to its sky high need of processing power. The most accurate and efficient model for our data is k-NN. It takes considerably less amount of time for the accuracy it gives. Moreover, the model is tuned to set the K parameter at such a value which avoids over-fitting and provides the best accuracy. The area under ROC curve and the confusion matrix also supports that this model is superior to most of the other algorithms used in this research. Decision tree is also a great model. The reason that it provides such a great accuracy is that it was allowed to run till its maximum height making the full classification. Since the training data and test data are not significantly different as they are mostly collected during the monsoon season and are of Bangladeshi context only, the data are similar and decision tree can easily classify them well. Random forests are always superior to Decision Trees. It makes ensembles, randomly takes a set of features, makes decision trees, and votes a final result. This gives it a better capability to generalize well and eliminate small inaccuracies of Decision Trees via the voting process. Thus, it achieves a greater accuracy than Decision Tree. Logistic Regression computes the probability of each instance of data. Logistic Regression also achieves a considerably accuracy due to the fact that data being used are subject to less variation. As a result most probabilities are well beyond or within the threshold, making it a clear distinction and thus the high accuracy. To conclude, after evaluation and justification from all perspectives, k-NN is the best model to be voted with a stellar accuracy of 96.5% and with an execution time of 0.18 s.

The target of the research is to continue gathering more data from all over Bangladesh and to make it a more robust system to prevent any sort of intrusion into data exchanging stations. In addition, more features should be added to the processed data to make sure all the elements that are needed to cover intrusion detection in a system exchanging environmental data. Finally, we are also interested to use deep learning algorithms when the data-set gets better with more added features, as in such a case, deep learning models would perform much better with the added complexity.

References

1. Abomhara M, Køien GM (2015) Cyber security and the internet of things: vulnerabilities, threats, intruders and attacks. J Cyber Secur 4(1):65–88
2. Alexander D, Finch A, Sutton D, Taylor A, Taylor A (2013) Information security management principles, 2nd edn. BCS, The Chartered Institute for IT, Swindon
3. Alghuried A (2017) A model for anomalies detection in internet of things (IoT) using inverse weight clustering and decision tree. Semantic scholar. https://doi.org/10.21427/D7WK7S
4. Aljawarneh S, Toth T, Yassein MB (2018) Anomaly-based intrusion detection system through feature selection analysis and building hybrid efficient model. J Comput Sci 25:152–160
5. Andreev S, Koucheryavy Y (2012) Internet of things, smart spaces, and next generation networking. LNCS, vol 7469. Springer, Berlin, pp 464
6. Atzori L, Iera A, Morabito G (2010) The internet of things: a survey. Comput. Netw. 54(15):2787–2805
7. Azeez NA, Ayemobola TJ, Misra S, Maskeliūnas R, Damaševičius R (2019) Network intrusion detection with a hashing based apriori algorithm using Hadoop MapReduce. Computers 8(4):86
8. Azeez NA, Bada TM, Misra S, Adewumi A (2020) Intrusion detection and prevention systems: an updated review. In: Data management, analytics and innovation, pp 685–696
9. Chaabouni N, Mosbah M, Zemmari A, Sauvignac C, Faruki P (2019) Network intrusion detection for IoT security based on learning techniques. IEEE Commun Surv Tutorials 21(3):2671–2701
10. Depren O, Lin MT, Anarim E, Ciliz MK (2005) An intelligent intrusion detection system (IDS) for anomaly and misuse detection in computer networks. Expert Syst Appl 29(4):713–722

11. Fu R, Zheng K, Zhang D, Yang Y (2011) An intrusion detection scheme based on anomaly mining in internet of things, pp 315–320
12. García-Teodoro P, Díaz-Verdejo J, Maciá-Fernández G, Vázquez E (2009) Anomaly-based network intrusion detection: techniques, systems and challenges. Comput Secur 28:18–28
13. Golmah V (2014) An efficient hybrid intrusion detection system based on c5.0 and SVM. Int J Database Theory Appl 7(2):59–70
14. Hasan M, Islam MM, Zarif MII, Hashem MMA (2019) Attack and anomaly detection in IoT sensors in IoT sites using machine learning approaches. Internet of Things 7:100059
15. Hassanalieragh M, Page A, Soyata T, Sharma G, Aktas M, Mateos G, Kantarci B, Andreescu S (2015) Health monitoring and management using internet-of-things (IoT) sensing with cloud-based processing: opportunities and challenges. In: 2015 IEEE international conference on services computing, pp 285–292
16. Jain A, Verma B, Rana JL (2014) Anomaly intrusion detection techniques: a brief review. Int J Sci Eng Res 5(7):1372–1383
17. Jain A, Verma B, Rana JL (2016) Low cost solution for temperature and humidity monitoring and control system using touch screen technology. Int. J Latest Res Eng Technol 2(1):10–14
18. Kazi SS, Kazi SS, Thite T (2018) Remote heart rate monitoring system using IoT. Int Res J Eng Technol 5(4):2956–2963
19. Kizza JM (2013) Guide to computer network security. Springer, Berlin
20. Kruegel C, Toth T (2003) Using decision trees to improve signature-based intrusion detection. In: International workshop on recent advances in intrusion detection, vol 2820, no 4, pp 713–722
21. Li C, Hu X, Zhang L (2017) The IoT-based heart disease monitoring system for pervasive healthcare service. Procedia Comput Sci 112:2328–2334
22. Liao HJ, Lin CHR, Lin YC, Tung KY (2013) Intrusion detection system: a comprehensive review. J Netw Comput Appl 36(1):16–24
23. Madden S (2004) Intel lab data. http://db.csail.mit.edu/labdata/labdata.html
24. Masdari M, Khezri H (2020) A survey and taxonomy of the fuzzy signature-based intrusion detection systems. Appl Soft Comput 92:106301
25. Middleton P, Tully J, Kjeldsen P (2013) The internet of things, worldwide. https://www.gartner.com/en/documents/2625419/forecast-the-internet-of-things-worldwide-2013
26. Moustafa N, Slay J (2016) The evaluation of network anomaly detection systems: statistical analysis of the UNSW-NB15 data set and the comparison with the KDD99 data set. Inf Secur J Glob Perspect 25(1–3):18–31
27. Muhic I, Hodzic MI (2014) Internet of things: current technological review. Period Eng Nat Sci 2(2):1–8
28. Mukherjee S, Sharma D (2012) Intrusion detection using naive Bayes classifier with feature reduction. Procedia Technol 4:119–128
29. Prabha DK, Sree SS (2016) A survey on IPS methods and techniques. Int J Comput Sci Issues 13(2):38–43
30. Sujan NI (2018) What is entropy and why information gain matter in decision trees. https://medium.com/coinmonks/what-is-entropy-and-why-information-gain-is-matter-4e85d46d2f01
31. Tan KMC, Killourhy KS, Maxion RA (2002) Undermining an anomaly-based intrusion detection system using common exploits. Recent Adv Intrusion Detect 2516:54–73
32. Tanpure SS, Jagtap J, Patel GD (2016) Intrusion detection system in data mining using hybrid approach. Int J Comput Appl 975:8887.
33. Timčenko V, Gajin S (2018) Machine learning based network anomaly detection for IoT environments. ICIST 1:196–201
34. Yassin W, Udzir NI, Abdullah A, Abdullah MT, Zulzalil H, Muda Z (2014) Signature-based anomaly intrusion detection using integrated data mining classifiers. In: 2014 International symposium on biometrics and security technologies (ISBAST), pp 232–237

System Level Knowledge Representation for Edge Intelligence

Paola Di Maio

1 Introduction

The Things' (IOT) [1] was created to describe objects connected to IP addresses with different levels of interactivity, from passive RDF tagging for the purpose of geo-location of physical objects to full interactivity of cyber physical devices capable of adaptive responses and autonomous intelligent behaviour, intended to deliver next generation capabilities like traffic control for self-driving cars and smart homes. In IoT scenarios, 'everyday objects', from intelligent suitcases to adaptive traffic lights, can be equipped with intelligence by using sensors, thanks to which these objects can be identified, located, monitored and operated remotely and thanks to which data about places things and people can be acquired and analysed to generate intelligence. The direction in which the data and information can flow (Fig. 1) is an important yet often underestimated characteristic of IoT architectures, which can be bidirectional [2] or multidirectional and configured according to variable parameters and criteria [3].

The term 'Web of Things' is also used to identify a more abstract layer that enables, in principle, some level of end-user interaction with IoT environments [4]. From a point of view of sociotechnical systems, however, data is not only about *things* but also about and related to *people*. For example, body sensors have already been in use in advanced health applications, and they are expected to become more widely integrated in consumer products especially in the field of health monitoring personal devices, telemedicine and neuroscience. The fundamental architecture of the IoT consists of data (audiovisual, geolocation, etc.) detected and collected by sensors, encrypted, bundled and broadcast together with network protocol data via

P. Di Maio (✉)
Center for Systems, Knowledge Representation and Neuroscience, Edinburgh, UK

Ronin Institute, Montclair, NJ, USA

S. Misra et al. (eds.), *Artificial Intelligence for Cloud and Edge Computing*,
Internet of Things, https://doi.org/10.1007/978-3-030-80821-1_12

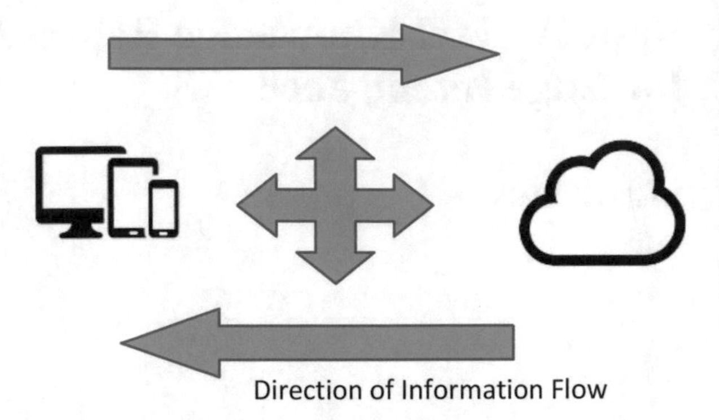

Fig. 1 Direction of information flow

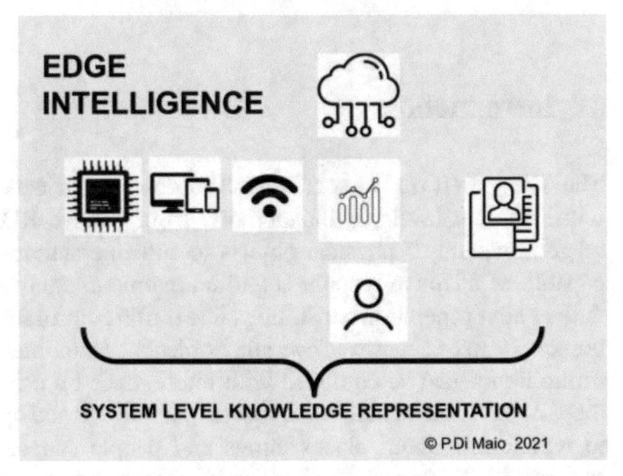

Fig. 2 System level KR for edge intelligence

standard telecommunication networks, to be later analysed and processed. In edge computing, the analysis and processing of the data increasingly takes place closer to the data source, contributing to the evolution of decentralised architectures where intelligence is at the edge of the IoT network (Fig. 2). AI capabilities can in principle be applied across the full spectrum of the IoT data lifecycle, from its acquisition to its routing and data crunching; however, AI technologies, in particular machine learning, come with their own set of challenges and risks [14]: no machine learning algorithm can guarantee the absence of bias, for example, and that the data in the network is going to be handled in respect of fairness, reliability and adherence to ethical policies and privacy laws. Adding AI layers to IoT results in increased fragility and higher risks and reduced transparency and accountability of the IoT stack. Furthermore, increased levels of complexity can be cognitively challenging for ordinary users to understand and interact with network information meaningfully and safely [5].

To achieve an acceptable level of cognitive awareness about the data and processes involved in IoT, it can be helpful to capture a 'system view' of the disparate layers, including infrastructure communication protocols, that support embedded intelligence both for things and humans. This can be achieved by leveraging one or more knowledge representation (KR) techniques. KR consists of knowledge modelling, and it is central to AI developments. In addition to encode knowledge and logic for the purpose of writing AI programs, the most abstract KR techniques can be used to simplify and detangle the conceptual representation of complexity, which is necessary to cognition and required for system development and use. A system level knowledge representation as an object is therefore proposed.

The scope of this paper is to present the model, while experimentation and further validation of the model are remanded to future work.

The paper contains an introductory overview of background issues (Sect. 2) and presents highlights from relevant research in AI, complexity and sensor-based architectures (Sects. 3 and 4).

The roles of KR and the system level are introduced (Sect. 5), and SLKR as an object is presented. Pointers to how the system level KR is being applied to address similar concerns in other fields are provided throughout the article.

2 Background

The Internet of Things (IoT) promised to make commercially accessible a new era of technological wonders: doors that open without keys, lights that switch on and off as people enter or leave rooms, ambient temperatures that auto-regulate, watches that monitor our healthcare and call the ambulance if something is wrong, coffee that makes itself, fridge that orders food directly when fresh supplies are needed, components that self-assemble and trucks that drive themselves and auto-park into the garage when they require maintenance. However, for most people, what really is happening instead is sensor-driven surveillance [6]. The electronic industry has managed to deliver sensors embedded in every gadget of ordinary use, from laptops to headphones to sensors that can detect pollution and traffic jams and monitor wildlife in public streets and at every road junction. Motion sensors are in laptops and mobile phones, but what these sensors detect and how this data is stored and handled happen without the knowledge, understanding and consent of the majority of ordinary users. While some level of sensing is already plausibly embedded everywhere, an 'intelligent future everywhere' is far from happening for most individuals, who, at best, are simply spied on. Ubiquity, which indicates the ability of technology to serve multiple purposes, also means that sensors equip devices with intelligence can actually be used to gather intelligence by unauthorised parties, i.e. break the privacy of users and even worse manipulate and distort reality itself by deconstructing and reconfiguring the data sets intended to represent reality.

This is currently the primary concern and a priority issue for IoT research. Given the diverse and fragmented nature of the possible scenarios, it is currently quite difficult for individuals and regulators and pretty much anyone to figure out in which direction the data is flowing according to what logic and policy. Regulatory safeguards that can ensure the legitimacy of IoT deployments and in particular privacy protection are still in development and very far from being enforceable in the majority of cases [7, 8]. In principle quite advanced IoT technologies are already in place; however, because of the complexity of IoT environments and the lack of transparency and auditability of data networks, they are quite hard to manage and query. Sensor data can be acquired fairly easily by anyone in every environment, thanks to inexpensive intelligent gadgets that can be purchased freely on Amazon.com. From airports and train stations to homes, using simply Internet connections and microphones built into devices, most innovation is currently being driven by an easily deployed ‘sensors grid’. However, data by itself, especially when it is bundled in large data sets, is meaningless and can be easily manipulated.

This is where AI and ‘embedded intelligence’ could be useful: everyday objects augmented with intelligence, connected to communication and information networks, can enable by default maximum information exchange and interaction between objects, agents and systems, as well as some level of autonomous behaviour, provided this takes place in ways defined and as intended by data owners and end users.

The ability of a system to generate intelligent behaviours depends solely on the configuration of the knowledge schemas used to interpret the otherwise rather meaningless data, as well as on the communication protocols adopted that should guarantee the integrity of such data when it is transported across networks [9]. These knowledge schemas and communication protocols must be explainable and accessible to intended users and made incorruptible by unintended users and unknown third parties. At the moment, no known AI fulfils this requirement.

3 AI as a Spectrum

Recent advancements in AI research have been focused largely on machine learning and on the application of neural networks to reasoning with terabyte size data sets, with great limitations. Intelligent functionalities delivered by ML are mostly related to pattern detection and autonomous learning, yet many of the underlying algorithmic processes are unreliable and unverifiable, since the logic of ML learning algorithms is not explicit. This has caused great alarm in the AI communities worldwide and has triggered a cascade of initiatives to advance AI ethics. Enabling automated identification of specific characteristics in certain images or sounds or text of information patterns, say sets of words, phrases or images that can be associated with features of interest to the user or to the researcher, does not deliver insights nor intelligence and comes with great risks of bias, misinterpretation and misrepresentation. AI as a whole is a highly sophisticated discipline consisting of a broad spectrum of technologies, of which ML and deep learning and neural networks are a subset of, as shown in Fig. 3.

Beyond machine learning, AI can therefore be used to supply intelligent capabilities. From language generation to various types of reasoning, including discrimination and inference but most importantly in the context of IoT, the level of knowledge and intelligence required by users to be able to visualise, query and modify the

Fig. 3 AI landscapes

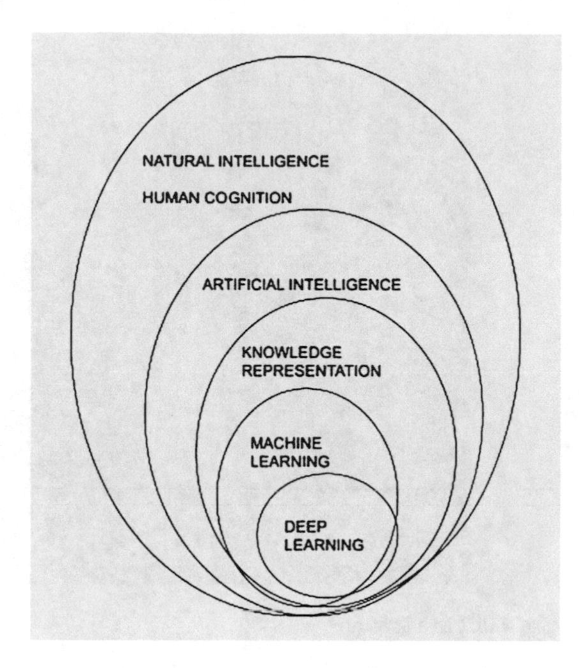

settings and configuration of their devices via interactive natural language interfaces Knowledge Representation (KR) consists of methods and techniques used in AI to capture and codify system logic, which can be expressed using different models and languages. In the same way as AI consists of a full spectrum of techniques of which ML, deep learning and NN are a subset of, knowledge representation (KR) also consists of a wide range of artifacts and techniques that can be developed and applied to support the engineering of autonomous intelligence. KR techniques such as ontologies can be particularly useful in addressing security and safety concerns in models of information systems underlying IoT [12].

KR can also be useful to capture and convey domain knowledge in complexity science and engineering as well as to model specific data and logical sets for the purpose of algorithmic encoding [13].

A complete catalogue of KR techniques developed in the context of AI research in the last fifty years is not available (and maybe not entirely feasible), but for simplification a spectrum of KR can be visualised as from the purely conceptual (ontology, taxonomy) to the more implementation-oriented techniques, such as logic programming, as shown in Fig. 4.

Ontologies, for example, can be considered the most complete and abstract form of KR use to capture and represent concepts, terms, axioms and relations. The figure below shows how ontologies can also be viewed as a spectrum from the more generalised, lightweight conceptual frameworks to the more formal, language-specific techniques.

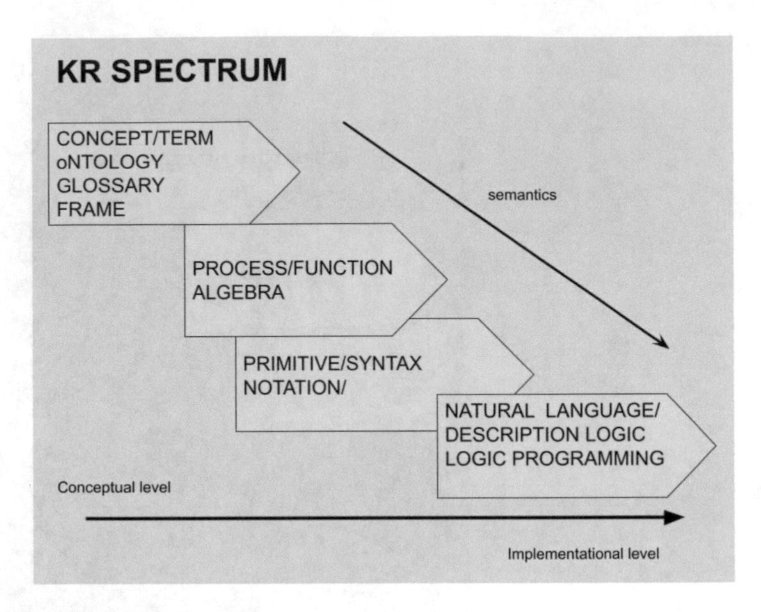

Fig. 4 KR as a spectrum

Corresponding to such a spectrum view of KR, different knowledge levels are identified in literature [15]:

Implementational: Including data structures such as atoms, pointers, lists and other programming notations

Logical: Symbolic logic propositions, predicates, variables, quantifiers and Boolean operations

Epistemic: Concept types with subtypes, inheritance and structuring relations

Conceptual: The level of semantic relations, linguistic roles, objects and actions

Linguistic: Deals with arbitrary concepts, words and expressions of natural languages

To tackle some of the emerging complexities observable in IoT, the system level knowledge representation is proposed that consider the entire spectrum of KR as applicable to represent a wider range of knowledge representation requirements for AI in IoT [16].

4 Complexity in Edge Intelligence

Sensing technologies are becoming more numerous and more powerful and capable of capturing increasing levels of detail, resulting in exponential amounts of data, each derived from a variety of sensory types, transmission protocols, encryption layers and data models. Complexity in IoT can be estimated by capturing factors, for example:

Sensors, types and quantities
Object model (of the object being sensed)
Data schema (representing the object)
Encryption method
Transmission protocol
Storage (location)
Availability (permission)
Data fusion method
Data analysis method

4.1 Seeing the Fog for the Cloud

Edge computing is considered by some as a 'brand new computational paradigm' allegedly based on the technical innovations of 5G communication networks. In reality, edge computing is a denomination for distributed, decentralised computing, for which equivalent definitions and specifications have been shifting over the last half a century to indicate more or less the same type of architecture, albeit with novel features becoming available for each new generation.

In this context, the edge stands for the periphery of the cloud (the geographically dispersed network of servers where data accessible via pubic network protocols such as the Internet is stored). It is also called FOG computing (named after its characteristic lack of visibility perhaps?) [17]. The complexity and the opacity of distributed network architectures increase as the network layers become more diverse and fragmented and more diverse AI and M technologies become embedded in them. The solution to address such opacity comes from the AI discipline itself, in the form of the KR spectrum of techniques. Edge intelligence therefore referred to as the AI at the edge of the cloud. According to research, it can be divided into AI for edge (intelligence-enabled edge computing) concerned mostly with network optimisation AI on edge (artificial intelligence on edge) concerned with data analysis [18, 19].

Novelty in edge intelligence is impacted by advancements in network technologies such as enhanced mobile broadband (eMBB) [20] millimetre-wave (mmWave) frequency adoption [21] and personal data storage architectures [22]. Technologies such as radio frequency spectrum regulation, data compression and transport protocol optimization also contribute to increase the complexity of IoT information networks. Additionally, complexity can be estimated including variables such as the amount of devices, antennas, sensors, and actuators, as well as sociotechnical factors noted earlier including privacy and data protection policies.

4.2 Networks of Things and People

The co-evolution of IoT and sociotechnical systems takes place either between people and things or between things and other things.

- Thing-to-person (and person-to-thing), where objects equipped with embedded sensors constantly communicate.
- Thing-to-thing, where there is no human intervention and communication and data exchange and interaction take place, including machine-to-machine communications including large-scale IT systems.

Although there are different types of sensors and networks, in the resulting mesh, the boundaries between what can be monitored (information flowing one way) and monitored and controlled (information flowing in two or more directions) are determined by the systems configuration, rather than by physical constraint.

Ideally, an optimal model of pervasive embedded intelligence can be described with the following attributes:

Pervasive – widespread
Mobile – portable, independent of location
Embedded – fully integrated
Adaptable – capable of change according to circumstance
Reliable – consistently good in quality or performance, that can be trusted
Scalable – that can meet increase or decrease in capacity
Intentional – as intended by the user

The system resulting from the combination of such complex interaction can be described as Networks of Things and People (Notap) [4] to help identify and characterise the dimensions that need to be represented in the requirements for system level KR for edge AI. The most notable advances in sensor technologies are extending the quantity, quality and granularity of signals that can be captured, by leveraging novel technologies that can sense a new spectrum of frequencies such as electromagnetic, infrared as well as others that emerge from research laboratories on a day-by-day basis.

4.3 Sensors

Despite technological advancements of the grid [23], there is still ambiguity in the use of the terminology. For example, sensors are also called transducers, indicating the low power devices capable of responding to physical stimuli, yet adoption of the terminology can vary greatly among countries and industries.[1] The most basic type

[1] https://www.electronics-tutorials.ws/io/io_1.html

of knowledge representation in AI is terminology defining and representing unambiguously specific concepts. This lack of terminology harmonization is common in technical standards and remains the cause of much misunderstanding, costs and potential risk, when considering that IoT are at global scale, for example, 'sensor', the preferred term used in the USA that literally means 'a device that detects a change in a physical stimulus and turns it into a signal which can be measured or recorded', from the corresponding Latin word sense which means perception. 'Transducers' by contrast are defined as 'devices that transfer power from one system to another in the same or in the different form' and it is the name preferred in Europe. Therefore, the word sensor generally indicates a device made up of a sensing element, and the word 'transducer' indicates both the sensing element and the data transmission component. All transducers contain at least one sensor, and most sensors require a transducer component. However, all this needs to be constantly reiterated in technical specifications and system documentation. Another type of components of a sensor network is the actuators, which translate signals into mechanical responses carried out by peripheral devices at the end of a process chain, such as valves. There are different criteria to categorise sensors [24] including industrial vs. commercial/consumer oriented based on different properties. In brief, and purely for the purpose of discussing the need for a system level KR, the following attributes are identified:

Accuracy: level of precision
Environmental conditions: temperature, humidity, pollution, wind speed
Range: upper and lower limits
Calibration: degree of adjustment among various parameters
Property: temperature pressure flow
Level sensors: volume measurements
Position: geolocation, proximity, displacement
Speed: acceleration
Signal: frequency

'Virtual sensors' consist of simulations that replace either in part or in full the actual readings from physical devices [25]. They are used when measurements cannot be taken physically, or when data readings from physical sensors are inaccessible or impractical or not possible for other reasons, their accuracy approximation limits their reliability and usefulness [26, 27].

New types of sensors are continually being developed, including contactless sensors for neuroscience and health monitoring, widening the potential field of applications and increasing the requirements for privacy and safety of the IoT [28].

Biosensors

Biosensors transform biological inputs into electronic signals, which from an information management viewpoint are like other data sets: they require processing to be converted to information and used to support decision making. Examples of biosensors include:

- Wearable, either embedded in devices or clothes
- Implantable, normally under the skin and require minor surgery as well as a higher conformity to health and safety
- Digestible, embedded in nanocapsules and swallowed to provide data readings from inside the body

Biosensing is the basis of all biometric technologies, which include, among others, fingerprinting, scanning (of the iris or the whole body) and voice recognition. They are used to detect subtle changes to physical conditions of the body or of the environment, for example, devices that can reproduce human sensorial perceptions such as smell and taste, also known as synthetic receptors. One notable example with critical applications is the 'electronic nose' capable of 'smelling'. Already widely used in airports, electronic noses can sniff the contents of passengers' luggage to detect dangerous or prohibited substances. In the food industry, they are used to test the appeal of certain aromas or can be deployed to detect gas leaks.

Biosensors data coming from individuals is part of a personal data network also known as body area network (BAN). Its use was initially specified in the standard protocol IEEE 802.15.6 and its corresponding wireless version, WBAN [29].

4.4 Interfaces

Sensors architectures interact with people and environments via interfaces. At the moment, the majority of systems use command-based interfaces typical of electronic controller devices, using machine-specific languages as determined by the operating system. Higher level user interfaces, known as GUIs, are necessary to provide a unified view and a single point of access and control of all the network components, and because they allow the configuration, manipulation and querying of said components do not require knowledge of the machine language nor of the system architecture.

Processing data from sensor networks requires algorithms capable of operating 'autonomously' the data acquisition, and that require minimal computer language and domain knowledge to perform information processing tasks which guarantee the control of end users over the IoT processes, including modelling the accuracy of the sensor readings, accuracy evaluation and prediction. Existing standard specifications for IoT interfaces include:

1. P1451.0 – Standard for a smart transducer interface for sensors, actuators, devices, and systems [30], which specifies a common functionality for the family of IEEE 1451 smart transducer interface standards, independently of the physical communications media and includes the common transducer services required to control and manage smart transducers and Transducer Electronic Data Sheet

(TEDS) formats. It defines a set of implementation-independent application programming interfaces (API).

2. The Generic Sensor API W3C Candidate Recommendation (published on12 December 2019) [31] – A framework for exposing consistently sensor data to the open web in a consistent way, by defining a blueprint for writing specifications of physical sensors along with virtual sensors interface that can be extended to accommodate different sensor types.

Interfaces are vital in the IoT because it is at interface level that the permissioning and privacy can be set, and the properties of the data sets which are relevant to the intended usage become visible and are made available for interaction and manipulation.

In other words, only GUI interfaces can enable the management and operations of fundamental functions such as selection, querying of the continuous streams of data into discrete chunks of information. Typical interface functions include real-time service interfaces providing time-sensitive information, diagnostic and management interfaces for configuration and planning.

4.5 The Sensor Mesh

A mesh resulting from sensor connectivity is called a sensor network, which can be either communicating within a physical boundary such as a building, a car or a city or have open bounds, in unconstrained environments, either physical such as roads, cities or even oceans or cyberspace, such as the Internet. The portion of space covered by a sensor network is referred to as a 'sensing field'. Network AI or edge AI can combine vast amounts of data of the same type, from different sources, such as the readings of acoustic data from different locations, or combine different types of data from one source, such as getting acoustic and visual readings from the same location. Although sensors tend to be low power, they do require some form of electricity to function, and power supply is the main physical constraints for sensor networks, as energy determines their lifespan, which in turn impacts the cost and economics of IoT.

'Energy harvesting' technologies are a key area of innovation enabling 'small but usable' amounts of electrical energy to be harnessed from the environment. They can leverage different sources, intentional, anticipated and accidental or unknown [32] as described below.

Intentional sources: dedicated transmitters enabling control based on application requirements

Anticipated ambient sources: identifiable based on expected reliability and are expected to be available as needed

Unknown ambient sources: sources of energy that are likely to occur accidentally and cannot be planned nor anticipated

4.6 Selection Automation

ML-based AI is used to support the automation of the management of sensors using 'selection schemes', execution patterns that cluster sensor data according to certain criteria or parameters, for example, physical attribute, task utility, as well as optimal power sources [33]. Some examples of how the schemes can be grouped according to coverage, target tracking and localization, single or multiple missions are as follows:

1. Coverage schemes: include selection schemes that are used to ensure sensing coverage of the location or targets of interest
2. Target tracking and localization schemes: include schemes that select sensors for target tracking and localization purposes
3. Single mission assignment schemes: include schemes that select sensors for a single specific mission
4. Multiple mission assignment schemes: include schemes that select sensors so that multiple specific missions are collectively accomplished

Sensors can detect the state of a particular object that they are designed to sense and will trigger communication with other sensors or given systems parameters according to rules, predetermined processes and instruction sets. The ability to merge data from different sources is called sensors fusion, particularly important especially when the readings are of different kinds – say, for example, acoustic and visual – and it consists of computing data from heterogeneous sets with advanced algorithms. It is also the main source of risk in machine learning algorithms because it is here that the integrity of the data and logic can easily become lost and distorted.

Techniques like Kalman filters are used to reduce noise (random variations) and other inaccuracies, to produce more accurate values [34]; however, the same can be used to completely scramble the signal and invalidate the data, without leaving any trail of how automation may have modified the data or the process in any way. This, together with the risks mentioned earlier, is the utmost concern for AI at the edge as well as anywhere else.

4.7 Physical Transport Layer

Initially generated by RFID pioneers, who considered radio frequency identification as one of the best possible mechanisms for enabling communication between devices and identifying unique items using radio waves, RFID tags typically consist of a reader – sometimes also called an interrogator – equipped with minimal memory, which communicates with a transponder. Chipless tags instead of memory use special reflective material to transmit the radio waves and constitute the latest generation of technologies of CRFID. Instead of being printed on a layer of silicon, CRFID are printed on soft, transient and more disposable materials.

RFID tags, chipless or otherwise, are embedded or attached to any object and can be used to enable their tracking and movements. Active and semi-passive tags have batteries; the former use power to transmit radio signals, and the latter depend on readers to power transmission, but they are both used in mid and long range. Passive tags rely on the reader as their power source, have short range, are cheaper and are designed to be disposable, for example, to be used on fast consumer goods. Tags come with three different types of memory: read-write, where data can be added to or overwritten; read-only, which contain only the data that is stored in them when they were manufactured; and worm (write once, read many), which cannot be overwritten.

4.8 Real-Time Data Streams

The ability to gather, query, interpret and make decisions based on real-time knowledge, information and data streams is being integrated with all types of software. AI-driven applications are being developed to merge, clean and map continuous data streams; however, much transformation can happen in real-time data processes to warp the data and introduce bias to the point that outcomes may not correspond to true values. In addition, values that may be true at any given point in time can also be modified at any stage of the transmission process. The accuracy of real-time (or near real-time) data warehousing applications that store real-time data from the IoT stream is a variable that can impact the quality of process execution in any field of IoT application. No AI nor ML learning algorithm can guarantee absolute accuracy of real-time data streams because they are not stored anywhere and thus cannot be checked.

As systems adapt to data coming from ‘sensing fields’, via the web or other communication networks, feeding straight into automated decision support systems, developing methods with the ability to evaluate and monitor accuracy of the outcomes, is also a high level priority.

4.9 Personal Storage Devices and the Privacy Layer

With explainability, transparency and privacy being the primary concerns in the IoT stack, emerging solutions typically try to create a fence that protects the individual from the undesired effects of hyperconnectivity [35].

Commercial and user services are being built around ML algorithms that require constantly updated personal data, alongside technology services that can address the growing ‘privacy preservation’ requirements. Current commercial solutions include VPNs (virtual private networks), peer-to-peer networks for anonymous messaging and mixnets. Each of these categories of solution offers advantages and disadvantages [36], and all of them can be manipulated ubiquitously to either preserve it or to infringe it.

5 The Role of KR in IoT

With the knowledge level [37], fifty years ago computer science started to acknowledge the role of knowledge, cognition and representation at the centre of computation. In the light of recent advances in science and technology including neuroscience which allow for better understanding of the physiological aspects of cognition, the system level is starting to become more cognizable. By computing, visualizing and representing end-to-end system processes and relations, the importance to design whole systems and processes is starting to become better understood. To be able to address the challenges of AI and IoT is now necessary to conceptualise and operationalise this new level of understanding to address the requirements for unification of cognition [38].

5.1 KR to Capture Complexity

There are no sufficiently comprehensive agreed upon models to represent complexity in all of its aspects, nor a single functional language that can capture its diversity of applicable multidisciplinary perspectives. In its early days, AI was conceived with the aim to understand and reproduce computationally intelligent behaviour observed in humans, but more recently AI has been broken down to target individual intelligent functions and has lost its ability to tackle and execute higher level cognitive functions. This is expected to change in the near future, as the shortfalls and limitations of ML-centred AI are overcome.

5.2 The Knowledge System Level

The system level KR results from a convergence between cognitive systems and KR, whereby intelligent processes start with perception, which is highly subjective, and transform progressively into cognition, explicit representation and theory formulation and then become encoded into algorithm and eventually power the AI system (Table 1).

The convergence between AI KR and cognitive systems has started to become visible in research. For example, a survey of cognitive architectures shows how these can be clustered and mapped to respective KR techniques according to their suitability to solve different categories of problems in AI. In addition, the many

Table 1 Complexity factors mapped to dimensions of system complexity

Input function	Data	Architectural component
Perception	Visual (text, images) Audio (sound) Tactile Other	Sensor
Cognitive	Thoughts Beliefs Mental models Reasoning Learning	Text Image
Representational	Symbol Facts Rules Encryption	Language Logic
Epistemic/ontological	Context model	Structure
Procedural/algorithmic	Type of encoding Syntax Programming language	Algorithm
Functional/systemic	Policy Integrity	Process Outcome

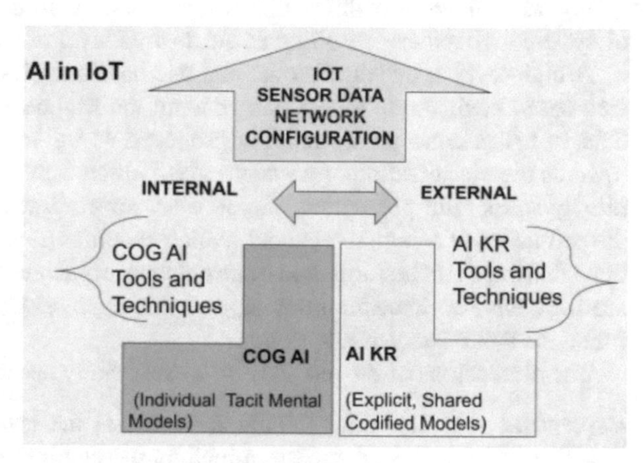

Fig. 5 Convergence between cognitive AI and KR in IoT

cognitive systems functions carried out by different architectures provide a way to group KR according to cognitive features [39].

Cognitive systems and KR constitute two different aspects of edge intelligence, one concerned with capturing the inner descriptions of intelligent agents and the other to make them explicit and shared for the purpose of engineering. Figure 5 shows that cognitive AI is concerned with internal cognitive representations (thoughts, mental states), while KR (knowledge representation) as practiced is concerned mostly with external knowledge which is explicitly codified for the purpose of systematization and serialization. It is expected that AI in IoT will evolve

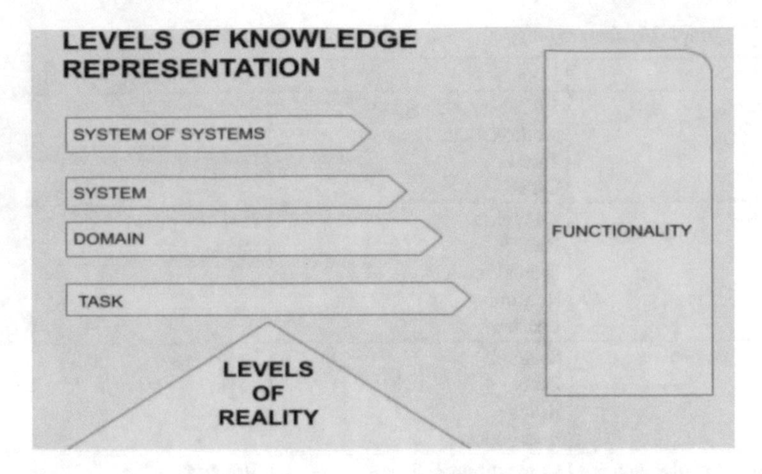

Fig. 6 Levels of knowledge representation

by [40] the explicit codification using KR methods of different types of human intelligent functions.

Figure 6 shows how different KR levels can be used to describe separate layers of system architecture, resulting in the desired level of functionality.

A high-level knowledge model can use heuristic classification that can support task-based navigation. Taking into account the fragmented state of knowledge and data in IoT, a conceptual model is proposed as top level representation as a step towards the visualization of a system view which captures and describes how complexity starts with perception, the core sensorial dimension, which contributes to the formation of a cognitive model, which becomes translated into a representation, that contributes to the formation of an epistemic and ontological levels which in turn are the basis for procedural and algorithmic expressions, that deliver the end-user functions that characterise systems.

The dimensions of the model (Fig. 7) are briefly described as follows:

Perceptual – perception, as defined, occurs at the level of the individual agent, either biological, such as determined by the configuration of the sense organs of the human, animal or vegetable species, or engineered (depending on the sensors types and configuration)

Cognitive – internal (mental models)

Representational – external (explicit, agreed models), caused by limitations in the ability to capture, represent and communicate the representation of knowledge [10]

Epistemic: relating to demarcation, validation of truth and belief and choice of paradigm

Ontological: a container for assumptions, axioms, boundary, definitions, relations and assertions

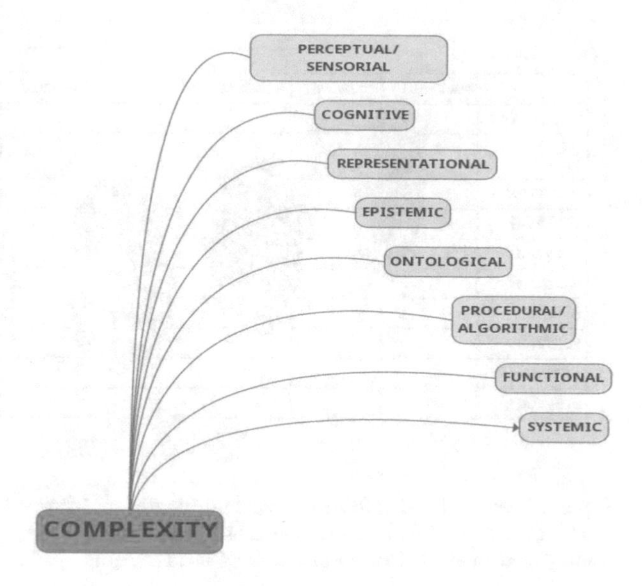

Fig. 7 Dimension of system complexity

Procedural: procedural bias relates to imperfections and flaws in the explicit representation of procedures or the incorrect implementation and management

Functional: a malfunction or dysfunction of the system, whether intentional or accidental

Systemic: any emergent bias that cannot be ascribed to a single cause and is generated by a combination of the other types of bias listed above

5.3 *System Level KR as an Object (KRO)*

Although primarily intended as a cognitive tool for the conceptualization of abstraction at the ontology/taxonomy/vocabulary level of the KR spectrum, the SLKR for edge intelligence (Fig. 8) can be implemented as a KRO, a knowledge representation object [41]. Object-based knowledge representation defines objects and classes in a system, which behaves according to the constrained specified in a model, standard or by a policy. It supports the encapsulation of certain data elements and embeds the mechanism of operation for the information they contain. A working example of such object-based knowledge representation system is found in TROPE [42], a general purpose knowledge repository whose basic entities are objects, i.e. entities modelled based on classes and fields. TROPE can serve as a useful model of KRO as it addresses the need to maximise bidirectional data flows as well as provide encapsulating of critical data, whereby objects are partitioned into concepts, which have attributes constrained under specific permissions that warrant integrity, resulting in principle in knowledge objects that contain all the required information for service identification and process execution and, at least in theory, cannot be

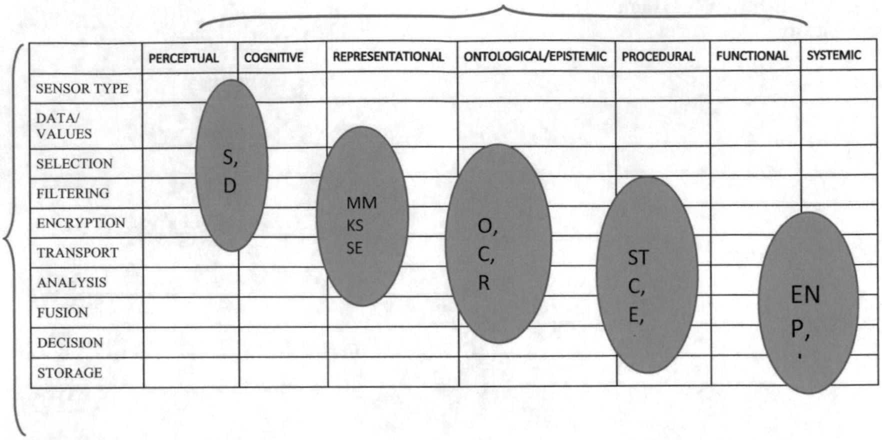

	PERCEPTUAL	COGNITIVE	REPRESENTATIONAL	ONTOLOGICAL/EPISTEMIC	PROCEDURAL	FUNCTIONAL	SYSTEMIC
SENSOR TYPE							
DATA/ VALUES							
SELECTION							
FILTERING							
ENCRYPTION							
TRANSPORT							
ANALYSIS							
FUSION							
DECISION							
STORAGE							

S, D

MM KS SE

O, C, R

ST C, E,

EN P,

Fig. 8 A view of the SL KRO for edge intelligence [43]. S sensor, D data, MM mental model, KS knowledge schema, SE selection criteria, O object, C class, R relation, ST stream, C configuration, E entity, A attribute, EN encapsulation, P policy, I integrity

modified without corrupting the object, therefore its integrity validity making it unusable by the system.

5.4 *Model Validation*

This resulting object is validated using a case-based approach and documented in related work.[2]System level KR is being adopted to capture and disentangle complexity in diverse use cases such as algorithmic bias, neuroscience and other use cases in science and engineering.

The system level knowledge representation object can be evaluated by applying to modelling the complexity in diverse fields of application, from supporting the development of algorithmic bias, (as in the p7003 IEEE standard) to research in systems neuroscience [43].

Combined and applied to the IoT stack, system level KR can be used as a diagnostic tool to detect gaps, faults and flaws that corrupt model integrity in IoT systems, as well as a reference model for cognitive support. The main intended benefit, however, is to support the development of management interfaces for personal data storage devices and other privacy solutions that enable users to access control and interact with the different layers of the IoT system, so that edge intelligence can finally be in the hands and minds of the intended users.

[2] https://ohbm.github.io/osr2020/speakers/paola_di_maio.html

6 Conclusion

AI applications for IoT are increasing the complexity of the scenarios, and corresponding risks associated with the lack of explainable, accountable AI are widely discussed.

In consideration of such risks, and in support of the explicitly represented shared knowledge that describes and specifies in sufficient detail the spectrum of technologies, possible configurations and modes of operations of IoT-enabled devices as well as of the AI layers therein embedded, a system level knowledge representation object is proposed that provides an end-to-end view of the components and processes involved in the IoT be it anywhere in the cloud, at the edge of the network or somewhere foggy in between. Future work includes the further refinement and continuous evaluation of adequate system level knowledge representation mechanism for the IoT and corresponding embedded AI technologies and the drilled down specification of its key components.

References

1. Ashton K (2009) That 'internet of things' thing. RFID J 22(7):97–114
2. Zhang J, Xu C, Gao Z, Rodrigues JJ, Albuquerque V (2021) Industrial pervasive edge computing-based intelligence IoT for surveillance saliency detection. IEEE Trans Ind Inform 17:5012–5020
3. Kiran WS, Smys S, Bindhu V (2020) Enhancement of network lifetime using fuzzy clustering and multidirectional routing for wireless sensor networks. Soft Comput 24(15):11805–11818
4. Azeta AA, Azeta VI, Misra S, Ananya M (2020) A transition model from web of things to speech of intelligent things in a smart education system. In: Data management, analytics and innovation. Springer, Singapore, pp 673–683
5. Radanliev P, De Roure D, Nurse J, Montalvo RM, Burnap P (2019) Standardisation of cyber risk impact assessment for the internet of things (IoT). arXiv preprint arXiv:1903.04428
6. Mihailescu RC, Davidsson P, Eklund U, Persson JA (2018) A survey and taxonomy on intelligent surveillance from a system perspective. Knowl Eng Rev 33. https://www.cambridge.org/core/journals/knowledge-engineering-review
7. Radanliev P, De Roure DC, Maple C, Nurse JR, Nicolescu R, Ani U (2019) Cyber risk in IoT systems. MPRA paper. https://mpra.ub.uni-muenchen.de/92569/ last retrieved online 1 October 2021
8. Moongilan D (2019) 5G internet of things (IoT) near and far-fields and regulatory compliance intricacies. In: 2019 IEEE 5th world forum on internet of things (WF-IoT). IEEE, pp 894–898
9. Li R, Mo T, Yang J, Jiang S, Li T, Liu Y (2020) Ontologies-based domain knowledge modeling and heterogeneous sensor data integration for bridge health monitoring systems. IEEE Trans Ind Inform 17(1):321–332
10. Müller VC (2020) Ethics of artificial intelligence and robotics. Stanford Encyclopedia of Philosophy. https://plato.stanford.edu/entries/ethics-ai/
12. Arogundade OT, Abayomi-Alli A, Misra S (2020) An ontology-based security risk management model for information systems. Arab J Sci Eng 45(8):6183–6198
13. Scharei K, Heidecker F, Bieshaar M (2020) Knowledge representations in technical systems–a taxonomy. arXiv preprint arXiv: 2001.04835

14. Yigitcanlar T, Desouza KC, Butler L, Roozkhosh F (2020) Contributions and risks of artificial intelligence (AI) in building smarter cities: insights from a systematic review of the literature. Energies 13(6):1473
15. Davis R, Shrobe H, Szolovits P (1993) What is a knowledge representation? AI Mag 14(1):17–33
16. Di Maio P (2021) In: Gruenwald L, Jain S, Groppe S (eds) Leveraging artificial intelligence in global epidemics. Academic Press, Elsevier (forthcoming). https://www.elsevier.com/books/leveraging-artificial-intelligence-in-global-epidemics/gruenwald/978-0-323-89777-8
17. Bonomi F, Milito R, Natarajan P, Zhu J (2014) Fog computing: a platform for internet of things and analytics. Springer International Publishing, Cham, pp 169–186
18. Deng S, Zhao H, Fang W, Yin J, Dustdar S, Zomaya AY (2020) Edge intelligence: the confluence of edge computing and artificial intelligence. IEEE Internet Things J 7(8):7457–7469. https://doi.org/10.1109/JIOT.2020.2984887
19. Park J, Samarakoon S, Bennis M, Debbah M (2019) Wireless network intelligence at the edge. Proc IEEE 107(11):2204–2239. https://doi.org/10.1109/JPROC.2019.2941458
20. Abedin SF, Alam MGR, Kazmi SA, Tran NH, Niyato D, Hong CS (2018) Resource allocation for ultra-reliable and enhanced mobile broadband IoT applications in fog network. IEEE Trans Commun 67(1):489–502
21. Sakaguchi K, Haustein T, Barbarossa S, Strinati EC, Clemente A, Destino G, Heath RW Jr (2017) Where, when, and how mmWave is used in 5G and beyond. IEICE Trans Electron 100(10):790–808
22. Buyle R, Taelman R, Mostaert K, Joris G, Mannens E, Verborgh R, Berners-Lee T (2019) Streamlining governmental processes by putting citizens in control of their personal data. In: International conference on electronic governance and open society: challenges in Eurasia. Springer, Cham, pp 346–359
23. Foster I, Kesselman C, Nick J, Tuecke S (2002) The physiology of the grid: an open grid services architecture for distributed systems integration. Technical report, Global Grid Forum
24. Haller A, Janowicz K, Cox SJ, Lefrançois M, Taylor K, Le Phuoc D et al (2019) The SOSA/SSN ontology: a joint W3C and OGC standard specifying the semantics of sensors, observations, actuation, and sampling. Semant Web-Interoper Usability Appl IOS Press J 56:1–19
25. Gupta A, Mukherjee N (2017) Can the challenges of IOT be overcome by virtual sensors. In: 2017 IEEE international conference on internet of things (iThings) and IEEE green computing and communications (GreenCom) and IEEE cyber, physical and social computing (CPSCom) and IEEE smart data (SmartData). IEEE, pp 584–590
26. Morgul EF et al (2014) Virtual sensors: web-based real-time data collection methodology for transportation operation performance analysis. Transp Res Rec 2442(1):106–116
27. Nguyen C, Hoang D (2020) Software-defined virtual sensors for provisioning IoT services on demand. In: 2020 5th international conference on computer and communication systems (ICCCS). IEEE
28. Valenzuela-Valdes JF, Lopez MA, Padilla P, Padilla JL, Minguillon J (2017) Human neuro-activity for securing body area networks: application of brain-computer interfaces to people-centric internet of things. IEEE Commun Mag 55(2):62–67. https://doi.org/10.1109/MCOM.2017.1600633CM
29. IEE.E http://ieeexplore.ieee.org/document/8244985
30. P1451.0 – Standard for a smart transducer interface for sensors, actuators, devices, and systems
31. Generic Sensor API W3C Candidate Recommendation. https://www.w3.org/TR/generic-sensor/
32. Zungeru AM, Ang LM, Prabaharan S, Seng KP (2012) Radio frequency energy harvesting and management for wireless sensor networks. In: Green mobile devices and networks: energy optimization and scavenging techniques, vol 13. CRC Press, New York, pp 341–368
33. Chen G, Huang J, Cheng B, Chen J (2015) A social network based approach for IoT device management and service composition. In: 2015 IEEE world congress on services. IEEE, pp 1–8

34. Yu T, Wang X, Shami A A novel fog computing enabled temporal data reduction scheme in IoT systems. In: GLOBECOM 2017: 2017 IEEE global communications conference, vol 2017. IEEE
35. Fernandes E, Paupore J, Rahmati A, Simionato D, Conti M, Prakash A (2016) FlowFence: practical data protection for emerging iot application frameworks. In: 25th USENIX security symposium (USENIX security 16), pp 531–548
36. Diaz C, Halpin H, Kiayias A (2021) The next generation of privacy infrastructure. Nym Technologies SA. https://nymtech.net/
37. Newell A (1982) The knowledge level. Artif Intell 18(1):87–127
38. Newell A (1990) Unified theories of cognition. Harvard University Press, Cambridge, MA
39. Kotseruba I, Tsotsos JK (2020) 40 years of cognitive architectures: core cognitive abilities and practical applications. Artif Intell Rev 53:17–94. https://doi.org/10.1007/s10462-018-9646-y
40. Clancey WJ (1985) Heuristic classification. Artif Intell 27(3):289–350
41. Patel-Schneider PF (1990) Practical, object-based knowledge representation for knowledge based systems. Inf Syst 15(1):9–19
42. Euzenat J (1993) Brief overview of T-tree: the tropes taxonomy building tool. In: 4th ASISSIG/CR workshop on classification research, Columbus, United States, pp 69–87. hal-01401186
43. Di Maio P (2021) Knowledge representation for algorithmic auditing to detangle systemic bias. In: ISIC 2021 proceedings, pp 149–157 (*Best Paper Award*)

AI-Based Enhanced Time Cost-Effective Cloud Workflow Scheduling

V. Lakshmi Narasimhan, V. S. Jithin, M. Ananya, and Jonathan Oluranti

1 Introduction

A new distributed and heterogeneous model of cloud computing described in this paper offers computing resources such as software, platform and infrastructure as a service over a network. The cloud model is more commercially driven than academically (or research) driven and is primarily based on "pay-as-you-go" approach [1]. In a cloud environment, resources and services can be dynamically deployed, allocated, or withdrawn at any given time. The concepts of on-demand resource conception and virtualization [2] is integrated in cloud computing. Scheduling tasks and developing cost models for operating the cloud are critically important. Pricing is generally inversely proportional to processing time [3], and when it comes to reducing processing time, the scheduling costs will almost always rise.

Representing task groups as workflows is a common practice to describe complex applications [4, 5]. Workflow is a logical sequence of the tasks of an application, which is commonly represented by directed acyclic graph (DAG). While typical examples of workflows include instance-intensive, transaction-intensive, and time-intensive, the time-intensive workflows have the need to be executed at the earliest

V. L. Narasimhan
University of Botswana, Gaborone, Botswana

V. S. Jithin
Srikar & Associates, Delhi, India

M. Ananya
Technical University of Munich, Munich, Germany
e-mail: ge25daj@mytum.de

J. Oluranti (✉)
Covenant University, Ota, Nigeria
e-mail: jonathan.oluranti@covenantuniversity.edu.ng

S. Misra et al. (eds.), *Artificial Intelligence for Cloud and Edge Computing*,
Internet of Things, https://doi.org/10.1007/978-3-030-80821-1_13

so that results can be evaluated within the given deadline. The time-intensive workflows also contain interdependent time critical tasks, and the cost in scheduling time critical tasks increases with decrease in completion time. Workflows represent tasks and all their interactions but do not specify any resource provisioning methods. In cloud computing, task scheduling relates to a variety of costs depending on resource consumption such as processor, memory, I/O, etc. [6].

Workflow scheduling algorithms concern the minimizing of the makespan [7–11], and [12] over a (finite) set of resources. A number of papers [13–18] have appeared in this regard on a variety of variants of workflow scheduling. Users of cloud services have various quality of service (QoS) requirements. For example, some users want to execute their workflow within the deadline without any cost constraints, while others would want to optimize the cost and are willing to pay higher price for the service they use. Yet another type of users may not always need to complete workflows earlier than they require; instead, they want QoS-based execution by considering both cost and deadline as constraints. In QoS-based execution, service cost depends on the level of QoS offered. Typically, service providers charge more as the QoS level increases; such users prefer to use lower QoS levels that are adequate to satisfy their needs. When the makespan (overall completion time) of tasks is minimized, the cost of computation increases as more resources will be needed.

The following are the paper's key contributions:

- Development of an AI-based time-constrained early-deadline cost-effective algorithm (TECA) for scheduling time-constrained workflows for completion in the earliest possible time within the deadline as well as cost optimization.
- Development of AI-enhanced versatile time-cost algorithm (VTCA) for QoS-based scheduling of the workflows that have both time and cost constraints. VTCA is versatile to keep the balance between completion time and cost.
- Definition of a QoS AI-based pricing plan which aids the user to select a QoS level.

As in the case of [10, 11], the TECA algorithm minimizes both the completion time of non-pre-emptive tasks and the number of virtual machines used. As and when resources are needed to finish a workflow within a time frame, TECA allocates as many virtual machines as possible and can auto-scale the capacity of existing virtual machines. In contrast, VTCA provides QoS-based scheduling by considering the cost and deadline constraints at the same time. It keeps a balance between cost and completion time. In the QoS-based scheme, a user can select the level of QoS needed based on the price list provided by the service provider.

The remainder of the paper is laid out as follows: Sect. 2 contains a description of the problem and the underlying mathematical models, while Sect. 3 details the QoS-based pricing plan. Section 4 provides our scheduling algorithms and the details on the experimentation and their analysis. Section 5 summarizes related research, while the conclusions and pointers for further work are provided in Sect. 6.

2 Problem Description

Just as in [10, 11], and [19], the NP-hard problem of scheduling-dependent tasks is formulated as rvm|prec|npmtn [4]. This results in the decomposition of tasks of workflow into levels (heights) according to their parallel and synchronization properties [3, 20]. Two categories of tasks in each level exist. The first category is critical tasks, which are usually allocated to high capacity VMs with a view in minimizing their time of execution. The second is the non-critical tasks, in which the execution times of critical tasks become deadlines for non-critical tasks. The aim of TECA algorithm is to reduce the number of VMs used for the allocation of tasks (thereby reducing the pay-for-use costs) [21], while the aim of VTCA algorithm is to exploit cost-effective VMs (i.e., by replacing the VMs with higher cost with lower cost VMs). Other related works in this area include those reported in [22–26].

2.1 Summary of the Mathematical Model – From [10]

Typically, a given set of virtual machines VMs are used to process a given number of workflows n given as the set $W = \{W_1, W_2 \ldots \ldots., W_n\}$ (see [21] for more details). On the other hand, a workflow W_i usually comprises a number of tasks given as the set $T = \{T_1, \ldots, T_k\}$.[2] ***Prec*** is a model describing interdependency among workflows, while **R** represents all the VMs that are not related and parallel to one another, and ***nptmn*** represents all scheduled non-pre-emptive tasks on these VMs.

Every task has a completion time T_i which is denoted CT_i, together with the related cost denoted, $f_i(CT_i)$. The overall model is dependent on the processing of a task T_i on a VM_i; denoted P_{ij}, which minimizes the maximum completion time denoted CT_{max} as well as cost written as $f(CT)$, is given in Eq. (1):

$$P_j = \sum_{i=1}^{n} p_{ij} \qquad j = 1, \ldots\ldots, m \tag{1}$$

Minimizing CT_{max},

$$\sum_{i=1}^{n} p_{ij} \quad \leq \quad CT_{max} \qquad j = 1, \ldots\ldots, m \tag{2}$$

$$P_j \leq CT_{max} \qquad j = 1, \ldots\ldots\ldots\ldots, m \tag{3}$$

Optimally,

$$CT_{max} = \left\{ \max_{i=1}^{n} CT_i, \max_{j=1}^{n} \sum_{i-1}^{n} P_{ij} \right\} \tag{4}$$

Minimizing $f(CT)$,

$$f(CT) = \sum_{i=1}^{n} f_i\,(CT_i) \tag{5}$$

$f_i(CT_i)$ is the cost of task T_i which completes in CT_i time, which needs to be minimized.

2.2 Summary of Definitions – From [10]

The formula for predecessor and successor tasks is as depicted in Eqs. (6) and (7). As can be seen, task N_j is preceded by task N_i and the other way in the case of successor [21].

$$ImPred\,(N_i) = \left\{N_j | N_j \rightarrow N_i\right\} \tag{6}$$

$$ImSuc\,(N_i) = \left\{N_j | N_i \rightarrow N_j\right\} \tag{7}$$

$$Pred\,(Ni) = \cup Nj \tag{8}$$

Quality of service (QoS) concerns meeting user demands, like the overall completion period, cost of service, precision and readiness, and so on. As a result, a service can be specified by its service QoS and functionality only and that the workflow completion time relates to the total completion period of wholly tasks whether parallel and/or serial, while the service cost represents the cumulative cost of virtual machines, network services, security services, etc. We therefore consider the cost of execution on virtual machines since this cost is much higher than other costs and directly depends on scheduling. Similar to the definition given in [10], the cost of implementation is the total service offered through the cloud service provider (i.e., sum of virtual machines) multiplied together with the cost for every unit service. Both deadline constrictions and cost must be fulfilled based on the scheduling scheme.

As described in [10, 11, 19], and [21], Fig. 1 represents a cloud computing workflow scheduling model for which the user inputs are workflow, resource specification and QoS requirements. The scheduling of workflows to the VMS is undertaken by the workflow management system. There are three aspects of the scheduling, namely, resource capacity estimator which approximates the capacities

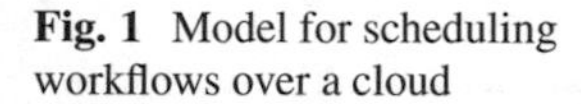
Fig. 1 Model for scheduling workflows over a cloud

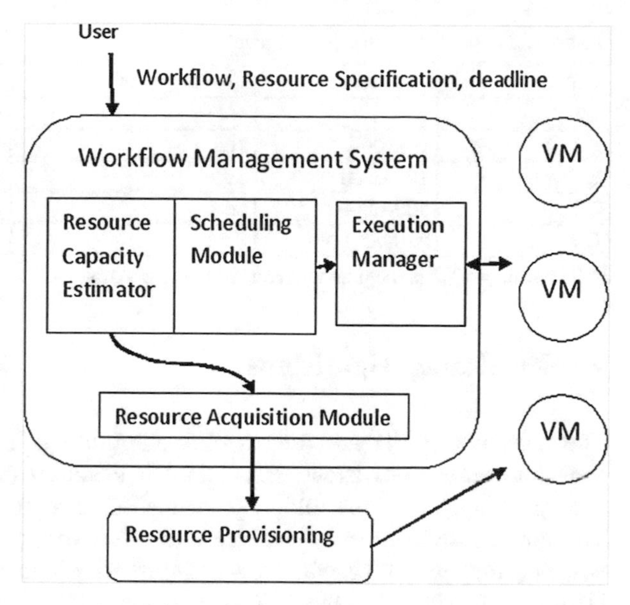

of the active VMs; the resource acquisition module which serves as an interface to both resource provisioning and capacity estimator; and resource provisioning module which detects the VMs to be assigned to the tasks. The module for scheduling determines the order of task to be performed, and at the same time, the execution manager delivers the tasks to the VM.

3 QoS-Based Pricing Plan

The cloud service providers generally propose different pricing plans which vary on their customer's requirements. The service provider requires a cost model with a view to building a pricing plan. AI-based cost model provides methods to evaluate the monthly cost for the cloud computing deployment [27] and [22–26]. The capital and operational expenditure can be used to compute the monthly cost. After the cost model has been developed, the provider can detect all the billable items and identify their atomic units. Forming a pricing plan involves linking costs to billable items and then combining these into various plans. For the workflow that has both cost and time constraints, we require a new pricing plan and scheduling strategy.

In QoS-based pricing, the service provider will analyze the workflow and present a pricing plan, while the client can opt for a QoS level that meets up their requirements. Each QoS level has a distinct price – a higher price for upper QoS levels and a lower price for lower QoS levels. Each QoS level has a different structure of resources, such as virtual machines, bandwidth, etc. Table 1 shows a sample pricing plan with arbitrary values for lower QoS levels.

Table 1 Sample QoS-based pricing plan

QoS level	Usage time	Cost per unit time	Resource configuration
Q1	1–10 t	50c	RC1
Q2	11–100 t	10c	RC2
Q3	101–500 t	5c	RC3
Q4	500–5000 t	1c	RC4
Q5	>10,000 t	0.5 c	RC5

In this example, "t" is the time unit and "c" is the cost unit

4 Scheduling Algorithms

The objective of the AI-based VTCA algorithm is to decrease the number of VMs for resource provisioning and maintain a balance (within user-defined values) among completion time and implementation cost for the selected QoS level. The time-constrained early-deadline cost-effective algorithm comprises steps for ensuring that workflow will be accomplished within the time limit. Similar to [10], our algorithm has the following inputs [21]: (1) a workflow outline, which is represented by an acyclic graph described as $G = (T, E)$, where T denotes the workflow tasks, while E represents the set of all dependencies between the tasks. The second input is the VMs (resources), while the third is QoS parameters such as deadline and cost.

As in [10], a task group refers to a group of tasks allocated to the virtual machines, while task length refers to the number of directives that have to be implemented on the VM. While critical tasks are characterized by their highest completion time as obtained on all machines for a given task group, non-critical tasks are measured by the following parameters:

Latest Completion Time (LCT)
As in [10], the latest completion time is the period upon which a non-critical task completes its implementation. The local critical completion time (LCCT) is the time before which every one of the tasks of a certain height should finish off their execution. The value of LCCT can be considered as the completion time of a critical task of a certain height or product of the completion period of a critical task and a scale factor. Scale factor is used to fix the LCCT so that workflow execution completes within the deadline. LCCT is defined in Eqs. (19) and (23).

$$LCT\ (T_{nc}) = LCCT, \tag{9}$$

where T_{nc} is non-critical task.

Latest Start Time (LST)
As in [10], the LST is the period in which the non-critical task begins its implementation, so as to ensure that it achieves execution on or prior to the completion of critical tasks [21]. The formula for this is as given in Eq. (10).

$$LST\left(T_{nc}, VM_i\right) = LST\left(T_{nc}\right) VM_i) - C\left(T_{nc}, VM_j\right) \tag{10}$$

The calculated value may result in a negative where the task completion time is higher than that of critical tasks [21]. The completion time is the completion time of a non-critical task at a VM_j.

Earliest Start Time (EST)
As in [10], the earliest start point is the period after performance of all predecessors of a task is thorough so that the task under consideration is set to execute.

$$EST\left(T_h\right) = \text{Max}\left\{C\left(T_{h-1}, VM_j\right)\right\} \tag{11}$$

where T_{h-1} is the predecessor of task T_h.

Earliest Completion Time (ECT)
As in [10], the ECT of a task is the earliest possible time to execute a task provided that all machines are prepared to execute.

$$ECT\left(T_{nc}\right) = \text{Min}\left\{C\left(T_{nc}, VM_j\right)\right\} \tag{12}$$

In algorithm, critical tasks are first assigned to a VM using the Max-Min model. Following the allocation of critical tasks, the outstanding non-critical tasks are allocated to the VMs in such a manner that for any VM the completion time of critical is always more than that of the non-critical tasks.

4.1 Versatile Time-Cost Algorithm (VTCA)

The AI-based VTCA algorithm is utilized for QoS-based scheduling wherein the service provider will scrutinize the workflow provided by the client in addition to formulate a price list. Based on the price list, a customer can opt for a QoS level with suitable time limit and price. The aim of VTCA algorithm is to minimize the completion period and implementation cost as much as possible but also maintain an equilibrium between completion time and execution cost for the selected QoS level. Initially an estimated cost per deadline (epd) is computed from user-provided deadline and estimated cost:

$$epd = {}^{e}\!/_{d} \tag{13}$$

where e is estimated cost and d is the deadline. The actual cost and elapsed time are determined after the scheduling of each task group TGh. The virtual machine configuration of each QoS level is different. The cost of the virtual machine declines from higher QoS levels to lower QoS levels – higher levels propose higher configuration of virtual machines.

Height Calculation Hierarchical sorting of tasks [28, 29] is conducted so that the height of a specific task is precisely higher than its predecessors, so that the entire tasks at a provided height can be interpreted as nonrelated tasks. The height ($H(T_i)$) of a node T_i can be labeled as follows:

$$H\left(T_i\right)=\begin{cases}0, & pred\left(T_i\right)=\text{null}\\ 1+\max\left(H\left(pred\left(T_i\right)\right)\right), & pred\left(T_i\right)!=\text{null}\end{cases} \tag{14}$$

where $Pred\ (T_i)$ is all the predecessors of task T_i.

Local Critical Task and Time-Cost Balancing In the scheduling of each task group prior to obtaining the critical task, the actual cost per elapsed time cpt is calculated as follows:

$$cpt={}^{c}\big/_{t} \tag{15}$$

where c is the actual cost and t is the elapsed time. The estimated cost per deadline is evaluated using Eq. (13) and can be applied to choose upper cost limit (*ucl*) and lower cost limit (*lcl*). Let *Va* be the set of available virtual machines and *Vc* be set of virtual machines that completed the billing cycle. *Vc* is the subset of *Va*. The billing cycle of a virtual machine is the period it can be applied for unit price.

If *cpt* is higher than upper cost limit, *ucl* implies that the cost is increased more than the balanced level. Therefore, with the aim of reducing the cost, some of the virtual machines that completed the billing cycle from the set *Vc* is substituted with virtual machines of lower configuration. If *cpt* is less than the lower cost limit, *lcl* implies that the elapsed time is more than the balanced level. Hence, to balance it, virtual machines of lower configuration from set *Vc* is replaced with virtual machines of higher configuration.

A VM is assigned a task only if its predecessor has been implemented in the course of workflow execution. Thus, the completion time of a given task T_i is the combination of execution time of task and ready time of VM. This is as shown in Eq. (16).

$$C\left(T_i,VM_j\right)=R\left(VM_j\right)+E\left(T_i,VM_j\right) \tag{16}$$

T_k is the parent task of T_i.

Now, the local critical task can be calculated in two steps based on the Max-Min algorithm.

(a) *Computing the minimum completion time*:

As stated in [10], the min-time (T_i) of every given group of tasks can be obtained using the formula given in Eq. (17).

$$MInTime\,(T_i) = \text{Min}\left\{C\left(T_i, VM_j\right)|\forall T_i \in TG_h\right\} \tag{17}$$

(b) *Computing maximum completion time T_c of task:*

The maximum completion time of a task can be computed using the formula given in Eq. (18).

$$T_c = \{T_c|\text{Max}\,\{MinTime\,(T_i)\}\} \tag{18}$$

If T_c is the local critical task, then the local critical completion time (LCCT) is as given in Eq. (19):

$$LCCT = \text{Min}\left\{C\left(T_c, VM_j\right)|\forall VM_j \in V_a\right\} \tag{19}$$

During this time-critical workflow for which local scheduling of task at height h must be done, step 3 in Sect. 4.1 achieves the scheduling solution which minimizes the cost.

Scheduling Non-critical Tasks As stated in [10, 11], and [19], LCCT is given as the highest time required by the non critical tasks to execute. Since the client is charged depending upon the number of VMs utilized for the calculation, extra capacity VMs can become available at any time. The task with the least LCT is chosen first for the allotment. Then, this task is allocated to the VM that has the completion time less than or equal to LCCT. Following the allocation, the task is eliminated from the group of tasks. Figure 2 demonstrates all phases of the VTCA algorithm.

Lemma 1 shows theoretically that VTCA can reduce the completion time better than Min-Min algorithm. In the similar manner, we can prove that VTCA is better than any other scheduling policy having different task order, since the completion period of a task group is determined in accordance with the critical task, and in our algorithm critical task is allocated to the highest capacity virtual machine.

Lemma 1 Scheduling using VTCA reduces the completion time than Min-Min algorithm.

Proof Let N be the total no. of tasks and T_i be ith task, where $1 \leq i \leq N$. Let M be the number of levels (heights) in which N-tasks are mapped onto. Let the number of levels follow Poisson distribution with mean μ. Let $CT(T_i, VM_j)$ be the completion time of task T_i on virtual machine where the number of tasks per level follow Poisson distribution with mean λ. x is the average no. of task per level, where $CT(TG_h)$ and $CT(_{wf})$ be the completion time of the task group of height h and workflow, respectively.

Inputs:

- DAG of all tasks W=(T,E) where $T=\{T_1,T_2,\ldots,T_n\}$ where n is the number of tasks and $T_i \rightarrow T_j \in E$ when $pred(T_j)=T_i$
- Set of VMs ready to execute tasks from the selected QoS level.
- Estimated cost per deadline (epd) calculated from user given estimated cost 'e' and deadline 'd'.

Output: Schedule for all tasks.

1: Height calculation

$$H(T_i) = \begin{cases} 0, & pred(T_i) = null \\ 1 + \max\left(H\left(pred(T_i)\right)\right), & pred(T_i) \mathrel{!}= null \end{cases}$$

2: For each h in H

$TG_h \leftarrow$ Tasks that are ready to execute after its parent execution.
$Va \leftarrow$ Set of available virtual machines.
$Vc \leftarrow$ Set virtual machines completed billing cycle.
Find $cpt = {}^{c}/_{t}$
Select *ucl* and *lcl* based on *epd.*
where *lcl* <*epd* <*ucl.*
If(cpt > *ucl*)

- Terminate n no of virtual machines in *Vc* and remove from *Va* and *Vc*.
- Add n virtual machines of lower configuration to *Va*.

If(cpt < *lcl*)

- Terminate n no of virtual machines in *Vc* and remove from *Va* and *Vc*.
- Add n virtual machines of higher configuration to *Va*.

If TG_h is not empty do
Find
$MinTime(T_i) = Min\{C(T_i, VM_j) | \forall\, T_i \in TG_h\}$
$T_c = \{T_c | Max\{MinTime(T_i)\}\}$
$LCCT = Min\{C(T_c, VM_j) | \forall\, VM_j \in V_a\}$
Assign T_c to the VM with Minimum completion time
Remove T_c from TG_h

3: For each task in TG_h

Take a task T_i which has minimum LCT
For each VM
If(C(T_i, S_j)<=LCCT)

- Allocate T_i to S_j
- Remove Ti from TG_h
- Take task T_i which has minimum LCT
- Update S_j ready time

Else
Continue
If $TG_h \neq \phi$
Request a higher capacity VM.
Goto step 3.

Fig. 2 Versatile time-cost algorithm

$CT(_{wf})$ is given by

$$CT(wf) = \sum_{h=1}^{M} CT\,(TG_h) \tag{20}$$

Let $VTCA\_CT(TG_h)$ and $MM\_CT(TG_h)$ be the completion time of task group of height h scheduled using VTCA and Min-Min algorithms, respectively. Let $VTCA\_CT(wf)$ and $MM\_CT(wf)$ be the completion time of workflow scheduled using VTCA and Min-Min algorithms, respectively. Let V_x be the total capacity of virtual machine x. Let $V_x(t)$ be the available capacity of virtual machine x at time t where $V_x(t) \leq V_x$. Let the available capacity of virtual machine follow normal distribution.

Case 1

$$CT\,(T_1) > CT\,(T_2) > \cdots > CT\,(T_x) \tag{21}$$

on VM_j

$$V_1(t) \geq V_2(t) \cdots \geq V_u(t) \tag{22}$$

at time t where $u \geq 1$

In VTCA-Based Scheduling

For case 1, completion time for T_1 is higher .Therefore, T_1 becomes critical task. All the non-critical tasks should be completed before the completion of critical task. Therefore,

$$CT\,(TG_h) = CT\,(T_1)\,, \tag{23}$$

where $0 < h \leq M$.

The critical task will be allocated to the virtual machine with the highest available capacity. Therefore in this case, T_1 is allocated to VM_1.

$$VTCACT\,(TG_h) = CT\,(T_1, VM_1) \tag{24}$$

All the non-critical tasks will be allocated to virtual machine VM_j, where $1 \leq j \leq u$ if and only if there is enough available capacity so that

$$CT\,\left(T_i, VM_j\right) \leq CT\,(T_1, VM_1)$$

In Min-Min-Based Scheduling
In Min-Min scheduling, task T_i with lowest completion time from unmapped tasks will be allocated with the virtual machine with highest available capacity and so on. Therefore in this case, the critical task T_1 will be allocated to the virtual machine with the lowest available capacity VM_x since all tasks are executed in parallel.

From Eqs. (21) and (22),

$$CT\left(T_x, VM_1\right) < CT\left(T_{x-1}, VM_2\right) < \cdots < CT\left(T_1, VM_x\right) \tag{25}$$

Therefore,

$$MMCT\left(TG_h\right) = CT\left(T_1, VM_x\right) \tag{26}$$

From Eqs. (21) and (22),

$$CT\left(T_1, VM_1\right) \leq CT\left(T_1, VM_x\right) \tag{27}$$

From Eqs. (24), (26), and (27),

$$VTCACT\left(TG_h\right) \leq MMCT\left(TG_h\right) \tag{28}$$

From Eq. (20),

$$VTCACT(wf) = \sum_{h=1}^{M} VTCACT\left(TG_h\right) \tag{29}$$

and

$$MMCT(wf) = \sum_{h=1}^{M} MMCT\left(TG_h\right) \tag{30}$$

From Eqs. (28), (29), and (30),

$$VTCACT(wf) \leq MMCT(wf)$$

For the demonstration task allocation, let us consider the same example as given in [10] – as in [21]. Figure 3 shows a workflow example with 12 tasks based on the assumption that all VMs are of the same capacity. Figure 4a, b represent the two different task allocations which are time-based. Figure 4a shows task allocations using our algorithms. As in [21], Max-Min algorithm has also been applied to task allocations as indicated in Fig. 4b. The same overall completion time was obtained since our approach followed the same Max-Min scheduling for critical tasks. But the number of VMs used is lesser than the algorithm of [10] which resulted in

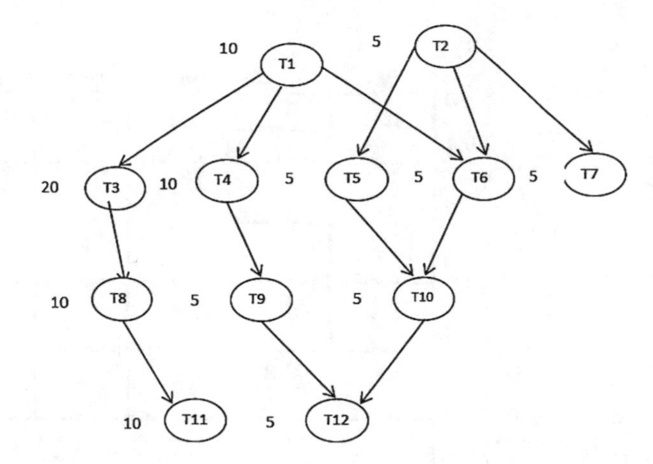

Fig. 3 Sample workflow with 12 tasks

reduced total cost. Our Max-Min algorithm offers the minimum possible time for the execution of non-pre-emptive tasks. This demonstration shows how our scheduling policy reduces the no. of virtual machines used.

4.2 AI-Based Time-Constrained Early-Deadline Cost-Effective Algorithm (TECA)

The AI-based TECA ensures workflow completion within the deadline, and it has three components, and Fig. 5 illustrates all phases of TECA algorithm:

Height Calculation The Height calculation is same as in VTCA.

Critical Task and Total Expected Completion Time In TECA critical task is found out to be using the same method that is used in VTCA. The critical completion time of a task group (CCT_h) is the expected completion time of the critical task of that task group on the virtual machine that gives the least time of completion.

$$CCT_h = \min\left\{C\left(T_C, VM_j\right) \mid \forall VM_j \in V_a\right\} \tag{31}$$

where is the set of available virtual machines.

The total expected completion time of the workflow can be calculated using critical completion time of each task group. Therefore, the total expected completion time,

$$TECT = \sum_{1}^{h} CCT_h \tag{32}$$

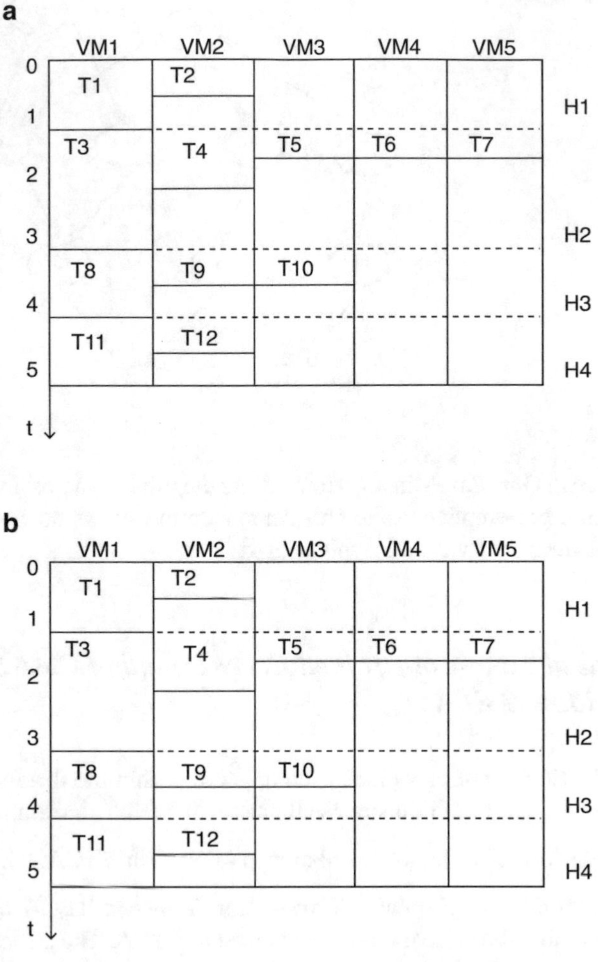

Fig. 4 (**a**) Allocation of tasks to VMs based on our algorithm. Only three VMs are used here. (**b**) Allocation of tasks to VMs based on Max-Min algorithm. Five VMs are used here

Calculate Scale Factor The total expected completion time of the workflow and deadline can be used to calculate the scale factor. Scale factor can be used to ensure that the workflow will be complete within its deadline.

$$\text{Scale factor,}\quad sf = \begin{cases} 1, & TECT \le d \\ d\big/_{TECT}, & TECT > d \end{cases} \tag{33}$$

Scaling of LCCT and Scheduling of Tasks If the total expected completion time of the workflow is higher than the deadline, the value of scale factor *sf* becomes less than one and workflow will not be completed within the deadline. Then, the LCCT

Inputs:

- DAG of all tasks W=(T,E) where $T=\{T_1,T_2,\ldots,T_n\}$ n is the number of tasks and $T_i \rightarrow T_j \in E$ when $pred(T_j)=T_i$
- Set of VMs ready to execute tasks from the selected QoS level.
- Deadline d

Output: Schedule for all tasks.

1: Height calculation.

$$H(T_i) = \begin{cases} 0, & pred(T_i) = null \\ 1 + \max\left(H\left(pred(T_i)\right)\right), & pred(T_i)! = null \end{cases}$$

2: For each h in H

$TG_h \leftarrow$ Tasks that are ready to execute after its parent execution.

If TG_h is not empty do

Find

$\mathrm{MinTime}(T_i) = \mathrm{Min}\{C(T_i, VM_j) | \forall\, T_i \in TG_h\}$

$T_c = \{T_c | Max\{MinTime(T_i)\}\}$

$CCT_h = Min\{C(T_c, VM_j) | \forall\, VM_j \in V_a\}$

$TECT = TECT + CCT_h$

3: Calculate Scale factor

$$sf = \begin{cases} 1, & TECT \le d \\ d/_{TECT}, & TECT > d \end{cases}$$

4: For each h in H

Find $LCCT = CCT_h \times sf$

If ($sf > 1$)

Auto scale the capacity of existing virtual machine or add virtual machines with higher capacity as needed.

Else

Continue

Assign T_c to the VM with Minimum completion time

Remove T_c from TG_h

For each task in TG_h

Take a task T_i which has minimum LCT

For each VM

If($C(T_i, S_j)$<=LCCT)

- Allocate T_i to S_j
- Remove Ti from TG_h
- Take task T_i which has minimum LCT
- Update S_j ready time

Else

Continue

If $TG_h \neq \varphi$

Request a higher capacity VM .

Goto step 4.

Fig. 5 Time-constrained early-deadline cost-effective algorithm

of each task group should be scaled down in order to finish within the deadline.

$$LCCT = CCT_h \times sf \tag{34}$$

Since LCCT is scaled down, and all the tasks of a task group should be finished within its LCCT, the available resources become insufficient. Then, the existing virtual machines should be auto scaled, or virtual machines with higher capacity should be added. The service provider such as Amazon EC2 support auto scaling of virtual machines. The procedure of scheduling critical and non-critical tasks are the same as in VTCA.

5 Experimental Analysis

Scheduling time-critical (non-pre-emptive) workflows is difficult by nature, particularly over a cloud computing system, because one has several hurdles to cross, such as load balancing on VMs, network flow, a group of clouds, scalability, and management of trust, along with others. Furthermore, cloud services have to handle additional issues such as temporal variation in demand, security, cost, speed, etc. Considering all these, experimentation of new techniques in real cloud computing operations becomes technically impossible, but a simulation environment, such as the CloudSim [3, 20, 30], could be of great use. CloudSim has a generalized simulation framework that allows experimenting, simulation, and modelling with the cloud computing infrastructure and application services. It accommodates quick and fast creation of different kinds of entities such as brokers and data centers along with their internal processing, communication mechanisms, and other features as described in [2].

The VTCA and TECA follow the same scheduling policy but have different resource provisioning policies for satisfying different QoS requirements. For a given configuration of resources, the completion time of the workflow and the total number of virtual machines used depend directly on the scheduling policy. We simulated our two scheduling policies over CloudSim and compared them with other familiar scheduling policies such as Min-Min, fair Max-Min, random selection, and queued selection. During simulation, we have assumed that the number of task groups in the workflow and the number of tasks in the task group follow the Poisson distribution. The available capacity of the virtual machine is assumed to be normally distributed. For comparison, consider a workflow W1 whose number of tasks per height follows a Poisson distribution with an average of 10,000, and the number of levels (heights) in the workflow is 5. The size of tasks in W1 is chosen randomly between 100 million instructions (mi) and 60,000 mi.

Figure 6 shows the total completion time of tasks at each height for the workflows W1, where the X-axis represents heights, and the Y-axis represents time. This shows that our scheduling policies perform better than other popular algorithms. Figure 7 shows the number of VMs used at each height for allocation of workflow W1, where the X-axis represents heights, and the Y-axis represents the number of virtual machines. It is again clear that our scheduling policies reduce the number of virtual machines used when compared with other scheduling algorithms. Note again that the cost is directly related to the number of virtual machines used in a scheduling system, i.e., as the number of VMs increases, the cost for scheduling also increases.

Figure 8 shows the overall time of completion used by five sample workflows. The number of tasks per each height of the sample follows a Poisson distribution

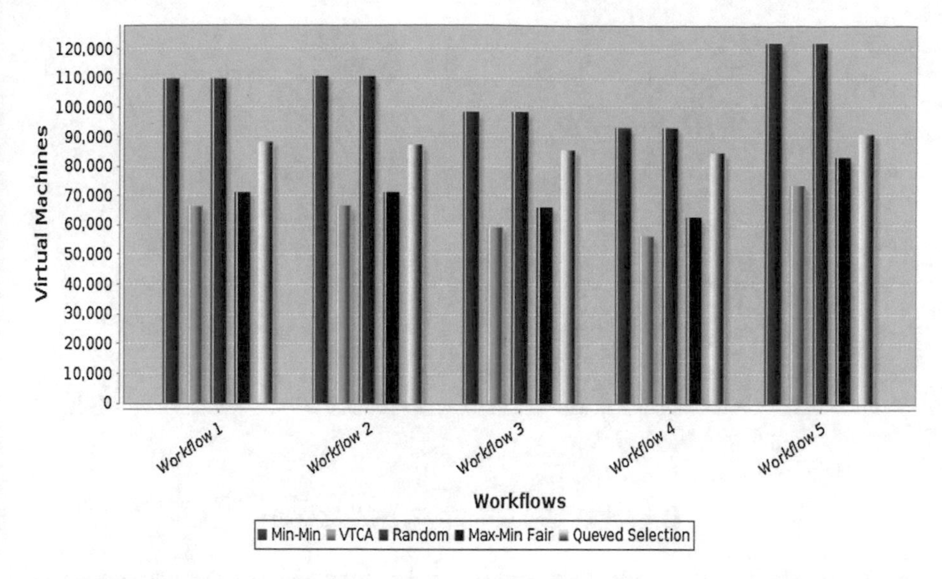

Fig. 6 Comparison of overall completion time of tasks at each height of the workflow W1 based on the proposed scheduling policy and other popular algorithms

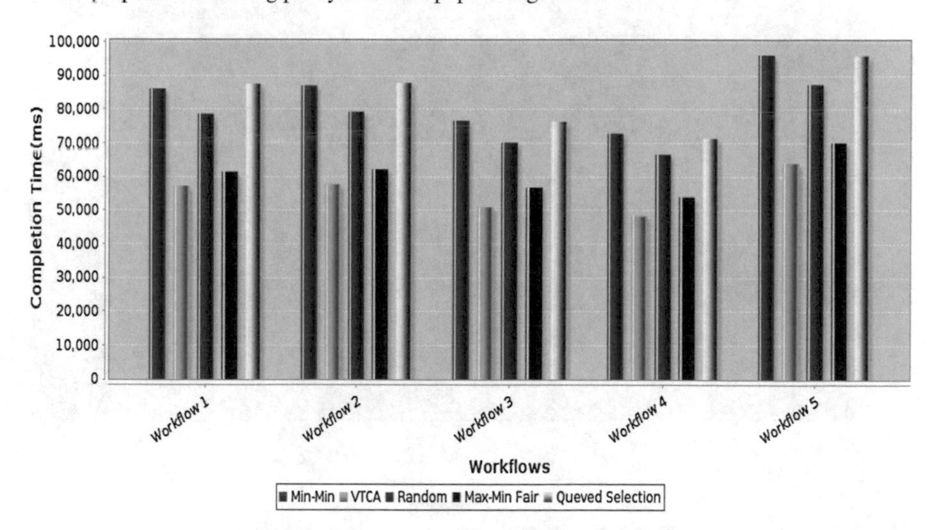

Fig. 7 Comparison of the number of virtual machines used at each height for allocation of workflow W1 based on proposed scheduling policy and other popular algorithms

with an average of 1000, and the number of levels also follows a Poisson distribution with an average of 100. The size of tasks in sample workflows follows a normal distribution with an average of 50,000 million instructions.

Figure 9 shows the total number of virtual machines (cost) involved in processing the tasks for the same sample of five workflows. The results indicate that using

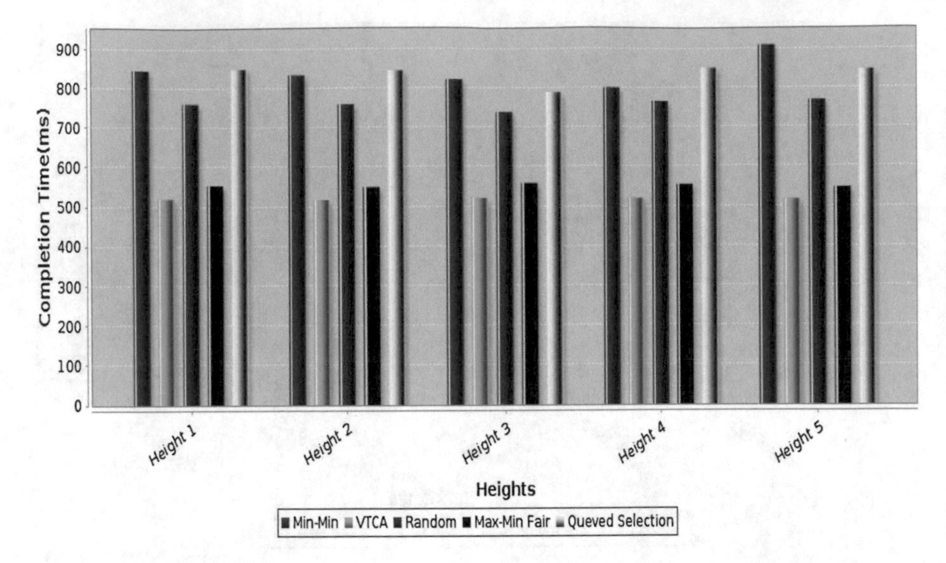

Fig. 8 Comparison of the overall completion time for five workflows using five different methods

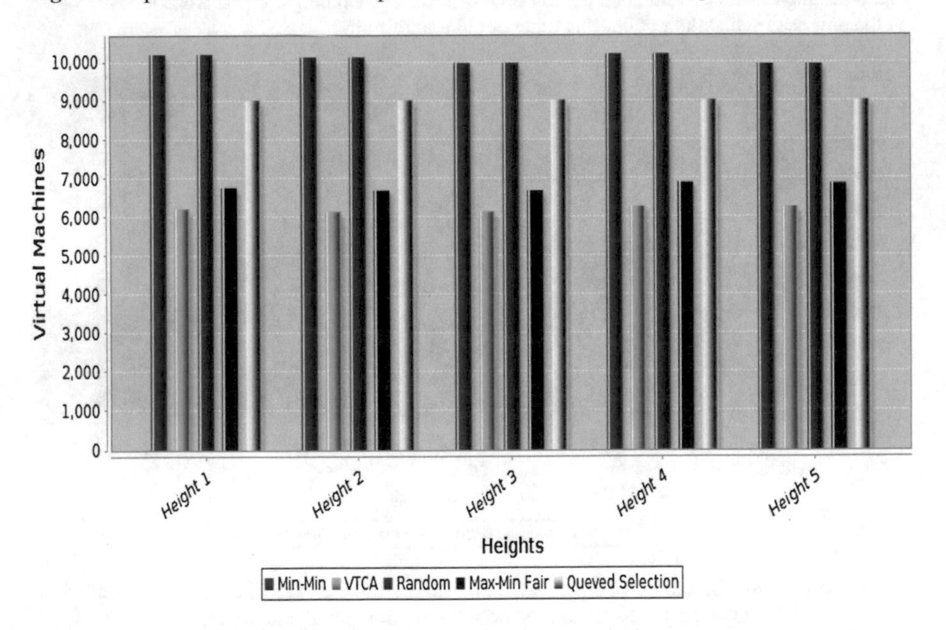

Fig. 9 Comparison of the total no. of virtual machines involved in processing a sample of 5 workflows between proposed scheduling policy and other popular algorithms

our proposed scheduling policies, it is possible to achieve the minimum possible completion time.

The above experimental results show that the VTCA algorithm reduces the completion time and cost when compared with other popular algorithms. Since the

TECA algorithm also uses the same task allocation policy that is used in the VTCA algorithm, TECA also outperforms all other popular algorithms.

6 Conclusions

To reduce the likelihood of scarcity of resources, cloud platforms dynamically allocate resources to customers. This requires scheduling of complex time-critical workflows which can be a daunting task. The reason is that workflow tasks are inter-dependent which gives preference to overall completion time as against individual task completion time. Some workflows have cost constraints along with time constraints. In this paper, the issues of both cost and time deadlines have been addressed. This paper addressed the issues of both cost and time deadlines by proposing two new variants of AI-based scheduling algorithms, named time-constrained early-deadline cost-effective algorithm (TECA) and versatile time-cost algorithm (VTCA); the original algorithms were proposed in [10]. The TECA algorithm reduces the overall time of completion based on height, wherein additional resources are sourced only when it cannot process with the already allocated resources. Further, the overall time of completion is also reduced. This modified algorithm uses less time to schedule tasks to the VMs in comparison with traditional algorithms. VTCA uses the same scheduling policy that is used in TECA, but the resource allocation is done based on the selected QoS level, and it imposes a time-cost controlling policy that balances the completion time and cost evolved. The results of the scheduling policy of both algorithms – as simulated using CloudSim – indicate that the proposed scheduling policies perform better in terms of allocating and managing of processing times of critical workflows. The time of completion of workflows (makespan) is much less than other methods and so are the total costs of consumed resources.

References

1. Buyya R, Pandey S, Vecchiola C (2009) Cloudbus toolkit for market-oriented cloud computing. In: CloudCom '09: proceedings of the 1st international conference on cloud computing, volume 5931 of LNCS. Springer, Berlin/Heidelberg, pp 24–44
2. Pandey S, Karunamoorthy D, Buyya R (2011) Workflow engine for clouds, cloud computing: principles and paradigms. Wiley Press, New York, pp 321–344. ISBN-13: 978-0470887998
3. Dhinesh Babu LD, Gunasekaran A, Venkata Krishna P (2014) A decision-based pre-emptive fair scheduling strategy to process cloud computing work-flows for sustainable enterprise management. Int J Bus Inf Syst 16(4):409–430
4. Plale B et al (2006) CASA and LEAD: adaptive cyberinfrastructure for real-time multiscale weather forecasting. IEEE Comput 39:56–64
5. Magistrale H, Day S, Clayton RW, Graves R (2000) The SCEC southern California reference three-dimensional seismic velocity model version 2. Bull Seismol Soc Am 90:S65–S76

6. Cao Q, Wei Z-B, Gong W-M (2009) An optimized algorithm for task scheduling based on activity based costing in cloud computing. In: 3rd International conference on bioinformatics and biomedical engineering, 2009. ICBBE 2009, 11–13 June 2009, pp 1–3
7. Yuan Y, Li X, Wang Q (2006) Time-cost tradeoff dynamic scheduling algorithm for workflows in grids. In: 10th International conference on computer supported cooperative work in design, 2006. CSCWD '06, pp 1–6, 3–5 May 2006
8. Yu J, Buyya R, Ramanohanarao K (2008) Metaheuristics for scheduling in distributed computing environments. Springer, Berlin
9. Dong F, Akl SG (2006) Scheduling algorithms for grid computing: state of the art and open problems. Technical report, School of Computing, Queen's University, Kingston, Ontario
10. Dhinesh Babu LD, Venkata Krishna P (2013) Versatile time-cost algorithm (VTCA) for scheduling non-preemptive tasks of time critical workflows in cloud computing systems. Int J Commun Netw Distrib Syst 11(4):390–411
11. Narendrababu Reddy G, Phani Kumar S Time and cost-aware method for scheduling workflows in cloud computing systems. In: 2017 international conference on inventive systems and control (ICISC)
12. Kazeem Moses A, Joseph Bamidele A, Roseline Oluwaseun O, Misra S, Abidemi Emmanuel A (2020) Applicability of MMRR load balancing algorithm in cloud computing. Int J Comput Math Comput Syst Theory:1–14
13. Byun E-K, Kee Y-S, Deelman E, Vahi K, Mehta G, Kim J-S (2008) Estimating resource needs for time-constrained workflows. In: Proceedings of the 4th IEEE international conference on e-science
14. Sudarsanam A, Srinivasan M, Panchanathan S (2004) Resource estimation and task scheduling for multithreaded reconfigurable architectures. In: Proceedings of the 10th international conference on parallel and distributed systems
15. Wieczorek M, Podlipnig S, Prodan R, Fahringer T (2008) Bi-criteria scheduling of scientific workflows for the grid. In: Proceedings of the 8th ACM/IEEE international symposium on cluster computing and the grid
16. Huang R, Casanova H, Chien AA (2007) Automatic resource specification generation for resource selection. In: Proceedings of the 20th ACM/IEEE international conference on high performance computing and communication
17. Sulistio A, Buyya R (2005) A time optimization algorithm for scheduling bag-of-task applications in auction-based proportional share systems. In: Proceedings of the 17th international symposium on computer architecture and high performance computing (SBAC-PAD), pp 235–242
18. Ma T, Buyya R (2005) Critical-path and priority based algorithms for scheduling workflows with parameter sweep tasks on global grids. In: Proceedings of the 17th international symposium on computer architecture and high performance computing (SBACPAD), Brazil, Oct 2005, pp 251–258
19. www.indersnline.com (17th Feb 2021)
20. Dhinesh Babu LD, Venkata Krishna P (2014) An execution environment oriented approach for scheduling dependent tasks of cloud computing workflows. Int J Cloud Comput 3(2):209–224
21. Foster I, Kesselman C (1999) The grid: blueprint for a new computing infrastructure. Morgan Kauffman, San Francisco
22. Symons A, Lakshmi Narasimhan V (1996) Development of a method of optimizing data distribution on a loosely coupled multiprocessor system. IEE Proc Comput Digit Tech 143(4):239–245
23. Tan RS, Lakshmi Narasimhan V (1997) Mapping of finite element grids over parallel computers using neural networks. IEE Proc Comput Digit Tech 145(3):211–214
24. Tan RS, Lakshmi Narasimhan V (1999) Time cost analysis of back-propagation ANNs over a transputer network. Eng Intell Syst J
25. Tan R, Lakshmi Narasimhan V (1994) Time cost analysis of back-propagation ANNs over a transputer network. In: Proceedings of the IEEE seventh international. conference on parallel and distributed systems (PDCS), 6–8 Oct 1994, IEEE Press, Los Vegas

26. Tan R (1996) Load balancing studies using ANNs for large scale FEM problems. MEngSc (research) thesis, Department of Electrical and Computer Engineering, The University of Queensland, Australia
27. Cisco Systems White Paper (2010) Managing the real cost of on-demand enterprise cloud services with chargeback models
28. Hou ESH, Ansari N, Ren H (1994) A genetic algorithm for multiprocessor scheduling. IEEE Trans Parallel Distrib Syst 5(2):113–120
29. Qu Y, Soininen J, Nurmi J (2007) Static scheduling techniques for dependent tasks on dynamically reconfigurable devices. J Syst Archit 53(11):861–876
30. Calheiros RN, Ranjan R, Beloglazov A, De Rose CAF, Buyya R (2011) CloudSim: a toolkit for modeling and simulation of cloud computing environments and evaluation of resource provisioning algorithms. Softw Pract Exp 41:23–50. https://doi.org/10.1002/spe.995

AI-JasCon: An Artificial Intelligent Containerization System for Bayesian Fraud Determination in Complex Networks

E. O. Nonum, K. C. Okafor, I. A. Anthony Nosike, and Sanjay Misra

1 Introduction

By definition, telecom frauds refer to any abusive activities on telecom-enterprise assets such as edge-to-cloud networks. In such networks, data characteristics generated through malicious attack vectors appear abnormal when compared with legitimate user behavior. Edge-to-cloud mobile networks that use voice and data algorithms are not free from fraudulent attacks. As such, to validate legitimate user (subscriber) actions, the conventional method used by existing telecom operators is to analyze subscriber usage call detail record (CDR) data. Through CDR, data classification via fraud detection rules is enforced. This approach suffers from complex computation processes and mostly has intrinsic delays in tracking fraudulent records that can mask business operations.

E. O. Nonum
Department of Electrical Electronic engineering, Federal Polytechnic Nekede, Owerri, IMO State, Nigeria
e-mail: enonum@fpno.edu.ng

K. C. Okafor (✉)
Department of Mechatronics Engineering (Computer Systems & Soft Development), Federal University of Technology, Owerri, Nigeria
e-mail: kennedy.okafor@futo.edu.ng

I. A. A. Nosike
Department of Electronic & Computer Engineering, Nnamdi Azikiwe University, Awka, Nigeria

S. Misra
Department of Computer Science and Communication, Ostfold University College, Halden, Norway
e-mail: sanjay.misra@hiof.no

S. Misra et al. (eds.), *Artificial Intelligence for Cloud and Edge Computing*,
Internet of Things, https://doi.org/10.1007/978-3-030-80821-1_14

Industries such as power, telecommunication, and financial institutions, among others, commonly experience frauds at various layers of the system interfaces. However, efforts are put in place to eliminate fraud-related losses [1–5]. In the vertical sectors, fraud and breaches are commonplace. The annual cost of cyber frauds is estimated at $945 billion exceeding 1% of global GDP [6]. Behavior-based fraud detection and threat landscape has continued to evolve [7, 8] within the edge-to-cloud networks.

Fraud is now normal in various mobile communication networks leveraging edge to cloud to process credit card fraud detection [9]. The telecom operators are already seeking ways of reducing this trend due to the economic loss impact factor. Deployment of anti-fraud detection systems to prevent losses requires a clear analysis of occurrence and trend patterns. These systems can analyze client's usage data in telco systems employing computational models and determine suspected fraud subscribers. Computational models with such capabilities are found in Bayesian network [10–12], artificial neural network (ANN) [13], rule inference, and data mining [14–18], including multi-agent systems [19–21]. Optimally, these systems can learn and analyze data generated by subscribers and detect illegal users via analytics. Such alerts can be used by telecom operators to minimize losses caused by frauds. Real-time fraud detection for voluminous telecom datasets fraud detection systems can be achieved to prevent various frauds. Modernized techniques could be applied to address fraud in network systems.

Fraud use case in revenue collection in the Nigerian telco industry offers two major challenges, namely, weak detection and account processing once detection has been completed. A computationally driven approach that ensures predictive response is necessary in this regard. For instance, Bayesian constructs can be applied in addressing such issues with good precision.

In this paper, AI-JasCon framework for fraud detection is presented. First, the related efforts on fraud contexts are presented; the architecture of the use-case operational system is discussed. Second, containerization set theory is introduced briefly. Third, AI-JasCon framework (Java containerized fraud detection method) for the telecom industry is described. The rule extracting algorithm is equally presented. A test case experimental prototype is presented to demonstrate AI-JasCon framework.

This paper is organized as follows. First, we present a taxonomy based on a survey of existing works on frauds within the edge-fog computing layers in Sect. 2. However, none of the works addressed computational containerization perspectives. In Sect. 3, we looked at the fraud detection system model concerning the global architecture. Section 4 discussed system implementation using the JAVA containerization approach. Section 5 highlighted the system design and its implementation. Section 6 concludes the work with future directions.

2 Literature Review

2.1 Existing Research Efforts

This section focused on various existing systems as well as their gaps within complex telecommunication applications. Efforts on fraud mitigation approaches in the telecom domain are increasing. This section then summarized the practices and efforts of telecom systems with their respective fraud detection procedures. Containerization perspectives are also examined.

In [22], the authors discussed anti-fraud measures available in the telecom industry, while establishing the architecture of the telecom system and using rough set theory to deal with anti-fraud detection procedure. The work in [23] discussed a fraud detection method driven by rules, which shared CDR data based on data characteristics, viz., average times of local call weekly and average calling duration. The work used a probabilistic model set up to describe user actions, while the parameters of the model were estimated by maximum likelihood estimation.

A threshold value was applied for each data group. This assisted in identifying the legitimate and illegitimate subscribers while monitoring the system to avoid losses. In [24], a fraud detection system based on fuzzy rules and ANN was used for classification and prediction according to history records: subscriber fraud, non-subscriber fraud, defaulting subscriber, and normal subscriber. Similarly, the work in [25] discussed a technique of fraud detection using an ANN training sample for error probability and subscriber classification. The authors in [26] proposed FraudDroid as a novel hybrid approach in detecting frauds in mobile Android apps for identifying fraudulent behaviors. In [27], a new fraud detection algorithm (using Gaussian mixed model (GMM) as a probabilistic model) was employed in speech recognition for fraud detection. Other works on fraud detection are studied in [28–31]. Table 1 highlighted contextual fraud systems in telecom environments.

In most excluded networks outside the edge and cloud zones, fraud and other attack payloads are prevalent. These attack payload-based signatures include denial of service (DoS), distributed denial of service (DDoS), and phishing, [46] among others. Fraud agents may use flooded traffic to disrupt service operations in telecom networks.

2.2 Limitations of Exiting Systems

Most works are still using computationally resource-intensive schemes to achieve fraud detection in non-edge-to-cloud domains. Early warning system for fraud is yet to be fully addressed. A containerized information system can be employed to deal with fraudulent tendencies.

Table 1 Existing work in fraud scenarios in telecommunication networks

References	Fraud scheme	Mitigation type	Risk data/effect	Domain
C. Cao et al. [32]	AdSherlock	Detection	URL tokenization/computationally intensive	Mobile/online
D. Cheng et al. [33]	Spatial-temporal attention-based graph network (STAGN)	Detection	3D convolution and detection networks/computationally intensive	Online detection system
C. Wang et al. [34]	Learning automatic window	Detection	Adaptive learning approach/less intensive	Online payment fraud
C. Wang et al. [8]	Network embedding schemes	Detection	Knowledge graph	Online payment
A. Dal Pozzolo et al. [35]	Learning algorithms	Detection	Computational intelligence algorithms	Credit card transactions
S. Ji et al. [36]	Multi-Range Gated Graph Neural Network (MRG-GNN)	Learning	Data mining and statistical techniques	Mobile communication technology
Tarmazakov et al. [37]	Fraud counteraction systems	Fraud detection and prevention	Data synchronization	Communications networks
Mohamed et al. [38]	Backpropagation neural network (BPNN)	Prediction	Fraud risk classification	Telecommunication network
Baharim et al. [39]	Naive Bayes approach posterior	Missing values method/detection	Rule-based classifier	Telecommunication network
Elrajubi et al. [40]	Bypass fraud	Detection	Speaker recognition	Telecommunication industry

Wei Xu et al. [41]	Rule-based system	Fuzzy detection	Fuzzy set-based approach	Telecommunication industry
Arafat et al. [42]	Wangiri telecom fraud	Learning/classification	Extreme gradient boosting algorithm/ensemble classifiers	Telecommunication industry
Özlan et al. [43]	Deep convolutional neural networks	Machine learning system/MLS	Speech recognition engine/text-categorization algorithm	Telecommunication industry
Niu et al. [44]	Manhattan distance/United Intelligent Scoring algorithm	Joint fraud probability/MLS	Lower computational complexity	Telecommunication industry
Zhong et al. [45]	Fraud identification systems	Broad Learning System (BLS)	Neural network	Telecommunication industry

In this case, the AI-JasCon software monitoring system obtains data records from non-homogeneous events in the edge-to-cloud domain while allowing for analysis and processing of records based on prior and posterior probability distributions. Processed event results are stored and can be flagged for alarms and error flags. The major issues in existing fraud detection schemes include:

(i) Data hugeness/complexity and data imbalance, for example, in credit card transactions making it very challenging to mine.
(ii) Weak accuracy in terms of computation for unstructured/imbalanced fraud datasets.
(iii) Most fraud prediction system is largely based on online auction sites which do not accurately represent telecommunication systems.
(iv) Most supervised fraud prediction classifications have the drawback that it requires "absolute certainty" that each event can be accurately classified as fraud or non-fraud.
(v) Statistical fraud prediction classification models (as an unsupervised scheme) do not prove that certain events are fraudulent but only suggest that these events should be considered as probably fraud suitable for further investigation.
(vi) Weak investigation on group activities in respect of fraudulent behavior within the context of network analysis. Accuracy, normalcy, and exact pattern matching in the specific graphic data structure are yet to be addressed.

In this work, AI-JasCon mining classification for fraud contexts will be considered. Light-weighted computational processing is used to gather attributes from the contextual fraud data. The second phase involves the identification of vulnerable cases for fraud detection through software routines. To achieve online detection of telecommunication frauds, this work created a JAVA containerized application (AI-JasCon) deployed on a customer's smartphone as a client interface. Anytime an ingress call is received, the application automatically analyzes the call contents to determine fraud instances.

3 Fraud Detection System Model

It is important to characterize predictive models for fraud elimination. The fraud detection model is a complex summation function of the Bayesian statistical model (predictive function), fuzzy function (sensitivity list), linear discriminant function, and critical value determination. The global system model is shown in Fig. 1. Mathematically, the SG model S_{g_m} is generally given by Eq. (1).

Let predictive function be Bsm_{V_α},
fuzzy function be F_{fm_β},
linear discriminant function be L_{df_μ},

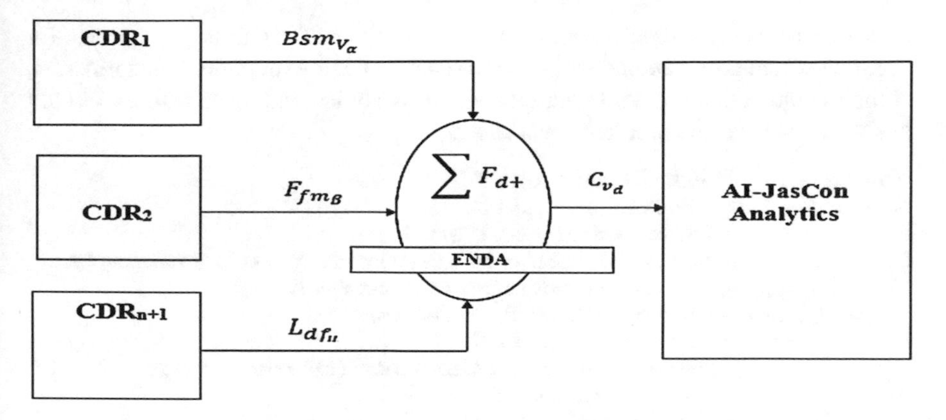

Fig. 1 Global architecture for telecommunication fraud ecosystem

and critical value determination be $C_{v_{dv}}$

$$S_{g_m} = CDR1g_{V_\alpha} + CDR2_{Cm_\beta} + CDRn + 1_{as_\mu} \tag{1}$$

where S_{g_m} are the combinations of several call record datasets relating to the traffic frauds in a communication ecosystem. The terminal point for overall load control is at the Internet analytic cloud network. From the global model equation represented in Eq. (1), the various subsystems are detailed in Eq. (2).

$$F_M = \sum_{ik_{=0}}^{n+j} \prod S_{g_m} * \left(Bsm_{V_\alpha} + F_{fm_\beta} + L_{df_\mu} + C_{v_{dv}}\right) \tag{2}$$

As depicted in Fig. 1, an enhanced neural discriminant algorithm was introduced and used to map all the input call records while ascertaining the call history. The prior and posterior moving average learning is derived for training the calls. Neural classification for AI-JasCon analytics was explored while monitoring the placement of calls with service level agreements (SLA) violations. The enforced linear discriminant controller (ELDC) was used for the classification of fraudulent entities based on their prior and posterior probabilities. The algorithm I is the engine of the elastic call classification which makes it possible for big data integration and security supports. The classification transform algorithm, predictor resource allocation, and class scaling are the novel features introduced to support the fraud optimization track. The call detail record (CDR) represents the exchanged data records for the telecommunications equipment. The record contains various attributes of the call, such as time, duration, completion status, source number, and destination number. In this case, algorithm I enable classification trusts on the AI-JasCon on possible fraud entities. With Algorithm I, fast transactions are coordinated by AI-JasCon making it difficult for fraudsters to execute their activities on a routine basis. It creates a pipeline structural view for CDRs. This makes for accurate schemas and automates processes such as CDR extraction, transformation, combination, validation, and data loading for analytics and visualization. In-built analytics manages the predictive workflow.

The Algorithm I: Fraud Prediction

```
Input: AI-JasCon (CDR1.........n+1)
Output: Bayesian statistical model Bsr ( )
1: {AI-JasCon class model with chaos fraudsters from AI trainingdata}
2: model_bayesain← trainSvmOneClassModel (trFraud)
3: pred_bayesian← predict (tested, modelsvm)
4:        fori = 1 to n_resamplesdo
5:              resample ← trFraudUresample (trNormal, ratio)
6:                      end for
7:                for all L in L_N do
8: if exists L^1in L_F listed up to 7 days before or after L then
9:      pred_final[L] ← 1
10:        return pred_final
11:   end for
12:  end for
```

4 System Implementation

In this section, component instantiation was explored while placing the fraud weighted position. From the edge through the fog layers, the main processes involved in the fraud prediction design include data acquisition, data cleansing and reconstruction, fraud detection algorithm selection, fraud analysis, and output of the results. In analyzing the fraud event, the system uses the acquired subscriber information, viz., the user registration area, ID, age, gender, phone number, etc. The behavioral features are examined. The user category frequently used in the end device is isolated. For the data cleansing and reconstruction, the data of mobile users include user registration ID, user phone number, IMEI number, calling time, calling duration, service category, up traffic and down traffic, and so on. Let us look at the containerization technique.

A. Java Containerization Approach (JCA)

Considering Fig. 1, the software method of fraud detection design is presented in this section. In this case, the foundation of the Java container framework is the Bayesian statistics. This captures the prior, posterior, and joint predictive distribution functions. Essentially, the JCA is a lightweight approach to the application design that removes the need for machine virtualization when deploying the fraud application in Fig. 1.

It entails encapsulating the fraud application in a Java container self-operating environment shown in Fig. 2. This offers merits in the area of loading the application into a virtual machine since the application runs on any suitable physical machine without any dependencies. Containerization with the open-source Docker Netbeans

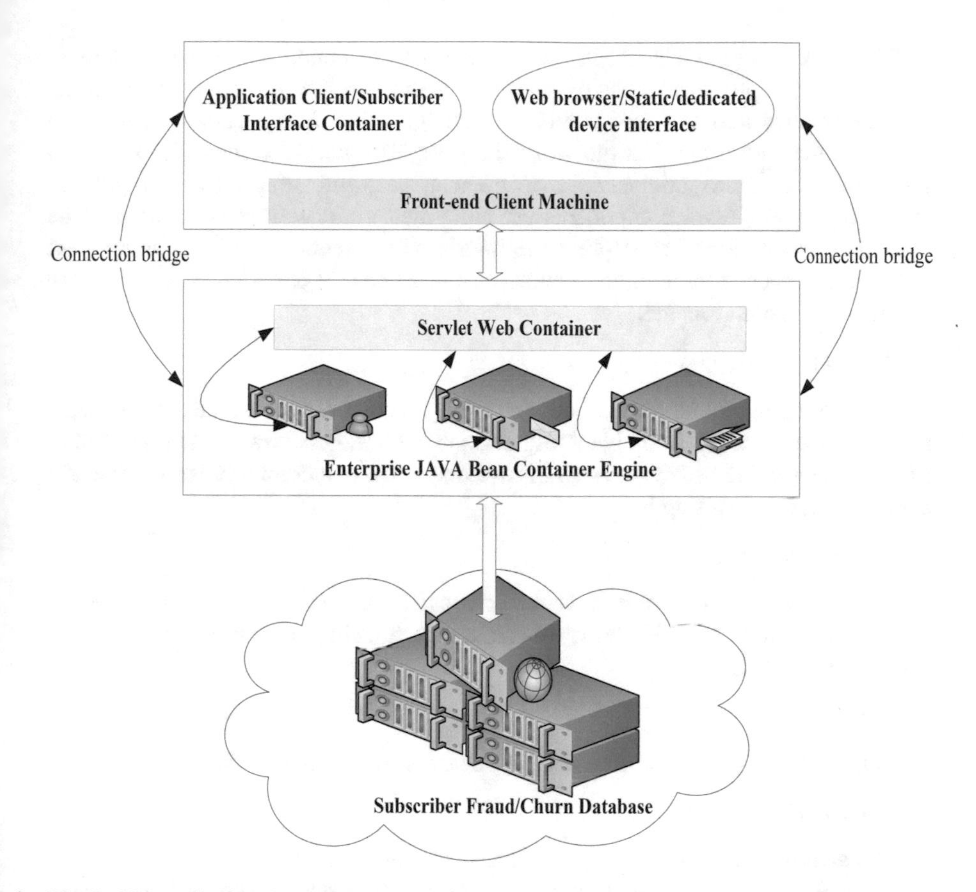

Fig. 2 JCF application deployment components

was used. The JAVA Docker container design is cloned to run on everything from physical computers to virtual machines, bare-metal servers, OpenStack cloud clusters, and public instances, among others. JSR-233 component of JAVA 6 offered an API scripting control which provides uniformity as a pluggable script AI engine.

With modularization, the class containers/data structure that have collections of instantiated objects provides an efficient tracking procedure. Hence, object storage in a coordinated fashion can comply with specific access rules. In the design, the size of the container depends on the number of objects (elements) it can contain. Underlying (inherited) implementations of various container types may vary in size and complexity and provide flexibility in choosing the implementation for the context scenario. The three established characteristic properties of JCA as depicted in Fig. 2 include:

- JCA *access* explains the method of accessing the objects of the JAVA container.
- JCA *storage* depicts container storage data model.
- JCA *traversal* depicts a method of traversing the objects of the container.

The JCA container classes in Fig. 2 were used to handle Bayesian implementation. It was explored to classify data models through the programming libraries, viz., associative and standard containers. The container-based data structures are implemented and cloned. In the work, JCF was explored as a reusable collection data structure. Transaction/state management, multithreading, resource polling, and complex low-level components are simplified via reusability containers. The container was explored to capture the interfaces between Java object component and related object functionality. Hence, the Java module container is deployed to capture all the components.

- *Java collection types*

Using the three major java collection types (viz., ordered lists, dictionaries/maps, and sets), the work realized object migration in the fraud/churn application stored in Fig. 2. Ordered lists solve the issue of inserting items in a certain order and retrieve accordingly, e.g., waiting list.

- *JCF list interface*

In the JCF, lists were realized through the *java.util.List* interface. It depicts a scaled flexible form of an array housing *java.util.ArrayList* and *java.util.LinkedList*.

- *JCF Stack class*

In the enterprise design, *java.util.LinkedList* is used to realize Stacks.

- *JCF queue interfaces*

The *java.util.The queue interface* depicts the queue data structure, which stores elements in the order in which they are inserted. Once the latest addition is made, it goes to the end of the line, and elements are removed from the front. It creates a first-in-first-out system. This interface is implemented using *java.util.LinkedList, java.util.ArrayQueue interface* and *java.util.PriorityQueue*.

5 System Design and Implementation

First, Bayesian computation was previously derived to determine fraud potentials through prior, posterior, and joint probability distributions. AI-JasCon framework was applied to generate predictive fraud detection. In this case, containerization and modularization are introduced using class models and data structures. At the network core layer (shown in Fig. 2), an enterprise management backend applied linear discriminant through fog controllers. These intrinsic processes identified fraud subscribers in the network. AI-JasCon framework provides a successful standard for determining fraudulent interactions while providing a pipeline application programming model for continuous integration and continuous delivery (CI/CD). To show the fraud classification implementation, a novel fraud detection container software tool was implemented with Java and SQL (Mysql), i.e., the JCF (software tool). JCF application deployment components are used. Let's briefly outline the system requirement before discussing the algorithms in Sect. 3.

A. *System Requirement*

For the JCF implementation context (Bayesian fraud detection), the system requirements for effective operation are to provide the following hardware and software components, viz., iCore 5, RAM: 10 GB, HDD: 1 TB, Windows 10, JAVA Netbeans, Apache Server.

B. *JCF Descriptions and Bayesian Computational Application*

The flowchart algorithms are described fully in this section. Essentially, before running the system from the edge network or the computer cloud servers, it is assumed that calls are going on between persons as depicted in Fig. 1. As such, when a user/subscriber logins, the views of those on-going calls by majorly the frequency of each caller and the conversations made by the callers are presented. By employing a determined classifier constant, it is feasible for the system to identify legitimate/non-churn entities as well as fraudulent callers. The software is a classification software designed with Java and SQL (Mysql).

Before executing the system, it is assumed that calls are going on between subscribers in Fig. 1. This is made possible by a call generator (simulation call) interface designed for the system's backend. Thus, it is possible to view the callers, the receivers, the call duration for each call, and the conversation during the call as shown in Fig. 3a, b. This shows the flowchart validation of the fraud detection predictive machine learning model. Using the Bayesian computational models in [19], the JCF was fully developed for seamless integration. The core of the JCF is the Bayesian statistics involving posterior predictive distribution. This was employed as the distribution of possible unobserved values conditional on the observed values (prior).

For a non-homogenous conjugate prior call record, the posterior predictive distribution is determined to belong to the same group as the prior predictive distribution. This is obtained by fixing the updated hyper-parameters associated with the posterior distribution into the model for depicting prior predictive distribution. In the design, the context form of the posterior update model for exponential distributions gives the posterior predictive distribution. This is then applied for fraud detection. Again, the frequency of calls is mapped for the inference frequency.

In this case, the design considered a likelihood function as a set function having parameters of the fraud statistical model within the context of a specific observed fraud data in Fig. 1. The coding design considered the likelihood functions in the call frequency inference for estimating fraud occurrence. For fraud detection, the use of Bayesian inference allowed for the determination of the likelihood of any random variable (given specified prior data or evidence). The features (attributes)

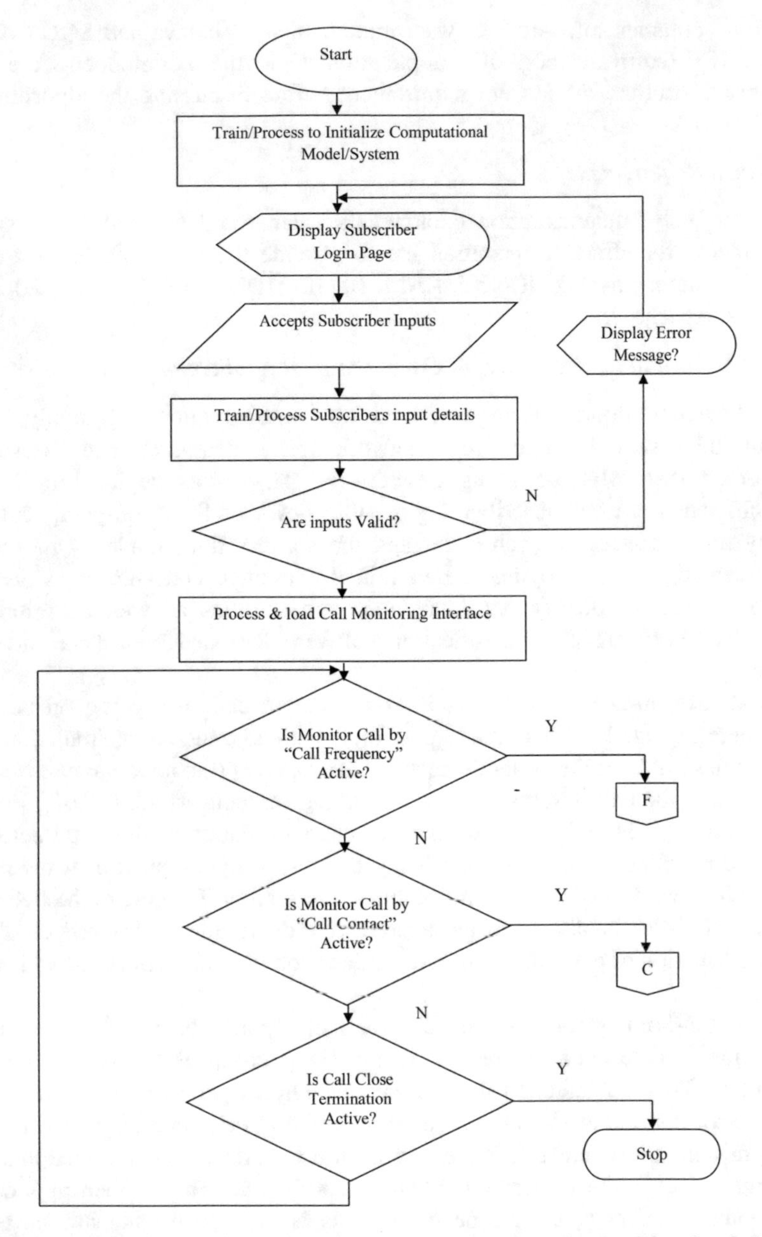

Fig. 3 (**a**) Bayesian Fraud detection container Implementation. (**b**) Bayesian fraud detection container validation flowchart

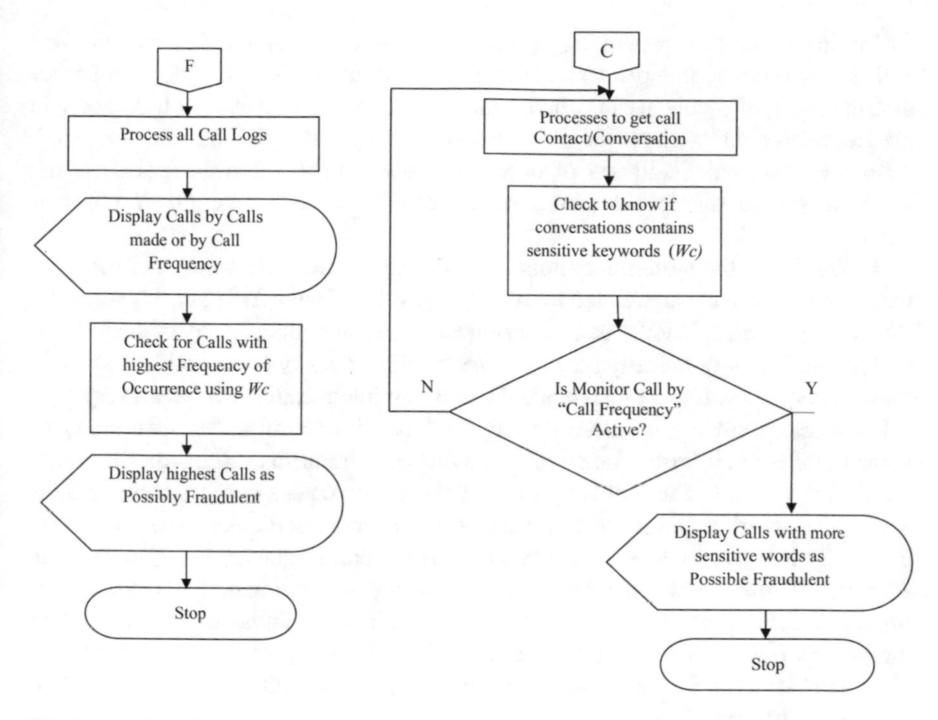

Fig. 3 (continued)

considered in the AI-JasCon emulation experiments include usage, connectivity, and interconnectivity.

Most of the attributes (i.e., usage attributes) are based purely on information extracted from simulated call data records. The second set of attributes (i.e., connectivity attributes) is based on the social ties of a (labeled) individual with existing (labeled) churners. Finally, the interconnectivity attributes are derived from the structural ties between these churners. This work used the JCF classifier to implement a part of the JAVA framework to obtain the predictions. In this case, the theories of Bayesian statistics were fully employed to determine the fraud instances. Figure 3a, b exploited the Bayesian likelihood function. This gives an idea of the possible fraud data in the call parameter. It gives a measure of information presented by the prior parameter value. Since a probability structure on the fraud parameter space can be introduced, it becomes feasible to ascertain a statistical scenario with a huge likelihood value for a given data. The case of low fraud probability is ignored during the detection phase. Using the Bayes rule, the conditional density likelihood can be multiplied by the prior probability density of the parameter and then normalized to give a posterior probability density. This demonstrates the joint predictive distribution for fraud determination.

In this case, the system aggregates the result as a joint distribution over a fixed number of independent identically distributed fraud samples with prior distribution while considering their shared parameters. For every observation in the Bayesian context, the prior and posterior predictive distributions are computed. Also, the marginal likelihood of observed fraud data is determined. This is the joint compound distribution for a set of call $\boldsymbol{X} = x_1 \ \ldots, x_N$ *for* N observations.

In Fig. 3a, b, the system monitors the call conversations for some very sensitive words or phrases that are related to fraud such as *"send your ATM pin," "free call," "browsing cheat," "BVN,"* etc. So whenever the system detects any of such sensitive phrases, it automatically runs the "check call frequency command" to ascertain the call frequency. It must be remarked that a fraudulent caller will have a very high call frequency since a fraudulent entity will like to maximize the opportunity of using the telecom infrastructure without paying or defraud an unsuspecting victim(s) within the period). The call frequency of the caller whose conversation contains a sensitive word or phrase is compared to the average call frequencies of all the callers for that day to observe any anomaly. If the call frequency is higher than the average call frequency [(maximum call frequency + minimum call frequency)/2] for all the callers for that day, as at the time the sensitive phrase was detected, then the system classifies the caller as "a caller with a high probability of fraudulent intension" or else the caller is classified as "a caller with a low probability of fraudulent intention."

Furthermore, using Fig. 3a, b, the Bayesian function demonstration context described above was implemented (AI-JasCon). The code scripts on JAVA NetBeans are demonstrated also. Standard Java middleware container for distributed transaction management, directory services, and messaging is used to test the application. Figure 4 shows the fraud detection container call setup/termination. Figure 5a, b illustrates the fraud detection container call content classification;

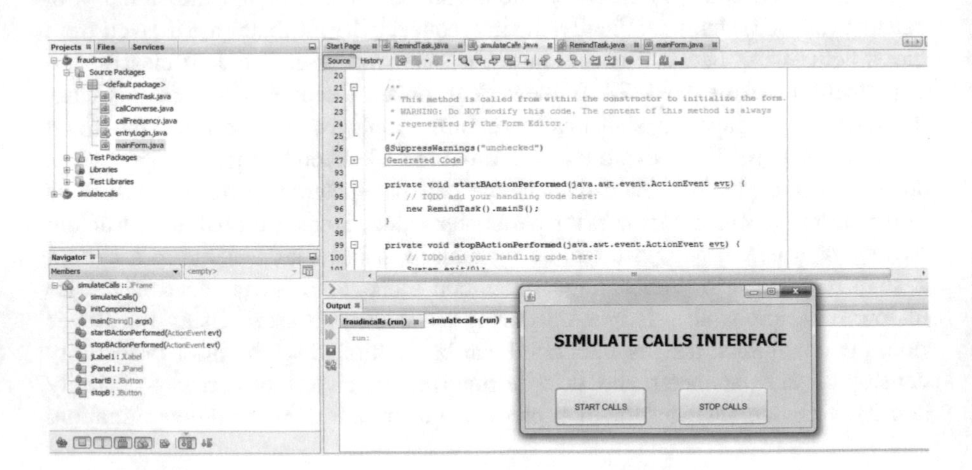

Fig. 4 Compiled code script for AI-JasCon fraud detection container (call setup/termination)

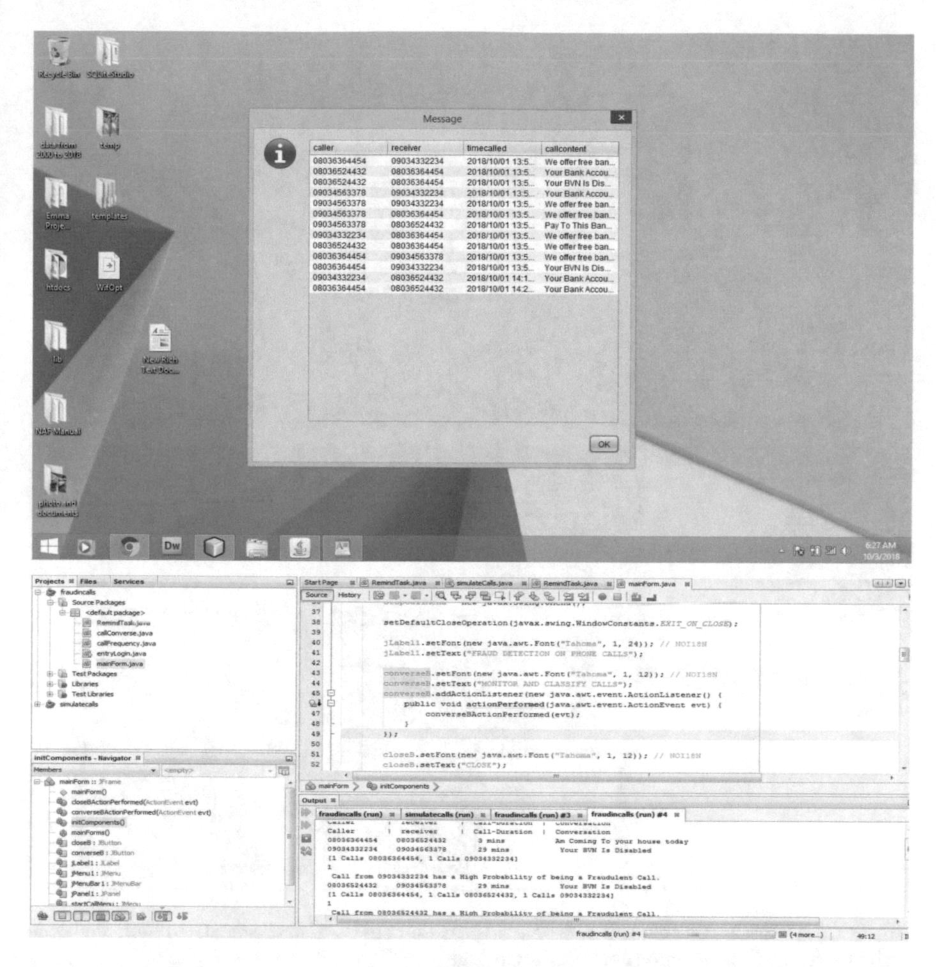

Fig. 5 (**a**) AI-JasCon fraud detection container call content classification. (**b**) Compiled code script for AI-JasCon detection container call content classification

Fig. 6a, b shows the fraud detection container login page for behavioral assessment. Figure 7, b shows the main form fraud detection container login page for call initialization and termination. Figure 8 shows the fraud detection container display page based on call frequency assessment, and Fig. 9a, b shows the fraud detection container call/SMS conversation for behavioral assessment/content form.

With computational Bayesian theory on the code scripts, an AI validation constructs on simulated but real financial transaction was dynamically tested. The machine learning scheme offered a significant detection of fraud. From the refactor mapping ratio, the highest rate of detection is 100% with 72% as the least rate. The mean rate offered 86% with the AI-JasCon algorithm. The proposed approach can be applied to cell signaling, system activation or disabling, logical script computation,

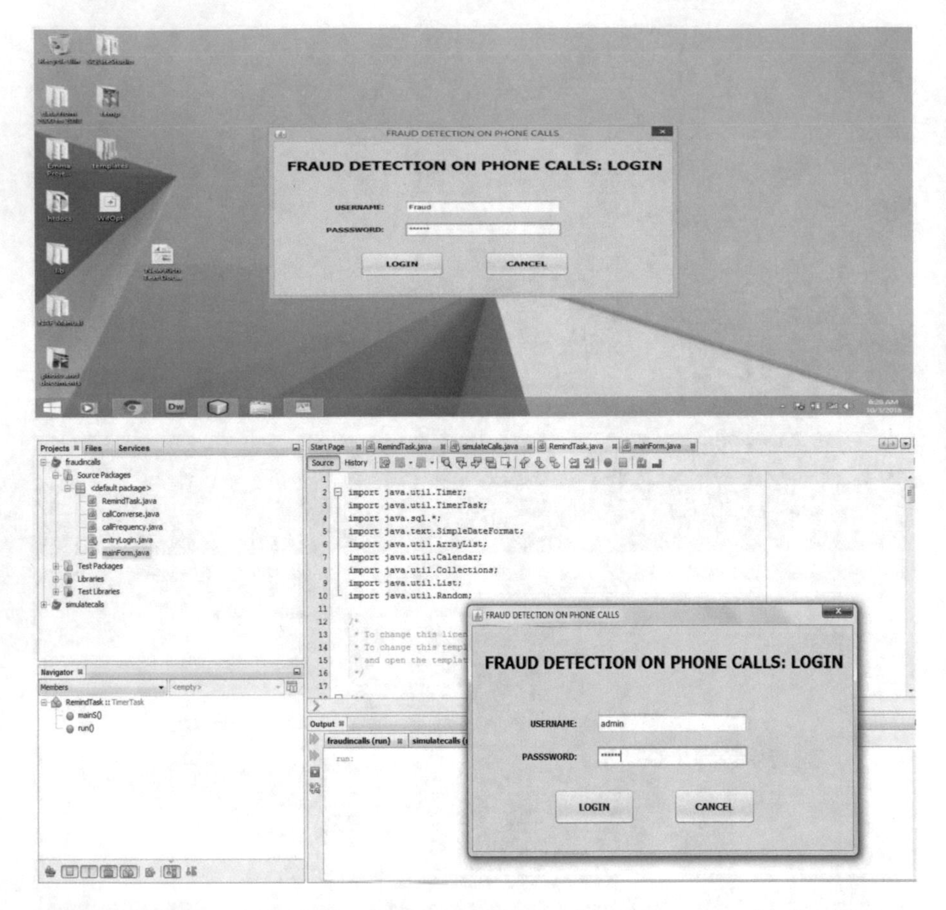

Fig. 6 (**a**) AI-JasCon detection container login page for behavioral assessment. (**b**) Compiled code script for AI-JasCon detection container login page

and non-equilibrium dynamics with hidden Markov model. The computational constraint is the large-scale CDR dimensionality which can be minimized by a summary statistic S. This is the frequency of call switching between endpoints. Besides, the main sources of error in this approach include non-zero tolerance, insufficient statistics summary, dimensionality scale, and model ranking, among others.

6 Conclusion

This paper has discussed the computational Bayesian theory used to achieve the Java containerization framework (AI-JasCon). This focused on the estimation of posterior distributions for fraud predictions, detection, and prevention. Java-

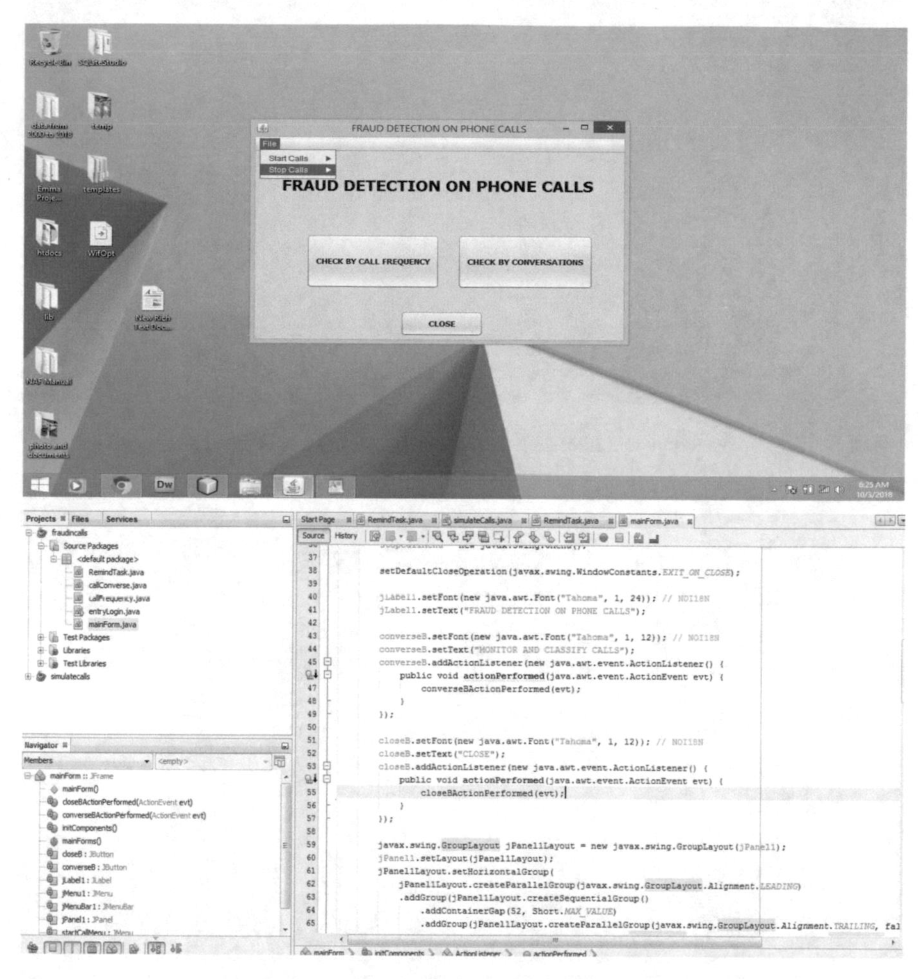

Fig. 7 (**a**) AI-JasCon detection container login page (main form) for call initialization and termination. (**b**)Compiled code scripts for AI-JasCon detection container login page

SQL containerization approach was explored in the context of API scripting for functional predictions. With the simulated prior and posterior distributions, a joint predictive distribution function for fraud detection was achieved. The technique of Java-SQL containerization framework for Bayesian fraud determination in telecommunication networks offers a flexible deployment context for tracking fraud-related activities. This approach is similar to a neural network data mining algorithm which illustrates a set of heuristics and calculations that creates the mining model from its generated data. The design creates the model that allows the Bayesian algorithm to generate and analyze available fraud data. AI-JasCon can identify specific types of patterns/trends. From the extracted information (huge sets of data), fraud data mining using the predictive classification technique is achieved. The system is useful

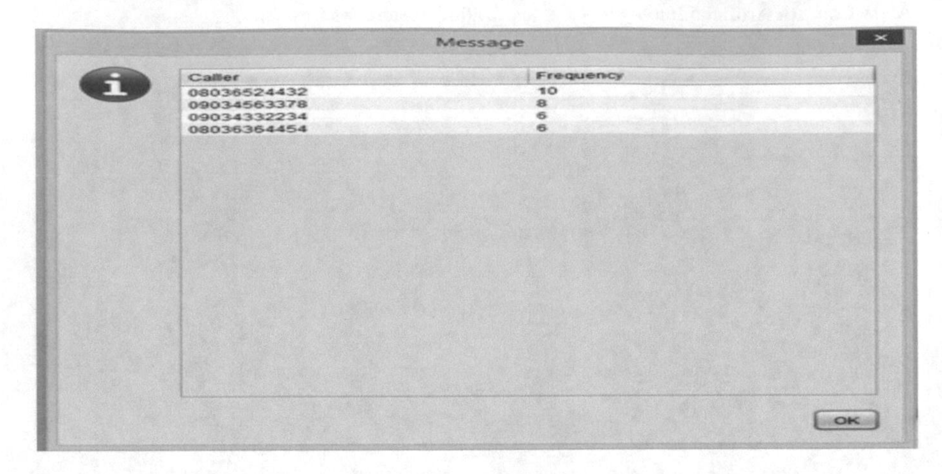

Message

Caller	Frequency
08036524432	10
09034563378	8
09034332234	6
08036364454	6

OK

Fig. 8 AI-JasCon detection container display page based on call frequency assessment

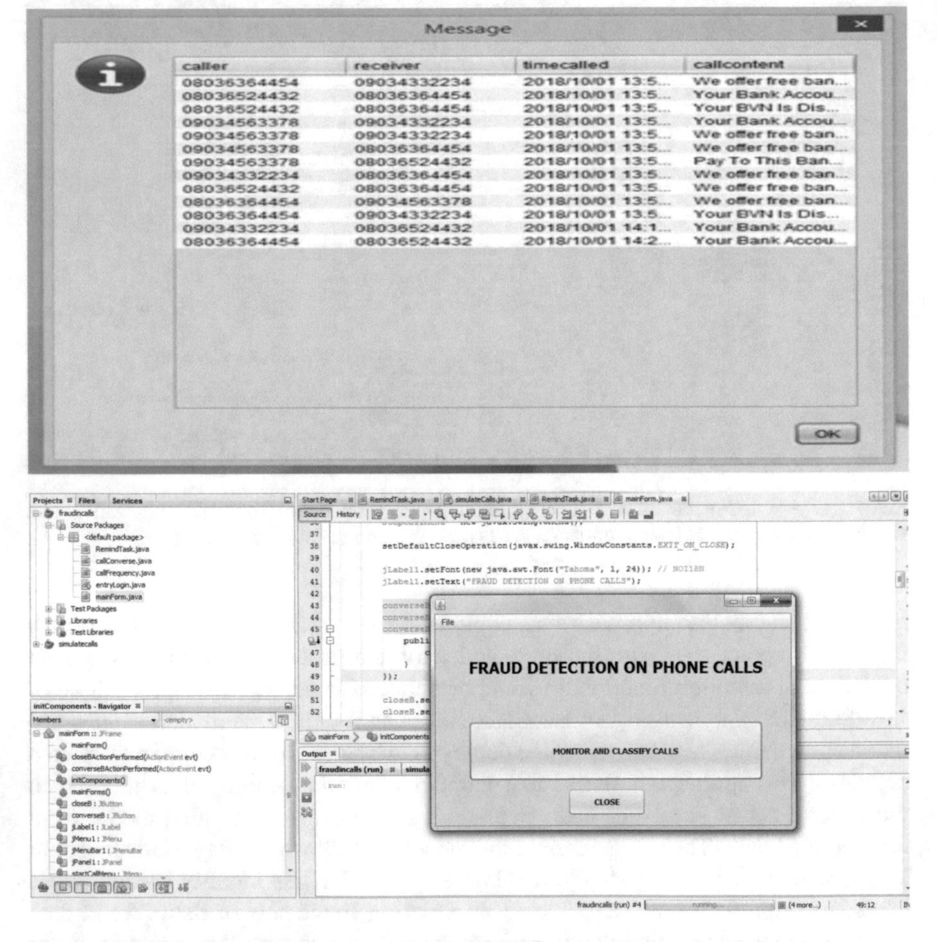

Message

caller	receiver	timecalled	callcontent
08036364454	09034332234	2018/10/01 13:5...	We offer free ban...
08036524432	08036364454	2018/10/01 13:5...	Your Bank Accou...
08036524432	08036364454	2018/10/01 13:5...	Your BVN Is Dis...
09034563378	09034332234	2018/10/01 13:5...	Your Bank Accou...
09034563378	09034332234	2018/10/01 13:5...	We offer free ban...
09034563378	08036364454	2018/10/01 13:5...	We offer free ban...
09034563378	08036524432	2018/10/01 13:5...	Pay To This Ban...
09034332234	08036364454	2018/10/01 13:5...	We offer free ban...
08036524432	08036364454	2018/10/01 13:5...	We offer free ban...
08036364454	09034563378	2018/10/01 13:5...	We offer free ban...
08036364454	09034332234	2018/10/01 13:5...	Your BVN Is Dis...
09034332234	08036524432	2018/10/01 14:1...	Your Bank Accou...
08036364454	08036524432	2018/10/01 14:2...	Your Bank Accou...

OK

Fig. 9 (**a**) AI-JasCon detection container call/SMS conversation for behavioral assessment/content form. (**b**) Code script for AI-JasCon detection container call/SMS conversation

for detecting fraud in any network system architecture based on the subscriber's behavioral synthesis. Future work will integrate the Bayesian statistics with neural network and compare with other computational models for detection reliability and accuracy.

References

1. Zhou C, Lin Z (2018) Study on fraud detection of telecom industry based on rough set. In: IEEE 8th annual computing and communication workshop and conference (CCWC), 8–10 Jan 2018, Las Vegas, NV, USA. https://doi.org/10.1109/CCWC.2018.8301651
2. DuHadway S, Talluri S, Ho W, Buckhoff T (2020) Light in dark places: the hidden world of supply chain fraud. IEEE Trans Eng Manag:1–14. https://doi.org/10.1109/TEM.2019.2957439
3. de Araujo B, de Almeida H, de Mello F (2019) Computational intelligence methods applied to the fraud detection of electric energy consumers. IEEE Lat Am Trans 17(01):71–77. https://doi.org/10.1109/TLA.2019.8826697
4. Yashchin E (2007) Modeling of risk losses using size-biased data. IBM J Res Dev 51(3.4):309–323. https://doi.org/10.1147/rd.513.0309
5. Almehmadi A, El-Khatib K (2017) On the possibility of insider threat prevention using intent-based access control (IBAC). IEEE Syst J 11(2):373–384. https://doi.org/10.1109/JSYST.2015.2424677
6. The Hindu Technology (2020) Cybercrime could cost the world almost $1 trillion in 2020, McAfee says – The Hindu. Online URL https://www.thehindu.com/sci-tech/technology/cybercrime-could-cost-the-world-almost-1-trillion/article33269047.ece
7. Allodi L, Hutchings A, Massacci F, Pastrana S, Vasek M (2020) WACCO 2020: the 2nd workshop on attackers and cybercrime operations co-held with IEEE European symposium on security and privacy 2020. In: 2020 IEEE European symposium on security and privacy workshops (EuroS&PW), Genoa, Italy, pp 427–427. https://doi.org/10.1109/EuroSPW51379.2020.00063
8. Wang C, Zhu H (2020) Representing fine-grained co-occurrences for behavior-based fraud detection in online payment services. IEEE Trans Dependable Secure Comput. https://doi.org/10.1109/TDSC.2020.2991872
9. Saheed YK, Hambali MA, Arowolo MO, Olasupo YA (2020) Application of GA feature selection on naive Bayes, random forest and SVM for credit card fraud detection. In: 2020 international conference on decision aid sciences and application (DASA), Sakheer, Bahrain, pp 1091–1097. https://doi.org/10.1109/DASA51403.2020.9317228
10. Jia X, Yang J, Liu R, Wang X, Cotofana SD, Zhao W (2020) Efficient computation reduction in Bayesian neural networks through feature decomposition and memorization. IEEE Trans Neural Netw Learn Syst. https://doi.org/10.1109/TNNLS.2020.2987760
11. Ye Q, Amini A, Zhou Q (2020) Optimizing regularized Cholesky score for order-based learning of Bayesian networks. IEEE Trans Pattern Anal Mach Intell. https://doi.org/10.1109/TPAMI.2020.2990820
12. Posch K, Pilz J (2020) Correlated parameters to accurately measure uncertainty in deep neural networks. IEEE Trans Neural Netw Learn Syst. https://doi.org/10.1109/TNNLS.2020.2980004
13. Yan S, Yao K, Tran CC (2021) Using artificial neural network for predicting and evaluating situation awareness of operator. IEEE Access. https://doi.org/10.1109/ACCESS.2021.3055345
14. Chiu C, Hsu P-L (2005) A constraint-based genetic algorithm approach for mining classification rules. IEEE Trans Syst Man Cybern Part C Appl Rev 35(2):205–220. https://doi.org/10.1109/TSMCC.2004.841919

15. Wang S, Cao L (2020) Inferring implicit rules by learning explicit and hidden item dependency. IEEE Trans Syst Man Cybern Syst 50(3):935–946. https://doi.org/10.1109/TSMC.2017.2768547
16. Zhang D, Zhou L (2004) Discovering golden nuggets: data mining in financial application. IEEE Trans Syst Man Cybern Part C Appl Rev 34(4):513–522. https://doi.org/10.1109/TSMCC.2004.829279
17. Díaz Vera JC, Negrín Ortiz GM, Molina C, Vila MA (2019) Extending knowledge based redundancy in association rules with imprecise knowledge. IEEE Lat Am Trans 17(04):648–653. https://doi.org/10.1109/TLA.2019.8891930
18. Charte F, Rivera AJ, del Jesus MJ, Herrera F (2014) LI-MLC: a label inference methodology for addressing high dimensionality in the label space for multilabel classification. IEEE Trans Neural Netw Learn Syst 25(10):1842–1854. https://doi.org/10.1109/TNNLS.2013.2296501
19. Mahmoud MS, Khan GD (2018) LMI consensus condition for discrete-time multi-agent systems. IEEE/CAA J Autom Sin 5(2):509–513. https://doi.org/10.1109/JAS.2016.7510016
20. Li H, Li X (2020) Distributed model predictive consensus of heterogeneous time-varying multi-agent systems: with and without self-triggered mechanism. IEEE Trans Circuits Syst I Regul Pap 67(12):5358–5368. https://doi.org/10.1109/TCSI.2020.3008528
21. Shi J (2020) Cooperative control for nonlinear multi-agent systems based on event-triggered scheme. IEEE Trans Circuits Syst II Express Briefs. https://doi.org/10.1109/TCSII.2020.3035075
22. Zhou C, Lin Z (2018) Study on fraud detection of telecom industry based on rough set. In: Proceedings of the IEEE 8th annual computing and communication workshop and conference (CCWC), 8–10 Jan. 2018, Las Vegas, NV, USA. https://doi.org/10.1109/CCWC.2018.8301651
23. Gopal RK, Meher SK (2007) A rule-based approach for anomaly detection in subscriber usage pattern. Int J Eng Appl Sci 3:7
24. Estévez PA, Held CM, Perez CA (2006) Subscription fraud prevention in telecommunications using fuzzy rules and neural networks. Expert Syst Appl 31:337–344
25. Awoyelu L, Adebomi A (2015) Fraud detection in telecommunications industry: bridging the gap with random rough subspace based neural network ensemble method. Comput Eng Intell Syst 6(10)
26. Dong F, Wang H, Li L, Guo Y, Bissyande TF, Liu T, Xu G, Klein J (2018) FraudDroid: automated ad fraud detection for android apps. In: Conference'17, July 2017, Washington, DC, USA 2018. ACM. ISBN 123-4567-24-567/08/06. https://doi.org/10.475/123v4. arXiv:1709.01213v4 [cs.CR] https://doi.org/10.1145/3236024.3236045
27. Yusoff MIM, Mohamed I, Bakar MRA (2013) Fraud detection in telecommunication industry using Gaussian mixed model. In: Proceedings of international conference on research and innovation in information systems (ICRIIS), 27–28 Nov. 2013, Kuala Lumpur, Malaysia. https://doi.org/10.1109/ICRIIS.2013.6716681
28. Filippov V, Mukhanov L, Shchukin B (2008) Credit card fraud detection system. In: Proceedings of the 7th IEEE international conference on cybernetic intelligent systems, 9–10 Sept. 2008, London, UK. https://doi.org/10.1109/UKRICIS.2008.4798919
29. Niu K, Jiao H, Deng N, Gao Z (2016) A real-time fraud detection algorithm based on intelligent scoring for the telecom industry. In: Proceedings of the international conference on networking and network applications (NaNA), 23–25 July 2016, Hakodate, Japan. https://doi.org/10.1109/NaNA.2016.11
30. Tarmazakov EI, Silnov DS (2018) Modern approaches to prevent fraud in mobile communications networks. In: Proceedings of the IEEE conference of Russian young researchers in electrical and electronic engineering (EIConRus), 29th Jan.–1st Feb. 2018, Moscow, Russia. https://doi.org/10.1109/EIConRus.2018.8317111
31. Hines C, Youssef A (2018) Machine learning applied to rotating check fraud detection. In: IEEE proceedings of the 2018 1st international conference on data intelligence and security (ICDIS), 8–10 April 2018, South Padre Island, TX, USA. https://doi.org/10.1109/ICDIS.2018.00012

32. Cao C, Gao Y, Luo Y, Xia M, Dong W, Chen C, Liu X (2020) AdSherlock: efficient and deployable click fraud detection for Mobile applications. IEEE Trans Mob Comput. https://doi.org/10.1109/TMC.2020.2966991
33. Cheng D, Wang X, Zhang Y, Zhang L (2020) Graph neural network for fraud detection via spatial-temporal attention. IEEE Trans Knowl Data Eng:1–1. https://doi.org/10.1109/TKDE.2020.3025588
34. Wang C, Wang C, Zhu H, Cui J (2020) LAW: learning automatic windows for online payment fraud detection. IEEE Trans Dependable Secure Comput:1–1. https://doi.org/10.1109/TDSC.2020.3037784
35. Dal Pozzolo A, Boracchi G, Caelen O, Alippi C, Bontempi G (2018) Credit card fraud detection: a realistic modeling and a novel learning strategy. IEEE Trans Neural Netw Learn Syst 29(8):3784–3797. https://doi.org/10.1109/TNNLS.2017.2736643
36. Ji S, Li J, Yuan Q, Lu J (2020) Multi-range gated graph neural network for telecommunication fraud detection. In: 2020 IEEE international joint conference on neural networks (IJCNN), Glasgow, United Kingdom, pp 1–6. https://doi.org/10.1109/IJCNN48605.2020.9207589
37. Tarmazakov EI, Silnov DS (2018) Modern approaches to prevent fraud in mobile communications networks. In: 2018 IEEE conference of Russian young researchers in electrical and electronic engineering (EIConRus), Moscow, pp 379–381. https://doi.org/10.1109/EIConRus.2018.8317111
38. Mohamed A, Bandi AFM, Tamrin AR, Jaafar MD, Hasan S, Jusof F (2009) Telecommunication fraud prediction using backpropagation neural network. In: 2009 International conference of soft computing and pattern recognition, Malacca, pp 259–265. https://doi.org/10.1109/SoCPaR.2009.60
39. Baharim KN, Kamaruddin MS, Jusof F (2008) Leveraging missing values in call detail record using Naïve Bayes for fraud analysis. In: 2008 International conference on information networking, Busan, pp 1–5. https://doi.org/10.1109/ICOIN.2008.4472791
40. Elrajubi OM, Elshawesh AM, Abuzaraida MA (2017) Detection of bypass fraud based on speaker recognition. In: 2017 8th International conference on information technology (ICIT), Amman, pp 50–54. https://doi.org/10.1109/ICITECH.2017.8079914
41. Xu W, Pang Y, Ma J, Wang S-Y, Hao G, Zeng S, Qian Y-H (2008) Fraud detection in telecommunication: a rough fuzzy set based approach. In: 2008 International conference on machine learning and cybernetics, Kunming, pp 1249–1253. https://doi.org/10.1109/ICMLC.2008.4620596
42. Arafat M, Qusef A, Sammour G (2019) Detection of Wangiri telecommunication fraud using ensemble learning. In: 2019 IEEE Jordan international joint conference on electrical engineering and information technology (JEEIT), Amman, Jordan, pp 330–335. https://doi.org/10.1109/JEEIT.2019.8717528
43. Özlan B, Haznedaroğlu A, Arslan LM (2019) Automatic fraud detection in call center conversations. In: 2019 IEEE 27th signal processing and communications applications conference (SIU), Sivas, Turkey, pp 1–4. https://doi.org/10.1109/SIU.2019.8806262
44. Niu K, Jiao H, Deng N, Gao Z (2016) A real-time fraud detection algorithm based on intelligent scoring for the telecom industry. In: IEEE 2016 international conference on networking and network applications (NaNA), Hakodate, pp 303–306. https://doi.org/10.1109/NaNA.2016.11
45. Zhong R, Dong X, Lin R, Zou H (2019) An incremental identification method for fraud phone calls based on broad learning system. In: 2019 IEEE 19th international conference on communication technology (ICCT), Xi'an, China, pp 1306–1310. https://doi.org/10.1109/ICCT46805.2019.8947271
46. Odusami M, Misra S, Abayomi-Alli O, Abayomi-Alli A, Fernandez-Sanz L (2020) A survey and meta-analysis of application-layer distributed denial-of-service attack. Int J Commun Syst., Wiley 33(18):e4603

Performance Improvement of Intrusion Detection System for Detecting Attacks on Internet of Things and Edge of Things

Yakub Kayode Saheed

1 Introduction

Edge computing (EC) is an emerging technology that allows computation at the network edge on behalf of cloud services on downstream data and on behalf of IoT services of upstream data [1]. In comparison with fog computing (FC) [2], they are interchangeable; EC focuses more on the things part, while the FC focuses on the side of infrastructure. The basis of EC is that computation should take place at the nearness of the data sources. Data storage, computing offloading, caching, and data processing can be performed by edge as well as deliver services and distribute requests from the cloud part to the user part. EC offers numerous benefits [3], such as that EC can be expanded flexibly from a house to different community or city. In applications that require low latency, for example, public safety and health emergency, EC is an appropriate model since it can shorten the network structure and save data transmission time. For applications that are geographically based like utility management and transportation, EC surpass cloud computing as a result of the location awareness.

In EC, rather than delegating data to the cloud, each of the end devices performs an active role in processing the data locally [4]. Consequently, the device, be it an actuator, sensor, or network device, relies on its own storage and computational power to perform processes on data. EC can produce services with greater quality and high-speed response as against cloud computing. EC is more appropriate to be inserted in the company of Internet of Things (IoT) to deliver secure and well-organized services for a huge figure of end users, and EC architecture can be used for the IoT future infrastructure [5].

Y. K. Saheed (✉)
School of Information Technology & Computing, American University of Nigeria, Yola, Nigeria
e-mail: yakubu.saheed@aun.edu.ng

S. Misra et al. (eds.), *Artificial Intelligence for Cloud and Edge Computing*,
Internet of Things, https://doi.org/10.1007/978-3-030-80821-1_15

IoT is a system of interrelated devices connected to the Internet to transform data from one form to the other. It is also a heterogenous device on networks that interlink other devices and exchange information without any form of intervention by human [6]. With IoT, all devices can be linked to the Internet and manage remotely. IoT is influencing the way we live our lives. It assists us in getting better insights into the working of things around us. A smart home is the best example of IoT. Home appliances like air condition, door bells, smoke detector, water heater, and security alarm can be interconnected to share data with the user over a mobile application. Nowadays, IoT is being utilized widely to reduce the burden on humans. It is deployed for wearables, smart homes, watches, smart cars and bracelets, smart retail, smart farming, smart cities, smart healthcare, and smart grids. The future of IoT looks more promising than ever before with such a wide area of applications. In the year 2018 [7], it was estimated that there were 23 billion Internet-connected devices which double the population of the world. Experts say there will be more than 80 billion connected devices by the year 2025 [7].

IoT performs an important role in improving the society like smart homes, smart cities, and smart health system. The omnipresence and being a large scale of IoT systems have brought present-day security issues [8–10]. Additionally, IoT applications and devices work in an unguarded setting; with malevolent intent, an attacker can substantially access these devices [11]. Besides, since IoT devices are connected over unsecure channel, such as wireless network, snooping of confidential information may be possible in network communication [12, 13]. Not only the security problem, but IoT also cannot support advanced security implementation features as a result of their limited computational resources and energy. Consequently, as compared to traditional computing environment system, IoT ecosystems are more susceptible to attacks [14]. This compels research in preventive and detective approaches for IoT environment systems to safeguard IoT devices threats. In order to protect IoT systems against cyber risk, IoT networks and devices should have another line of defense. This purpose is fulfilled by IDSs [15, 16]. There are several approaches based on machine learning for protection against IoT networks proposed in past studies. These studies cover work on IDSs for WSN [17–19], mobile ad hoc networks (MANETs) [20–22], and cyber physical systems [23]. The smart IoT devices can be linked through a wireless or wired linking. The wireless presents security issues, as numerous diverse protocols and wireless connection techniques can be utilized to connect IoT tools. These technologies include ZigBee, Z-Wave, Bluetooth, Near-Field Communication, and Low-Power Wireless Personal Area Networks [24]. The authors in [25] gave a detailed IoT architecture environment with different layers where different attacks can take place. The IoT ecosystem is made up of three (3) layers which are the application layer, network layer, and perception layer [26].

In order to design EC built on IoT, the attributes of IoT and its concepts should be studied firstly. IoT can link facilities and ubiquitous appliances with numerous networks to produce secure and efficient services for all the applications anywhere and anytime [27, 28]. Cloud computing (CC) is an emerging technology that is mature utilized by top information technology companies; for instance, International

Business Machine (IBM), Google, and Amazon are hosting cloud services for providing data storage and computing services. CC delivers the aids of efficiency, flexibility, and ability to use and store data. However, when CC is utilized in IoT, new problems will emerge. At the network edge, EC can produce computing service and storage to devices. An EC node may be network appliances that have the capacity of computing or network link connectivity such as routers, switches, VSC, and servers as explained in the study [5]. These appliances can be positioned where there is network connection to gather data coming from IoT appliances connected with the IoT-based applications.

EC system brings the routine administrations, obtainable by CC, closer to the client end; a big percentage of its protection and security problems are gotten from the cloud. Precisely, many privacy and security problems from cloud are now migrated to diverse layers of the edge architecture [29, 30]. So, spotting attacks is challenging in such an edge-cloud that is distributed in nature. The best method to solve this problem is the addition of IDSs to break down and screen the network traffic and the behavior of the device, in the edge-cloud environment [31–33]. Edge device attacks that are malignant in nature can be the main shortcoming in any edge IoT environment. There are many advantages of edge computing; however, the research on edge computing is still at the infancy stage. There are many aspects that needed to be improved upon, such as the defense strategy and intrusion detection [34, 35]. If the attack is not detected earlier, it can amount to immeasurable losses to the industry [34]. Hence, in the process of protecting data security and IoT devices, it is sacrosanct to construct an intelligent intrusion detection technique at the edge environment. To be precise, IDS is a significant security resistance technique that can be able to spot intrusion efficiently.

Intrusion detection systems (IDSs) is a technology that has been implemented for conventional networks, and consequently, the present IDS approaches for IoT are inadequate to spot different attack types for some reasons [36]. Firstly, the reason is that the present IDS defends against known attacks, that is, they can be defeated by new attacks as they can conceal traditional intrusion detection system [16, 37]. For example, the increase in volume of distributed denial-of-service (DDoS) attacks utilize methods that can spoof the source internet protocol addresses to conceal attacks [38]; consequently, it becomes unnoticed by conventional IDSs. Secondly, the specific attributes of IoT present an issue for IDS creation. IoT needs to host IDS agents, and IoT devices are enormous in number. On the other hand, the IoT environment connects with many actuators and sensors to realize numerous control and monitoring responsibilities [25]. Hence, applying conventional IDS to IoT environment is difficult as a result of its specific attributes, like protocol stacks, network requirements, and limited resource. However, EoT is a novel computing paradigm oriented by the IoT that enables storage, service, and data preprocessing to be moved from the cloud to systems or edge devices like routers, base stations, and smartphones that are nearby the IoT model. Nevertheless, this architectural change causes the privacy and security problems to move to dissimilar layers of the edge architecture. Therefore, spotting intrusion attacks in an environment that is distributed as such is problematic. In this situation, an IDS is required. For these reasons, lightweight IDS models were proposed in this paper to detect both known and zero-day attacks for IoT and EoT. The central contributions of this research are as follows.

- Utilization of minimum-maximum (min-max) normalization technique to ensure all the values of the features in UNSWNB15 dataset are in the same scale.
- Development of dimensionality reduction with principal component analysis (PCA) to select features that are relevant for the classifiers.
- Development of several lightweight IDS models for IoT and EoT that are effective and efficient which achieved 100% accuracy of detection.

This paper is structured as follows. Section 2 is the related work. We highlight the methodology employed in Sect. 3 and present in Sect. 4 the results and discussion. The conclusion is presented in Sect. 5.

2 Related Work

Authors [39] presented random forest (RF) algorithm for recognizing denial-of-service (DoS) attacks in WSN which is a type of networks that are part of IoT, and it consists of sensor nodes. The proposed technique achieved the best score for F1; results are 99% for blackhole, 96% for flooding, 98% for grayhole, 96% for scheduling, and 100% for normal attacks.

The study in [40] presented a supervised learning method for IDS using KNN. The results gave high accuracy with huge total amount of memory requirement. The shortcomings of their proposed model are that accuracy lessens when the size of dataset increases. Also, the dataset used in the work does not reflect contemporary attacks.

In a research by [41], the author introduced a novel dataset of diverse DoS attacks in WSN which is a part of IoT referred to as WSN-DS. The dataset comprises four attacks, namely, grayhole attack, blackhole attack, flooding, and scheduling together with the normal class. Neural network (NN) algorithm was utilized for the classification with tenfold cross-validation technique as the hold-out method. The result obtained in the work suffers from the problem of imbalance with detection rate of 75.6% for grayhole attack.

In the research of [42], PCA was employed as a dimensionality reduction method and SVM as a classification algorithm to classify the attacks in the data. The running time observed in the study is high which actually suggests that the proposed model cannot be adopted for IDS at the network layers of Internet of Things.

The researcher in [43] presented an IDS for WSN utilizing SMOTE and RF algorithm. The experimental findings gave an accuracy of 92.39% for RF, and SMOTE increased the accuracy to 92.57%.

In a similar study of [44], they presented a model based on PCA and LDA to extract attributes in the dataset. The work is two-layer dimensionality reduction and two-stage classification technique for IDS in IoT. The classification algorithms employed are NB and k-nearest neighbor for classification of the attack types. However, the computing complexity is very low likewise the memory resource utilized. Also, the proposed study was evaluated on the NSLKDD dataset which does not represent the modern-day attacks.

The author [45] proposed an IDS with multi-SVM algorithm using mutual information as the feature selection strategy in smart grid. The results performed better than other machine learning methods such as ANN.

3 Materials and Methodology

This section discusses about the proposed IDS-IoT models for detecting IoT application attacks and EoT. The data in the cloud is not safe as this data is vulnerable to attack. The attack can be launched directly on the cloud or via transmission from one IoT sensor to the others. The dataset used in this work is the UNSW-NB15 dataset. This dataset contains up-to-date attack types and was released recently [46]. Data preprocessing was the first analysis performed after the acquisition and loading of the dataset. The data preprocessing is very vital as it helps in eliminating outliers and removing redundant attributes. Min-max normalization technique was used for data preprocessing. The output of the min-max is fed into the feature selection algorithm known as PCA. The PCA selected ten (10) important components out of the forty-nine attributes in the dataset. The reduced dataset is then trained by the LightGBM, decision tree (DT), Gradient Boosting Machine (GBM), k-nearest neighbor (KNN), and Extreme Gradient Boosting (XGBoost) classifiers.

3.1 Data Preprocessing

This is the proposed method's initial step. The aim of preprocessing data is to convert raw data into a format that is simpler and more convenient to use for subsequent processing stages [47]. In this initial step, we normalize the data using minimum-maximum (min-max) technique.

3.2 Normalization Technique

Minimum-maximum (min-max) normalization is an approach where data can fit specifically in a pre-determined limit [48]. It is an approach that affords linear conversion on original array of the data [49, 50]. Assume a group of corresponding scores $\{z_i\}$, $i = 1, 2, \ldots..n$; the normalized scores can be expressed as:

$$zi = \frac{zi - \min}{\max - \min} \tag{1}$$

where min and max are the minimum and maximum values of the attributes.

Min-max technique keeps the original scores of the distribution excluding scaling factor and changes all scores into a mutual array [1,0] [51]; the attributes are normalized in the range [0,1] [52].

3.3 Feature Selection

The feature selection tries to pick the right features in the dataset. The data can be grouped into a set of class features and class target by machine learning algorithm. The usage of irrelevant attributes may lead to low-level performance. Feature selection or simply dimensionality reduction is very essential in machine learning IDSs [53].

3.4 Principal Component Analysis

PCA is a mathematical approach that is used to convert a number of interrelated attributes into a number of unrelated attributes known as principal components [54]. The aim of PCA is to lessen the dimensionality of the dataset. PCA is popularly employed in machine learning algorithms to address the multicollinearity issue and also to decrease the number of attributes utilized as inputs for the machine learning algorithms [55]. It reduces the feature count by means of the orthogonal linear combinations with the important variance [56]. The PCA can be used for both extraction and selection of data attributes [57]. In this research, we use PCA to select the attribute where ten (10) components are selected to lower the dimensionality of the UNSW-NB15 dataset.

PCA is the bedrock for dimensionality reduction for probability and statistics and very useful in data science. It is used to overwhelm redundancy of features in a dataset. These features are low dimensional. Several studies such as in [58, 59], and [60] utilized a mixture of PCA with several classification algorithms to detect anomalies behavior in IoT architecture networks.

3.5 *K-Nearest-Neighbor*

An instant learning approach for classifying objects based on the feature space's closest training instances is known as KNN [61]. KNN is a supervised machine learning classifier that classifies a test sample by first getting the class of the *k* trials close to the test sample [55]. The majority part of these nearest samples is returned as the prediction for sample test. For that test sample, the majority class of those closest samples is returned as the predictor. It is based on non-parametric classification method [62]. The Euclidean distance is a common easy to compute metric popularly used in KNN [63] that work based on the idea of measure of similarity [64]. The Euclidean distance function can be described as in Eq. (2).

$$\text{The Euclidean distance} = \sqrt{\sum_{j=1}^{k} (xj - yj)^2} \tag{2}$$

KNN has been used on diversity of datasets for computer security field with outstanding results. The authors [65] used the DARPA dataset with different values of k and achieved 100% detection accuracy. In a related study, the authors [66] used NSL-KDD dataset with a revised version of KNN which was referred to as fast k-nearest neighbor. Some studies [67–71] have employed KNN classification algorithm generally for anomaly IDS and network IDS for IoT mainly in [44, 72].

3.6 *Decision Tree*

The decision tree (DT) is one of the prevalent prediction and classification algorithms in the machine learning field [73]. The DT is one of the most often used classification algorithm that functions in divide and conquer method [74]. The structure of DT is given in the work [14]. The DT consists of two (2) procedures, referred to as inference and induction, intended for classification and building the model [75]. The DT contains edges, nodes, and leaves. Each attribute is designated by the tree node, and its estimates are designated as the branches initiating from that node. Any feature node that perfectly splits tree into two is seen as the node origin of the tree. Several measures are used for the origin node identification, which perfectly splits training datasets such as the information gain and Gini index [76]. The gain is considered using the formula [74]:

$$\text{Gain } (K, D) = \text{Entropy } (K) - \sum\nolimits_{i=1} Fs(Di) \times \text{Entropy (KDi)} \tag{3}$$

Here, the Gain (*K*, *D*) is the gain of set K after a division over the *D* feature; Entropy (*K*) is the information entropy of set *K*; *D* is the proportion of items possessing *Di* as the value for *D* in *K*; *Di* is the *i*th possible value of *D*; and *KDi* is a subset of K comprising all items where the value of D is Di. The Entropy also referred to as Shannon entropy [77] is given as:

$$\text{Entropy}\ (K) = \sum_{k=1}^{n} Gs(k) \times \log 2\ Gs\ (k) \tag{4}$$

where n is the number of the different attributes in S and $GS(k)$ is the proportional value of k in the set K.

In IDSs, DTs have been used in few studies [78–80]. In the IoT systems, the study in [81] employs DT to spot distributed denial-of-service (DDoS) attacks via network traffic.

3.7 Gradient Boosting Machine

The Gradient Boosting Machines (GBMs) procedure of learning successively fits novel models to produce a more precise approximation of the class response variable feature. The concept behind the principle of this classification algorithm is to set up novel base learners to be significantly connected with the gradient negative of its loss function that is related with the entire ensemble [82]. An example of GBMs is the Friedman gradient boosting algorithm described by the authors [83]. It uses forward stepwise algorithm to understand the optimization learning process [84]. The GBMs have the ability to learn with different loss functions. The high flexibility features of GBMs made it suitable for any specific data-intensive task. GBMs have shown significant results in various data mining and machine learning challenges [85–88]. GBMs were used specifically for Intrusion Detection Systems for IoT in [89].

3.8 Extreme Gradient Machine

XGBoost is an adaptive gradient boosting tree framework that generates sequential decision model trees [56]. In all computing environments, it could compute calculations quicker. With implementation in structured and tabular datasets, the usage of XGBoost has obtained significant popularity. XGBoost was recognized as a supporting technique to enhance the gradient boosting classification algorithm by eradicating values that are missing, removing overfitting problem utilizing the idea of parallel processing technique. The optimization system in the XGBoost is attained by applying tree pruning, parallelization, and optimization of hardware as revealed by the authors [56]. Recently, the authors [90] utilized Genetic Extreme Gradient Boosting (GXGBoost) and fisher score feature selection technique for IoT botnet attack detection.

3.9 *Light Gradient Machine*

The gradient boosting decision tree (GBDT) [91] is a popularly well-known effective algorithm used for classification and regression problems [92–95]. Recently, the author proposed a new GBDT classification algorithm referred to as Light Gradient Boosting Machine (LightGBM) [96]. Light GBM is built on decision tree idea and boosting [97]. Various researches have utilized LightGBM for IDS such as [98, 99], and [100]. The LightGBM has never been used for classification in IDSs for IoT. This paper would serve as the novel study to introduce LightGBM in IDSs for IoT network attacks and EoT.

4 Results and Discussion

4.1 *Evaluation Dataset*

The evaluation of the proposed study is done utilizing the UNSW-NB15 dataset which comprised a hybrid of attack vectors and genuine modern normal. The size of the network packets is around a hundred (100) gigabytes, generating 2,540,044 observations which are recorded in four (4) comma-separate value (CSV) files. Each record includes one (1) class label and forty-seven (47) features. The dataset consists of ten (10) different classes, one (1) normal and nine (9) events of security breaches and malware: Backdoors, Analysis, Exploits, DoS, Reconnaissance, Generic, Fuzzers for anomalous activities, Shellcode, and Worms. The proposed model is evaluated with ten (10) features selected utilizing PCA as dimensionality reduction from the UNSW-NB15 dataset as presented in Table 1.

Table 1 PCA features selected

#	Features name	Description of the feature
46	ct_dst_sport_ltm	Number of rows of the same dstip (3) and sport (2) in 100 rows
33	tcprtt,	Setup round-trip time, the sum of "ackdat" and "synack"
20	Dwin	The destination Transmission Control Protocol (TCP) window value advertisement
45	ct_src_dport_ltm,	Number of rows of the same srcip (1) and the dsport (4) in 100 rows
47	ct_dst_src_ltm	Number of rows of the same scrip (1) and the dstip (3) in 100 records
43	ct_dst_ltm	Number of rows of the same dstip (3) in 100 rows
23	Smeanz	The mean of packet size conveyed by the srcip
24	Dmean	The mean of the packet size conveyed by the dstip
14	Service	Like Hypertext Transfer Protocol (HTTP), ssh, FTP, dns, SMTP
5	Proto	Type of protocol such as UDP and Transmission Control Protocol (TCP)

The UNSW-NB15 was formed by using IXIA perfect tool to extract mixture of contemporary attack and normal activities of network traffic [46]. The dataset is split into two, where 75% of the dataset was used for training of the model and 25% was used for testing the model. The hold-out method used in the analysis was done on sixfold and tenfold cross-validation techniques. The experiments were carried out on a 64-bit Windows 10 Professional operating system, x64-based processor with 8.00 GB of RAM and Intel (R) Core (TM)i5-8250U CPU @1.60 GHz 1.80GHz.

4.2 *Performance Metrics of the Models on Sixfold Cross-Validation*

The performance of the proposed models PCA-LightGBM, PCA-DT, PCA-GBM, PCA-KNN, and PCA- XGBoost on sixfold cross-validation is given in Table 2. The performance of the model was evaluated in terms of the accuracy, area under curve (AUC), recall, precision, F1, kappa and Matthews correlation coefficient (MCC) utilizing sixfold cross-validation as the hold-out method. The experimental results of our findings of the proposed models are presented in Table 5.

As revealed in Fig. 1, the LightGBM outperformed all other algorithms in terms of the accuracy, AUC, recall, precision, F1, kappa, and MCC. The decision tree (DT) performed also better with an accuracy of 99.97, AUC of 99.98, recall of 99.95, precision of 100, F1 of 99.98, kappa of 99.93, and MCC of 99.93. The GBM gave an accuracy of 99.97, AUC of 100, recall of 99.95, precision of 100, Fl of 99.98, kappa of 99.93, and MCC of 99.93. The KNN gave accuracy of 99.97, AUC of 100, recall of 99.95, precision of 100, F1 of 99.98, kappa of 99.93, and MCC of 99.93. Lastly, the XGBoost gave an accuracy of 99.93, AUC of 100, recall of 99.90, precision of 100, F1 of 99.95, kappa of 99.85, and MCC of 99.85.

Table 2 Performance evaluation on sixfold cross-validation method

Algorithms	Accuracy	AUC	Recall	Precision	F1	Kappa	MCC
KNN	99.97	100	99.95	100	99.98	99.93	99.93
DT	99.97	99.98	99.95	100	99.98	99.93	99.93
GBM	99.97	100	99.95	100	99.98	99.93	99.93
XGBoost	99.93	100	99.90	100	99.95	99.85	99.85
LightGBM	100	100	100	100	100	100	100

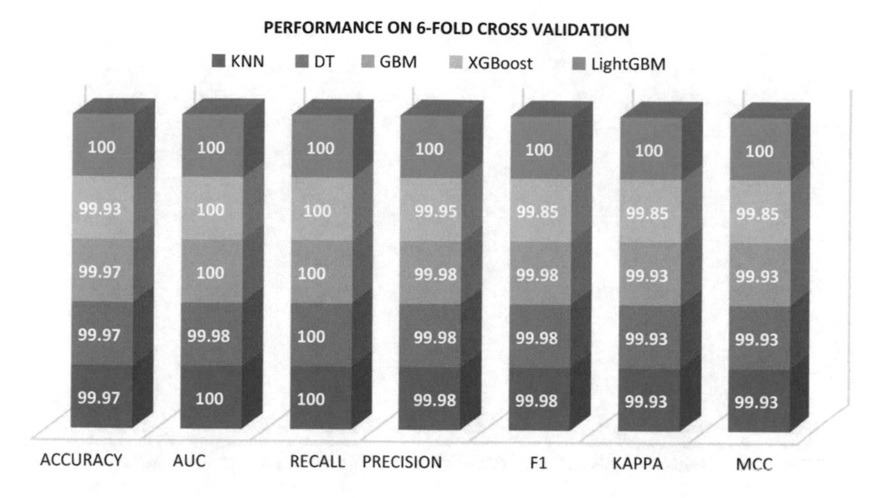

Fig. 1 Performance of the proposed models on sixfold cross-validation method

Table 3 Performance evaluation on tenfold cross-validation method

Algorithms	Accuracy	AUC	Recall	Precision	F1	Kappa	MCC
KNN	99.98	100	99.98	99.99	99.99	99.96	99.96
DT	99.99	99.99	99.99	99.99	99.99	99.97	99.97
GBM	99.99	100	99.99	100	99.99	99.98	99.98
XGBoost	99.98	100	99.98	100	99.99	99.96	99.96
LightGBM	100	100	100	100	100	100	100

4.3 *Performance Metrics of the Models on Tenfold Cross-Validation*

The performance of the proposed models PCA-LightGBM, PCA-DT, PCA-GBM, PCA-KNN, and PCA-XGBoost on tenfold cross-validation is given in Table 3. The performance of the model was done in terms of the accuracy, area under curve, recall, precision, F1, kappa, and Matthews correlation coefficient utilizing tenfold cross-validation as the hold-out method. The experimental results of our findings of the proposed models are presented in Table 3.

As revealed in Fig. 2, the LightGBM outperformed all other algorithms in terms of the accuracy, AUC, recall, precision, F1, kappa, and MCC. The decision tree (DT) performed also better with an accuracy of 99.99, AUC of 99.99, recall of 99.99, precision of 99.99, F1 of 99.99, kappa of 99.97, and MCC of 99.97. The GBM gave an accuracy of 99.99, AUC of 100, recall of 99.99, precision of 100, F1 of 99.99, kappa of 99.98, and MCC of 99.98. The KNN gave accuracy of 99.98, AUC of 100, recall of 99.98, precision of 99.99, F1 of 99.99, kappa of 99.96, and MCC of 99.96. Lastly, the XGBoost gave an accuracy of 99.98, AUC of 100, recall of 99.98, precision of 100, F1 of 99.99, kappa of 99.96, and MCC of 99.96.

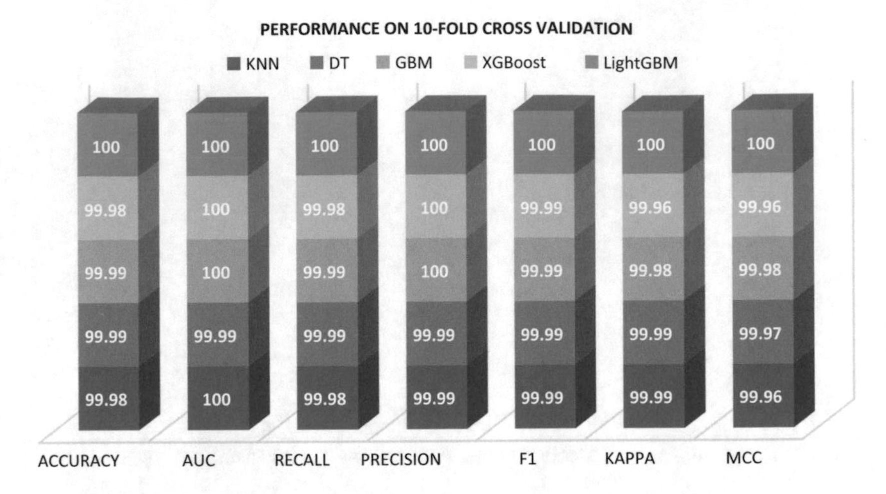

Fig. 2 Performance of the proposed models on tenfold cross-validation method

Table 4 Comparison of the proposed models on sixfold with the State-of-the-art

References	IoT IDS techniques	IoT datasets	Accuracy	F1	MCC
[103]	ELM-FCM	NSLKDD	86.53	X	x
[102]	DM	NLSKDD	99.20	X	x
[45]	M-SVM	ADFA-LD	100	X	x
[104]	ANN	NSLKDD	99.40	X	x
[105]	SVM-ELM	KDDCUP 99	95.75	X	x
Proposed model 1	PCA-LightGBM	UNSWNB15	100	100	100
Proposed model 2	PCA-DT	UNSWNB15	99.97	99.98	99.93
Proposed model 3	PCA-GBM	UNSWNB15	99.97	99.98	99.93
Proposed model 4	PCA-KNN	UNSWNB15	99.97	99.98	99.93
Proposed model 5	PCA-XGBoost	UNSWNB15	99.93	99.95	99.85

4.4 Comparison of the Proposed Models Utilizing Sixfold Cross-Validation with the State-of-the-Art Methods

The proposed models 1–5 was compared with the state-of-art methods as shown in Table 4. The authors in [85, 86] and [87, 88] employed a dataset that has many shortcomings as pointed out in [101]. We maintain that this dataset ought not to be utilized as a benchmark dataset in IoT and EoT, since it was obtained in a traditional network setting [102]. Thus, majority of the state-of-art methods adopt this dataset that do not reflect present-day attacks. This guides us to develop Intrusion Detection System models that support modern-day attacks that satisfied the requirements of IoT protocol and EoT. Additionally, our accuracy, F1, and MCC are superior to the state-of-art results.

Table 5 Comparison of the proposed models on tenfold with the state-of-art techniques

References	IoT IDS techniques	IoT datasets	Accuracy (%)	F1	MCC	Validation technique
[106]	GXGBoost	WSN-DS	99.5	X	X	Tenfold
[41]	ANN	WSN-DS	98.5	X	x	Tenfold
[44]	CA-LDA	NSLKDD	94.43	X	x	x
[44]	NB-KNN	NSLKDD	94.57	X	x	x
[58]	PCA – -softmax regression	KDDCup99	84.40	X	x	Tenfold
[58]	PCA-KNN	KDDCup99	85.24	X	x	Tenfold
[25]	C5-SVM	BoT-IoT	99.97	X	x	x
[59]	PCA-NTAD	Kyoto 2006	95.9	X	x	x
[59]	PCA-AD	Kyoto 2006	99.9	X	x	x
Proposed model 6	PCA-LightGBM	UNSW NB15	100	100	100	Tenfold
Proposed model 7	PCA-DT	UNSW NB15	99.99	99.99	99.96	Tenfold
Proposed model 8	PCA-GBM	UNSW NB15	99.99	99.99	99.98	Tenfold
Proposed model 9	PCA-KNN	UNSW NB15	99.98	99.99	99.96	Tenfold
Proposed model 10	PCA-XGBoost	UNSW NB15	99.98	99.99	99.96	Tenfold

4.5 *Comparison of the Proposed Models Utilizing Tenfold Cross-Validation with the State-of-the-Art Techniques*

The proposed models 6–10 was compared with the state-of-art techniques as shown in Table 5. The evaluation was done on WSN-DS, NSLKDD, BoT-IoT, and Kyoto 2006 datasets. These datasets are suitable for IoT except for NSLKDD which suffers the same dataset limitation to support IoT. The accuracy, F1, and MCC of our proposed models as shown in Table 5 outperformed the reported state-of-the art techniques.

5 Conclusion and Future Work

In this paper, we presented an improved performance of IDSs for IoT and EoT attacks detection. We utilized ten different lightweight models that satisfied IoT requirements and EoT. The min-max method was used for normalization and PCA utilized for dimensionality reduction. Ten features were selected that are important for both IoT and EoT applications. The LightGBM, DT, GBM, KNN, and XGBoost algorithms were used for classification. The tenfold cross-validation and sixfold cross validation were used as hold-out methods. The LightGBM attained 100% accuracy on the two hold-out strategies used in this study. The findings were also compared with the state-of-the-art results, and our results outperformed in terms of accuracy, F1, MCC, and validation datasets. The future work would be to introduce a deep learning approach for IDSs in improving IoT and EoT attack classification. The BoT-IoT dataset can be utilized to analyze the deep learning model in future and compare with the UNSWNB-15 dataset.

References

1. Zeng M, Yadav A, Dobre OA, Vincent Poor H (2019) Energy-efficient joint user-RB association and power allocation for uplink hybrid NOMA-OMA. IEEE Internet Things J 6(3):5119–5131. https://doi.org/10.1109/JIOT.2019.2896946
2. Bonomi F, Milito R, Zhu J, Addepalli S (2012) Fog computing and its role in the internet of things. In: MCC'12 — Proceedings of the 1st ACM mobile cloud computing workshop, pp 13–15. https://doi.org/10.1145/2342509.2342513
3. Wang Y, Xie L, Li W, Meng W, Li J (2017) A privacy-preserving framework for collaborative intrusion detection networks through fog computing. In: Lecture notes in computer science (including subseries Lecture notes in artificial intelligence and Lecture notes in bioinformatics), vol. 10581 LNCS, pp 267–279. https://doi.org/10.1007/978-3-319-69471-9_20
4. Mukherjee M et al (2017) Security and privacy in fog computing: challenges. IEEE Access 5:19293–19304. https://doi.org/10.1109/ACCESS.2017.2749422
5. Lin J, Yu W, Zhang N, Yang X, Zhang H, Zhao W (2017) A survey on internet of things: architecture, enabling technologies, security and privacy, and applications. IEEE Internet Things J 4(5):1125–1142. https://doi.org/10.1109/JIOT.2017.2683200
6. Tahsien SM, Karimipour H, Spachos P (2020) Machine learning based solutions for security of internet of things (IoT): a survey. J Netw Comput Appl 161. https://doi.org/10.1016/j.jnca.2020.102630
7. Cisco Annual Internet Report – Cisco Annual Internet Report (2018–2023) White Paper – Cisco. https://www.cisco.com/c/en/us/solutions/collateral/executive-perspectives/annual-internet-report/white-paper-c11-741490.html. Accessed 11 Feb 2021
8. Chaqfeh MA, Mohamed N (2012) Challenges in middleware solutions for the internet of things. In: Proceedings of 2012 international conference on collaboration technologies and systems. CTS 2012, pp 21–26. https://doi.org/10.1109/CTS.2012.6261022
9. Torkaman A, Seyyedi MA (2016) Analyzing IoT reference architecture models. Int J Comput Sci Softw Eng 5(8) ISSN 2409-4285, [Online]. Available: www.IJCSSE.org
10. Atzori L, Iera A, Morabito G (2010) The internet of things: a survey. Comput Netw 54(15):2787–2805. https://doi.org/10.1016/j.comnet.2010.05.010
11. Asharf J, Moustafa N, Khurshid H, Debie E, Haider W, Wahab A (2020) A review of intrusion detection systems using machine and deep learning in internet of things: challenges, solutions and future directions. Electronics 9(7):1177. https://doi.org/10.3390/electronics9071
12. Sicari S, Rizzardi A, Grieco LA, Coen-Porisini A (2015) Security, privacy and trust in internet of things: the road ahead. Comput Netw 76:146–164. https://doi.org/10.1016/j.comnet.2014.11.008
13. Kolias C, Kambourakis G, Stavrou A, Voas J (2017) DDoS in the IoT: Mirai and other botnets. Computer (Long Beach California) 50(7):80–84. https://doi.org/10.1109/MC.2017.201
14. Asharf J, Moustafa N, Khurshid H, Debie E, Haider W, Wahab A (2020) A review of intrusion detection systems using machine and deep learning in internet of things: challenges, solutions and future directions. Electron 9(7). https://doi.org/10.3390/electronics9071177

15. Marsden T, Moustafa N, Sitnikova E, Creech G (2017) Probability risk identification based intrusion detection system for SCADA systems. Springer International Publishing
16. Azeez NA, Ayemobola TJ, Misra S, Maskeliūnas R, Damaševičius R (2019) Network intrusion detection with a hashing based apriori algorithm using Hadoop MapReduce. Computers 8(4). https://doi.org/10.3390/computers8040086
17. Gould WA (1992) Spoilage of canned tomatoes and tomato products. Tomato Prod Process Technol 16(1):419–431. https://doi.org/10.1533/9781845696146.3.419
18. Varma PRK, Kumari VV, Kumar SS (2018) Progress in computing, analytics and networking, vol 710. Springer, Singapore
19. Shamshirband S et al (2014) Co-FAIS: cooperative fuzzy artificial immune system for detecting intrusion in wireless sensor networks. J Netw Comput Appl 42:102–117. https://doi.org/10.1016/j.jnca.2014.03.012
20. Şen S, Clark JA (2009) Intrusion detection in mobile ad hoc networks, pp 427–454. https://doi.org/10.1007/978-1-84800-328-6_17
21. Zarpelão BB, Miani RS, Kawakani CT, de Alvarenga SC (2017) A survey of intrusion detection in internet of things. J Netw Comput Appl 84:25–37., ISSN 1084-8045. https://doi.org/10.1016/j.jnca.2017.02.009
22. Kumar S, Dutta K (2016) Intrusion detection in mobile ad hoc networks: techniques, systems, and future challenges. Secur Commun Networks 9(14):2484–2556. https://doi.org/10.1002/sec.1484
23. Mitchell R, Chen IR (2014) A survey of intrusion detection techniques for cyber-physical systems. ACM Comput Surv 46(4). https://doi.org/10.1145/2542049
24. Jāmiʻat al-Zaytūnah al-Urdunīyah, Universiti Sains Malaysia, and Institute of Electrical and Electronics Engineers. In: ICIT 2017: the 8th international conference on information technology : internet of things IoT: conference proceedings: May 17–18, 2017, Amman, Jordan, pp 697–702, 2017
25. Khraisat A, Gondal I, Vamplew P, Kamruzzaman J, Alazab A (2019) A novel ensemble of hybrid intrusion detection system for detecting internet of things attacks. Electron 8(11). https://doi.org/10.3390/electronics8111210
26. Sonar K, Upadhyay H (2014) A survey on ddos in internet of things. Int J Eng Res Dev 10(11):58–63
27. Athreya AP, Tague P (2013) Network self-organization in the internet of things. In: 2013 IEEE international workshop on internet-of-things networking and control. IoT-NC 2013, pp 25–33. https://doi.org/10.1109/IoT-NC.2013.6694050
28. Lopez P, Fernandez D, Jara AJ, Skarmeta AF (2013) Survey of internet of things technologies for clinical environments. In: Proceedings of the 27th international conference on advanced information networking and applications workshops. WAINA 2013, pp 1349–1354. https://doi.org/10.1109/WAINA.2013.255
29. Raponi S, Caprolu M, Di Pietro R (2019) Intrusion detection at the network edge: solutions, limitations, and future directions, LNCS, vol 11520. Springer International Publishing
30. Wang Y, Meng W, Li W, Liu Z, Liu Y, Xue H (2019) Adaptive machine learning-based alarm reduction via edge computing for distributed intrusion detection systems. Concurr Comput 31(19):1–12. https://doi.org/10.1002/cpe.5101
31. Alam MGR, Hassan MM, Uddin MZ, Almogren A, Fortino G (2019) Autonomic computation offloading in mobile edge for IoT applications. Futur Gener Comput Syst 90:149–157. https://doi.org/10.1016/j.future.2018.07.050
32. Roman R, Lopez J, Mambo M (2018) Mobile edge computing, Fog et al.: A survey and analysis of security threats and challenges. Futur Gener Comput Syst 78:680–698. https://doi.org/10.1016/j.future.2016.11.009
33. Hosseinpour F, Vahdani Amoli P, Plosila J, Hämäläinen T, Tenhunen H (2016) An intrusion detection system for fog computing and IoT based logistic systems using a smart data approach. Int J Digit Content Technol Appl 10(5):34–46
34. Wu D, Yan J, Wang H, Wang R (2019) Multiattack intrusion detection algorithm for edge-assisted internet of things. In: Proceedings of IEEE international conference on industrial internet cloud. ICII 2019, pp 210–218. https://doi.org/10.1109/ICII.2019.00046

35. Wu D, Si S, Wu S, Wang R (2018) Dynamic trust relationships aware data privacy protection in mobile crowd-sensing. IEEE Internet Things J 5(4):2958–2970. https://doi.org/10.1109/JIOT.2017.2768073
36. Khraisat A, Gondal I, Vamplew P, Kamruzzaman J (2019) Survey of intrusion detection systems: techniques, datasets and challenges. Cybersecurity 2(1). https://doi.org/10.1186/s42400-019-0038-7
37. Huda S, Abawajy J, Alazab M, Abdollalihian M, Islam R, Yearwood J (2016) Hybrids of support vector machine wrapper and filter based framework for malware detection. Futur Gener Comput Syst 55:376–390. https://doi.org/10.1016/j.future.2014.06.001
38. Odusami M, Misra S, Adetiba E, Abayomi-Alli O, Damasevicius R, Ahuja R (2019) An improved model for alleviating layer seven distributed denial of service intrusion on webserver. J Phys Conf Ser 1235(1):2–8. https://doi.org/10.1088/1742-6596/1235/1/012020
39. Le TTH, Park T, Cho D, Kim H (2018) An effective classification for DoS attacks in wireless sensor networks. In: International conference on ubiquitous and future networks, ICUFN, July 2018, pp 689–692. https://doi.org/10.1109/ICUFN.2018.8436999
40. Khraisat A, Gondal I, Vamplew P et al (2019) Survey of intrusion detection systems: techniques, datasets and challenges. Cybersecur 2:20. https://doi.org/10.1186/s42400-019-0038-7
41. Almomani I, Al-Kasasbeh B, Al-Akhras M (2016) WSN-DS: a dataset for intrusion detection Systems in Wireless Sensor Networks. J Sensors 2016. https://doi.org/10.1155/2016/4731953
42. Praneeth NSKH, Varma NM, Naik RR (2017) Principle component analysis based intrusion detection system using support vector machine. In: 2016 IEEE international conference on recent trends in electronics, information and communication technology, RTEICT 2016 – proceedings, pp 1344–1350. https://doi.org/10.1109/RTEICT.2016.7808050
43. Tan X et al (2019) Wireless sensor networks intrusion detection based on SMOTE and the random forest algorithm. Sensors (Switzerland) 19(1). https://doi.org/10.3390/s19010203
44. Pajouh HH, Javidan R, Khayami R, Dehghantanha A, Choo KKR (2019) A two-layer dimension reduction and two-tier classification model for anomaly-based intrusion detection in IoT backbone networks. IEEE Trans Emerg Top Comput 7(2):314–323. https://doi.org/10.1109/TETC.2016.2633228
45. Vijayanand R, Devaraj D, Kannapiran B (2017) Support vector machine based intrusion detection system with reduced input features for advanced metering infrastructure of smart grid. In: 2017 4th International conference on advanced computing and communication systems. ICACCS 2017. https://doi.org/10.1109/ICACCS.2017.8014590
46. Moustafa N, Slay J (2017) The significant features of the UNSW-NB15 and the KDD99 data sets for network intrusion detection systems. In: Proceedings of 2015 4th international workshop on building analysis datasets and gathering experience returns for security. BADGERS 2015, pp 25–31. https://doi.org/10.1109/BADGERS.2015.14
47. Ahmad T, Aziz MN (2019) Data preprocessing and feature selection for machine learning intrusion detection systems. ICIC Express Lett 13(2):93–101. https://doi.org/10.24507/icicel.13.02.93
48. Patro SGK, Sahu KK (2015) Normalization: a preprocessing stage. Iarjset:20–22. https://doi.org/10.17148/iarjset.2015.2305
49. Panda SK, Nag S, Jana PK (2015) A smoothing based task scheduling algorithm for heterogeneous multi-cloud environment. In: Proceedings of 2014 3rd international conference on parallel, distributed and grid computing. PDGC 2014, pp 62–67. https://doi.org/10.1109/PDGC.2014.7030716
50. Panda SK, Jana PK (2015) Efficient task scheduling algorithms for heterogeneous multi-cloud environment. J Supercomput 71(4):1505–1533. https://doi.org/10.1007/s11227-014-1376-6
51. Jain A, Nandakumar K, Ross A (2005) Score normalization in multimodal biometric systems. Pattern Recogn 38(12):2270–2285. https://doi.org/10.1016/j.patcog.2005.01.012
52. Jain S, Shukla S, Wadhvani R (2018) Dynamic selection of normalization techniques using data complexity measures. Expert Syst Appl 106:252–262. https://doi.org/10.1016/j.eswa.2018.04.008

53. PE Project (2020) The performance of intrusion detection system 14(12):1217–1223. https://doi.org/10.24507/icicel.14.12.1217
54. Bouzida Y, Cuppens F (2004) Efficient intrusion detection using principal component analysis. In: 3éme Conférence sur la Sécurité Archit. Réseaux (SAR), La Londe, Fr., no. October, pp 381–395
55. Howley T, Madden MG, O'Connell ML, Ryder AG (2006) The effect of principal component analysis on machine learning accuracy with high-dimensional spectral data. Knowl-Based Syst 19(5):363–370. https://doi.org/10.1016/j.knosys.2005.11.014
56. Bhattacharya S et al (2020) A novel PCA-firefly based XGBoost classification model for intrusion detection in networks using GPU. Electron 9(2). https://doi.org/10.3390/electronics9020219
57. Granato D, Santos JS, Escher GB, Ferreira BL, Maggio RM (2018) Use of principal component analysis (PCA) and hierarchical cluster analysis (HCA) for multivariate association between bioactive compounds and functional properties in foods: a critical perspective. Trends Food Sci Technol 72(2018):83–90. https://doi.org/10.1016/j.tifs.2017.12.006
58. Zhao S, Li W, Zia T, Zomaya AY (2018) A dimension reduction model and classifier for anomaly-based intrusion detection in internet of things. In: Proceedings of 2017 IEEE 15th international conference on dependable, autonomic and secure computing/2017 IEEE 15th international conference on pervasive intelligence and computing/2017 IEEE 3rd international conference on big data intelligence and computing, January 2018, pp 836–843, https://doi.org/10.1109/DASC-PICom-DataCom-CyberSciTec.2017.141
59. Hoang DH, Nguyen HD (2018) A PCA-based method for IoT network traffic anomaly detection. In: International conference on advanced communication technology. ICACT, February 2018, pp 381–386. https://doi.org/10.23919/ICACT.2018.8323766
60. Hoang DH, Duong Nguyen H (2019) Detecting anomalous network traffic in IoT networks. In: International conference on advanced communication technology. ICACT, February 2019, no 1, pp 1143–1152. https://doi.org/10.23919/ICACT.2019.8702032
61. Saheed YK, Hamza-usman FE (2020) Feature Selection with IG-R for improving performance of intrusion detection system. Int J Commun Netw Inf Secur 12(3):338–344
62. Serpen G, Aghaei E (2018) Host-based misuse intrusion detection using PCA feature extraction and kNN classification algorithms. Intell Data Anal 22(5):1101–1114. https://doi.org/10.3233/IDA-173493
63. L. Greche, M. Jazouli, N. Es-Sbai, A. Majda, and A. Zarghili, "Comparison between Euclidean and Manhattan distance measure for facial expressions classification," 2017 International conference on wireless technologies, embedded and intelligent systems. WITS 2017, pp. 2–5, 2017, doi: https://doi.org/10.1109/WITS.2017.7934618
64. Singh A, Pandey B (2016) An euclidean distance based KNN computational method for assessing degree of liver damage. 2016 International Conference on Inventive Computation Technologies (ICICT):1–4. https://doi.org/10.1109/INVENTIVE.2016.7823222
65. Liao Y, Vemuri VR (2002) Classifier for intrusion. Comput Secur 21(5):439–448
66. Rao BB, Swathi K (2017) Fast kNN classifiers for network intrusion detection system. Indian J Sci Technol 10(14):1–10. https://doi.org/10.17485/ijst/2017/v10i14/93690
67. Wazirali R (2020) An improved intrusion detection system based on KNN Hyperparameter tuning and cross-validation. Arab J Sci Eng 45(12):10859–10873. https://doi.org/10.1007/s13369-020-04907-7
68. Padigela PK, Suguna R (2020) A survey on analysis of user behavior on digital market by mining clickstream data. In: Proceedings of the third international conference on computational intelligence and informatics, Advances in intelligent systems and computing, vol 1090. Springer, Singapore
69. Rena R (2008) The internet in tertiary education in Africa: recent trends. Int J Comput ICT Res 2(1):9–161
70. Su MY (2011) Real-time anomaly detection systems for denial-of-service attacks by weighted k-nearest-neighbor classifiers. Expert Syst Appl 38(4):3492–3498. https://doi.org/10.1016/j.eswa.2010.08.137

71. Li L, Zhang H, Peng H, Yang Y (2018) Nearest neighbors based density peaks approach to intrusion detection. Chaos Solitons Fractals 110:33–40. https://doi.org/10.1016/j.chaos.2018.03.010
72. Li W, Yi P, Wu Y, Pan L, Li J (2014) A new intrusion detection system based on KNN classification algorithm in wireless sensor network. J Elect Comput Eng 1:2014. https://doi.org/10.1155/2014/240217
73. Sarker IH, Kayes ASM, Watters P (2019) Effectiveness analysis of machine learning classification models for predicting personalized context-aware smartphone usage. J Big Data 6(1). https://doi.org/10.1186/s40537-019-0219-y
74. Kim G, Lee S, Kim S (2014) A novel hybrid intrusion detection method integrating anomaly detection with misuse detection. Expert Syst Appl 41(4 Part 2):1690–1700. https://doi.org/10.1016/j.eswa.2013.08.066
75. Kotsiantis SB (2013) Decision trees: a recent overview. Artif Intell Rev 39(4):261–283. https://doi.org/10.1007/s10462-011-9272-4
76. Du W, Du W, Zhan Z, Zhan Z (2002) Building decision tree classifier on private data. In: Proceedings of the IEEE international conference on privacy, security and data mining, vol 14, pp 1–8. [Online]. Available: http://portal.acm.org/citation.cfm?id=850784
77. Haque MA, Verma A, Alex JSR, Venkatesan N (2020) Experimental evaluation of CNN architecture for speech recognition. In: First international conference on sustainable technologies for computational intelligence, vol 1045. Springer, Singapore
78. Nancy P, Muthurajkumar S, Ganapathy S, Santhosh Kumar SVN, Selvi M, Arputharaj K (2020) Intrusion detection using dynamic feature selection and fuzzy temporal decision tree classification for wireless sensor networks. IET Commun 14(5):888–895. https://doi.org/10.1049/iet-com.2019.0172
79. Goeschel K (2016) Reducing false positives in intrusion detection systems using data-mining techniques utilizing support vector machines, decision trees, and naive Bayes for off-line analysis. In: Conferences proceedings – IEEE SOUTHEASTCON July 2016. https://doi.org/10.1109/SECON.2016.7506774
80. Ahmim A, Maglaras L, Ferrag MA, Derdour M, Janicke H (2019) A novel hierarchical intrusion detection system based on decision tree and rules-based models. In: Proceedings of 15th international conference on distributed computing in sensor systems. DCOSS 2019, pp 228–233. https://doi.org/10.1109/DCOSS.2019.00059
81. Alharbi S, Rodriguez P, Maharaja R, Iyer P, Subaschandrabose N, Ye Z (2018) Secure the internet of things with challenge response authentication in fog computing. In: 2017 IEEE 36th international performance computing and communications conference. IPCCC 2017, January 2018, pp 1–2. https://doi.org/10.1109/PCCC.2017.8280489
82. Natekin A, Knoll A (2013) Gradient boosting machines, a tutorial. Front Neurorobot 7. https://doi.org/10.3389/fnbot.2013.00021
83. ส. ไทรทับทิม, “No Titleการนำสาหร่ายที่ผลิตน้ำมันไบโอดีเซลมาบำบัดน้ำเสียของ โรงงานอุตสาหกรรมรีไซเคิล,” 2554, [Online]. Available: http://library1.nida.ac.th/termpaper6/sd/2554/19755.pdf
84. Tian D et al (2018) An intrusion detection system based on machine learning for CAN-Bus. Lect Notes Inst Comput Sci Soc Telecommun Eng LNICST 221:285–294. https://doi.org/10.1007/978-3-319-74176-5_25
85. Bissacco A, Yang MH, Soatto S (2007) Fast human pose estimation using appearance and motion via multi-dimensional boosting regression. In: Proceedings of the IEEE Computer Society Conference on Computer Vision and Pattern Recognition. https://doi.org/10.1109/CVPR.2007.383129
86. Hutchinson RA, Liu LP, Dietterich TG (2011) Incorporating boosted regression trees into ecological latent variable models. Proc Natl Conf Artif Intell 2:1343–1348
87. Pittman SJ, Brown KA (2011) Multi-scale approach for predicting fish species distributions across coral reef seascapes. PLoS One 6(5). https://doi.org/10.1371/journal.pone.0020583
88. Title P, Fulton B, Energy R, Award P, Renewables B, Location P, Final technical report
89. Verma A, Ranga V (2020) Machine learning based intrusion detection systems for IoT applications. Wirel Pers Commun 111(4):2287–2310. https://doi.org/10.1007/s11277-019-06986-8

90. Alqahtani M, Mathkour H, Ben Ismail MM (2020) IoT botnet attack detection based on optimized extreme gradient boosting and feature selection. Sensors 20(21):6336. https://doi.org/10.3390/s20216336
91. Friedman J (2001) Greedy function approximation : a gradient boosting machine. Ann Stat 29(5):1189–1232. Published by: Institute of Mathematical Statistics Stable URL : https://statweb.stanford.edu/~jhf/ftp/trebst.pdf
92. Pan Y, Liu D, Deng L (2017) Accurate prediction of functional effects for variants by combining gradient tree boosting with optimal neighborhood properties. PLoS One 12(6):1–20. https://doi.org/10.1371/journal.pone.0179314
93. Kuang L et al (2018) A personalized QoS prediction approach for CPS service recommendation based on reputation and location-aware collaborative filtering. Sensors (Switzerland) 18(5). https://doi.org/10.3390/s18051556
94. Fan C, Liu D, Huang R, Chen Z, Deng L (2016) PredRSA: a gradient boosted regression trees approach for predicting protein solvent accessibility. BMC Bioinform 17(1). https://doi.org/10.1186/s12859-015-0851-2
95. Li C, Zheng X, Yang Z, Kuang L (2018) Predicting short-term electricity demand by combining the advantages of ARMA and XGBoost in fog computing environment. Wirel Commun Mob Comput 2018. https://doi.org/10.1155/2018/5018053
96. Ke G et al (2017) LightGBM: a highly efficient gradient boosting decision tree. Adv Neural Inf Process Syst (NIPS) 2017:3147–3155
97. Fan J, Ma X, Wu L, Zhang F, Yu X, Zeng W (2019) Light gradient boosting machine: an efficient soft computing model for estimating daily reference evapotranspiration with local and external meteorological data. Agric Water Manag 225(August):105758. https://doi.org/10.1016/j.agwat.2019.105758
98. Khafajeh H (2020) An efficient intrusion detection approach using light gradient boosting. J Theor Appl Inf Technol 98(5):825–835
99. Rai M, Mandoria HL (2019) Network intrusion detection: a comparative study using state-of-the-art machine learning methods. In: IEEE International conference on issues and challenges in intelligent computing techniques. ICICT 2019, pp 0–4. https://doi.org/10.1109/ICICT46931.2019.8977679
100. Calisir S, Atay R, Pehlivanoglu MK, Duru N (2019) Intrusion detection using machine learning and deep learning techniques. In: UBMK 2019 – proceedings of 4th international conference on computer science and engineering, pp 656–660. https://doi.org/10.1109/UBMK.2019.8906997
101. Mchugh J (2000) Testing intrusion detection systems: a critique of the 1998 and 1999 DARPA intrusion detection system evaluations as performed by Lincoln Laboratory. ACM Trans Inf Syst Secur 3(4):262–294. https://doi.org/10.1145/382912.382923
102. Diro AA, Chilamkurti N (2017) Distributed attack detection scheme using deep learning approach for internet of things. Futur Gener Comput Syst. https://doi.org/10.1016/j.future.2017.08.043
103. Rathore S, Park JH (2018) Semi-supervised learning based distributed attack detection framework for IoT. Appl Soft Comput J 72:79–89. https://doi.org/10.1016/j.asoc.2018.05.049
104. Hodo E et al (2016) Threat analysis of IoT networks using artificial neural network intrusion detection system. In: 2016 International symposium on networks, computers and sommunications. ISNCC 2016, pp 4–9. https://doi.org/10.1109/ISNCC.2016.7746067
105. Al-Yaseen WL, Othman ZA, Nazri MZA (2017) Multi-level hybrid support vector machine and extreme learning machine based on modified K-means for intrusion detection system. Expert Syst Appl 67:296–303. https://doi.org/10.1016/j.eswa.2016.09.041
106. Alqahtani M, Gumaei A, Mathkour H, Ben Ismail MM (2019) A genetic-based extreme gradient boosting model for detecting intrusions in wireless sensor networks. Sensors (Switzerland) 19(20). https://doi.org/10.3390/s19204383

Index

S. Misra et al. (eds.), *Artificial Intelligence for Cloud and Edge Computing*, Internet of Things, https://doi.org/10.1007/978-3-030-80821-1

W

X

Printed in the United States
by Baker & Taylor Publisher Services